Control Systems: Design and Engineering

Control Systems: Design and Engineering

Edited by Ashley Potter

CLANRYE
INTERNATIONAL
www.clanryeinternational.com

Clanrye International,
750 Third Avenue, 9th Floor,
New York, NY 10017, USA

ISBN: 978-1-63240-655-2

Cataloging-in-Publication Data

Control systems : design and engineering / edited by Ashley Potter.
    p. cm.
Includes bibliographical references and index.
ISBN 978-1-63240-655-2
1. Automatic control. 2. Control theory. 3. Engineering design. I. Potter, Ashley.
TJ213 .C65 2018
629.8--dc23

For information on all Clanrye International publications
visit our website at www.clanryeinternational.com

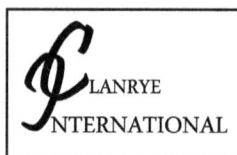

_C_LANRYE
_I_NTERNATIONAL

Contents

Preface

In recent years, control systems have come to play an important role in the development and in the progress of technology. Control systems help in the management and in the regulation of other systems. They can be found in appliances such as refrigerators, air conditioners, geysers, etc. as well as in industrial plants. They mainly consist of two classes of systems, open loop control systems and closed loop control systems. The ever growing need of advanced technology is the reason that has fueled the research in the field of control systems. The objective of this book is to give a general view of the different areas of this field and its applications. It is appropriate for students seeking detailed information in this area as well as for experts. It will help the readers in keeping pace with the rapid changes in this field.

After months of intensive research and writing, this book is the end result of all who devoted their time and efforts in the initiation and progress of this book. It will surely be a source of reference in enhancing the required knowledge of the new developments in the area. During the course of developing this book, certain measures such as accuracy, authenticity and research focused analytical studies were given preference in order to produce a comprehensive book in the area of study.

This book would not have been possible without the efforts of the authors and the publisher. I extend my sincere thanks to them. Secondly, I express my gratitude to my family and well-wishers. And most importantly, I thank my students for constantly expressing their willingness and curiosity in enhancing their knowledge in the field, which encourages me to take up further research projects for the advancement of the area.

Editor

The Function Requirement Analysis and Implementation of the Intelligent Fitting System

Ruan Jinhua, Zheng Jingjing, Hu Shouzhong

Fashion Design and Engineering, Shanghai University of Engineering Science, shanghai, China

Email address:

hushzh@sues.edu.cn (Hu Shouzhong)

Abstract: The intelligent fitting system is the product of the development of science and technology, at the same time, a new consumption experience method appears with the increasing demands of consumer. The intelligent fitting system not only increase the customer' shopping experience by the interactive fitting system, but also set up the brand image for merchants and increase the competitiveness. Therefore, the research of the fitting system is a looming thing. This article, through the investigation and expert scoring, etc, describes its role in the process of sales. Via analyzing the value of the intelligent fitting system through the comprehensive analysis method to find out the exist problems and a method to improve the fitting system and a way to its implementation, hoping it have an promoting effect on the development of the intelligent fitting system.

Keywords: Fashion, The Intelligent Fitting System, Value Analysis

1. Introduction

Along with the development of science and technology, the model of clothing retail is changing constantly. The fierce social competition, forcing the retailers to reform in retail models and the experience consumption which is changing from the initial drape to now the humanization designed intelligent fitting room, has always been changing. People's pursuit of a nice and comfortable fitting room has never stopped. The design style is changing with the change of the demand of people. Intelligence is the development trend of the fitting room.

The reality of how to balance the relationship between consumer demand and development cost of the intelligent fitting room, maximizing the value of the system, has become an important problem of the research of modern clothing retail enterprise.

The article, through comprehensive research methods, analyzed the element, influencing the "consumer experience", of customers, the need of retailers to the functional requirements of intelligent fitting system and the condition of function realization technology of the fitting system. Based on the original fitting system in the market, study and improve the design of the intelligent fitting system.

2. The Intelligent Fitting System

The intelligent fitting system, a kind of intelligent machines, provide "consumer experience" for customers during the process of the clothing retail. In the radio frequency identification technology (RFID), users, standing in front of the screen, do not need to touch screen only needs to control the system via hand gestures which can achieve interaction with the fitting mirror. The users just waving, the choice of clothes will be amazing natural wear on the user's body. Users, on the way, can easily replace different clothes by switching on the page. In addition, the virtual fitting system also provide high-definition cameras, developing close composite image between trying clothes, in a timely manner, and users. The user experience unprecedented shopping pleasure.

The intelligent fitting system is a set of integration platform composed by hardware and software, mainly including the fitting terminal, content management server and the Internet management. Besides the common functions of fitting room and the release of multimedia information, it also specially integrate the clothes attributes display, a discount promotion, the multi-machine networking management, and other functions, as well as access to the internet and emerging media such as Micro-Blog. The intelligent fitting system, through perfect union of the human body induction technology, network and digital display technology ,give strong visual impact, customer attraction and a sense of science and technology to the terminal clothing stores, shopping malls and other public places. It also provide the customer with high quality consuming experience and simpler management services. If chose their favorite clothes, the customer can scan to share to the micro letter, Micro-Blog and other social software. If determine to buy, the consumer can direct scanning to afford it.

But the smart design and the manufacture process of the intelligent fitting is complex, which have many technical difficulties such as how to realize the reality function simulation, how to balance the relationship between the customer demand and the cost, etc. The key to the intelligent fitting system design is improving it and making it conform to the principle of value engineering analysis, both economic benefit and satisfying the function of experience.

3. Mental Experience of Clothing Retail Fitting

The intelligent fitting system，through the change of people psychological experience, stimulate the desire of consumers. Consumer experience refers to that, under certain social economy condition and in a particular consumption environment, consumer personally complete a mode of special consumption ,which has a strangeness, freshness and novelty consumption object ,in order to obtain a novel stimulus, profoundly memorable consumer experience, improving the customer delivered value. Experience consumption subject, experience consumption object and consumption experience environment constitute the three elements of the consumer experience.

Market survey statistics show that every consumer spend about seven, eight minutes, on average, on the fitting room, more than fifty percent of the whole process of the clothing purchase. The female consumers in the fitting room was as high as seventy percent of the purchasing time. It is clear that the fitting room plays a critical role in the retail system. The purchase decision of consumer is made in the fitting room in a large percentage. In the act of the fitting, the customer, mainly through whether the clothing fit, it is to be improved and it can modify body type, to determine whether the clothing can satisfy the demands of their own wear. Therefore many consumers will be more carefully observe on its fabrics content, its feel, the price and the cost performance in the fitting room. Thus the fitting behavior will directly lead to the purchase behavior.

4. The Research of the Consumer Requirements of the Function of Intelligent System

The main method of collecting study materials is questionnaire survey, supplemented by interview method, to verify and understanding the analysis data in question. To the aspect of company employees, the research's object is mainly the person who understand the fitting system, who shop more and retired people; The analysis results indicate that different consumer groups demand different function of the intelligent fitting system. Through the analysis of the data, researchers summarizes the functional requirements of different consumer groups. This research mainly uses SPSS20.0 to analysis on the data obtained (see table1).

Table 1. Statistics of the problems existed in the fitting room at present.

		response		Case
		N	percentage	percentage
Inconvenience	Difficult in finding proper clothing	269	27.60%	65.00%
	Troublesome in Change clothes	298	30.60%	72.00%
In the fitting process	The body' privacy is leaked Protection	188	19.30%	45.40%
	Destruction of makeup look	142	14.60%	34.30%
	others	77	7.90%	18.60%
Total		974	100.00%	235.30%
	The release of privacy, unsafe	218	20.50%	52.50%
The problems in the fitting process reduce your purchase enthusiasm of clothing	Lack of necessary equipment	216	20.30%	52.00%
	The fitting room small or disorder	274	25.80%	66.00%
Others	Unable to quickly find proper clothes	283	26.60%	68.20%
		71	6.70%	17.10%
Total		1062	100.00%	255.90%
What kind of	Necessary equipment complete	294	33.00%	71.40%
Services should	Privacy is protected	203	22.80%	49.30%
Shopping malls and clothing store services in terms of the fitting in your piont	Customer self-service fitting	322	36.10%	78.20%
	Others	72	8.10%	17.50%
Total		891	100.00%	216.30%

Use structure validity to analysis on the consumer demand for the intelligent functions of the fitting system. The mainly measurement method of the structure validity is the factor analysis method. This study will, therefore, identify the

functional requirements as three parts. In the rotating component matrix table, the joint degree, less than 0.650, will be deleted and then getting the final three common factor: "show clothes details", "but the proposed amendments to the clothing" and "save the customer database data". The first factor is the display module and the later both is background processing module which are mainly operate through database management. Therefore this article named it as "data management"; The second common factor is "show trends and putting forward modified opinions". The factors make customers jointly with stylist participate in the popular trend forecast. So, this article named it as "resource sharing"; The third public factor is "simulate human". This article named it as" function of visual experience".

5. The Determination of "Objective Function" Based on Value Engineering System

In the consumer research, three functional requirements were extracted: the functions of database management, resource sharing and visual experience function. In the functional requirements of consumers, the database management include displaying clothing details, possible proposed amendments to the clothing and keep customer database information; Resource sharing function include showing trends and putting forward modified opinions; Visual experience function include possible simulating the reality in the action, etc. Three functions, database management, promoting enterprise image and resource sharing, are retained in the enterprise research. Database management function include storing customer' database, displaying clothing details and recording customer's preferences; Propaganda enterprise capabilities include recommending customers on different clothing collocation; Resource sharing feature include displaying the sale record of each piece of clothing in each branch and the customer review. Investigation on comprehensive consumer and corporate demand for functions of the intelligent fitting system, the final results as shown in table2.

Table 2. *The determination of functional requirements in research.*

The general functional requirements	
The database management function	Display clothes details, possible proposed amendments to the clothes, store customer database, record customer's preferences
Resource sharing feature	Showing trends and putting forward modified opinions and the customer review
Visual experience function	Simulate the reality in the action, etc
Propaganda enterprise capabilities	Recommended customers different clothing collocation

Table 3. *The functional requirements in investigation determine the value of the function.*

Function	Programming cost (yuan)	Data cost (yuan)	Cost (yuan)	Cost index C (%)	Function index F （%）	Value （F/C）
Save the customer size	180	0	180	1.98	13.89	7.02
Record customer's preferences	360	360	720	7.92	13.19	1.67
Proposed amendments to the clothing	540	0	540	5.94	13.89	2.34
Display clothes details	720	90	810	8.91	12.5	1.4
Showing trends and putting forward to modify opinions	900	900	1800	19.8	11.11	0.56
Showing trends and putting forward modified opinions and the customer review	1080	540	1620	17.82	10.42	0.58
Recommended customers different clothing collocation	1260	720	1980	21.78	12.5	0.57
Simulate the reality in the action, etc	1440	0	1440	15.84	12.5	0.79
Total	9090yuan			100	100	

As can be seen from the consumer and the market demand, the more comprehensive of the function, the ranger the system will be used. However, the more functions, the higher the system development will cost, which will decrease system business value and it may not be accepted by the market.

Based on the main ideas of value engineering theory: achieve certain necessary functions at the lowest cost. The goal is to realize the necessary function of products or services reliably and improve the value of the analysis object with the lowest cost. In value engineering, there are three parts: the Value(V),the Function(F),and the Cost(C).The relations between them is: V = F/C; Expressed in mathematical proportion is as follows: value = function/cost.

5.1. The Determination of Index of Function F

In this study, the method of quantitative is to replace functions with their important degree. We can use expert scoring method to determine the important degree of each function. This method is to organize5-10experts and the relevant personnel as judges. The judges will compare the products or components according to the importance of them by one-to-one, important for 1 score, not important to 0.

Taking down the results of the products or components, and then calculating score average of the contestant personnel for the same product or component. With the sum divided by the average, obtaining the function coefficient. In this study, we will organize eight experts and relevant personnel to mark and calculate the function index of various components.

5.2. The Determination of Index of Function C

The table3 shows storing customers' size, recording customers' preferences, putting forward amendments to the clothing and the details of garments; the functions' value is higher than 1, so this few functions will be retained. But to the displaying clothing fashion trend by putting forward modified opinions, displacing clothing sales, customer evaluation at each branch, and recommending customer different collocation, all the functions' value is lower than 1, so these features can be deleted. The value of these features of simulating the real people in action is lower than 1, but higher than 0.7, so after appropriate for cost reduction, the function can be retained.

After the rule out through survey and value engineering analysis, the five functions, saving the customers' size, the database record customers' preferences, possible putting forward amendments to the clothing and display clothing details and simulate the reality in action, are determined functions. According to the five functions, improve the intelligence of the original intelligent fitting system which is blend in these five functions. The "target" function of the system development in the enterprise.

6. Improvement Design of the Intelligent Fitting System

6.1. The Customer Using Process of the Original System

The figure 1 shows the customer' using process of the intelligent fitting system .

As shown in figure 2, the improved system of customers' using process as follows: three point need to be improved on the basis of the original system.

1) save the customer information database
2) possible proposed amendments to the clothing and evaluation
3) simulate the reality in action, scene

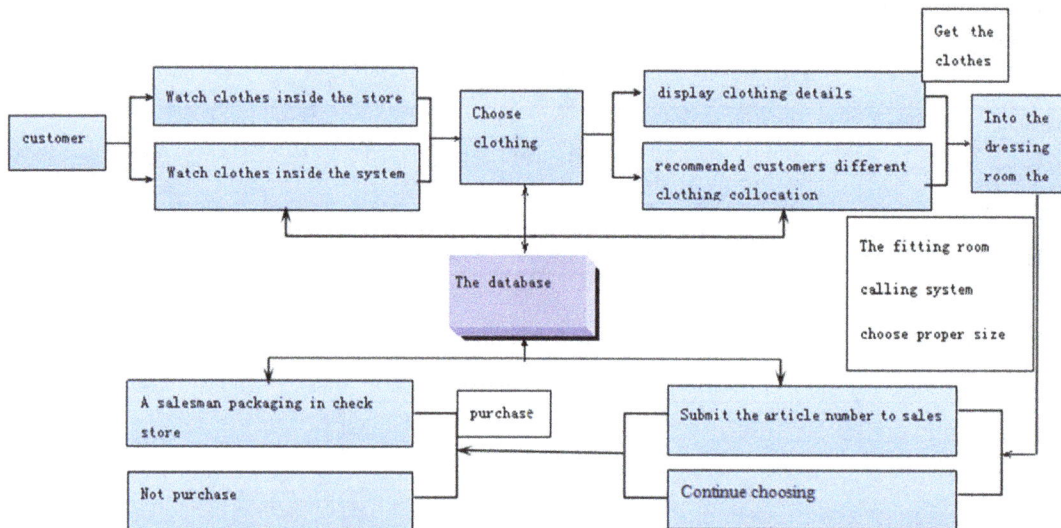

Figure 1. The customer' using process of the original system.

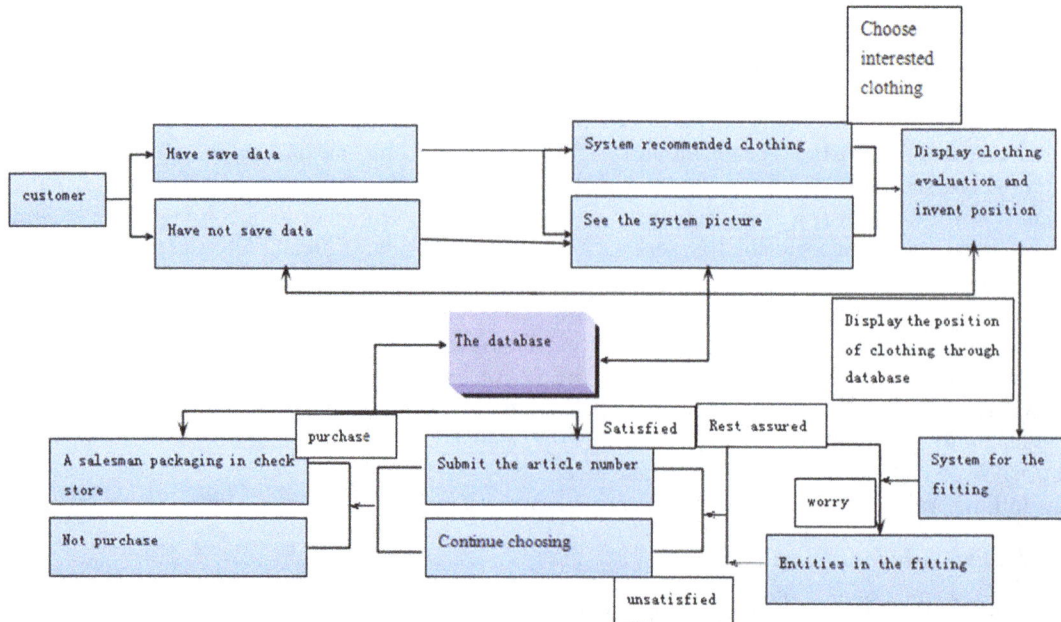

Figure 2. The improved system of customers' using process.

6.2. The Implementation of Function of the Modified Intelligent Fitting System

The realize of saving the customer information database and the evaluation of the proposed amendments to the clothing are mainly completed through the form of web and database. The system access to customer's size information and buying and trying information by client and effective integration and updating of background database information, directly showing the three-dimensional human body model and 3d fitting effect to the customer which help customers make buying decisions. After the customer determined to buy it, the purchasing information and trying information of customers, by automatic processing and format conversion, are sent directly to the background management system, so that the next time the customer purchase it could be generated by the customers' recommended list, designing the "o2o" practically intelligence fitting system.

To simulate reality in the action, our major build a 2d frame in Virtools to display all of the displaying clothing pictures. At the same time create a 2d frame used to display information, giving the corresponding materials. Clothing pictures in the gallery is used to control to change clothing and display status information area. Establish an array under level manager, which is named "clothing details", used to store each clothing style, price, material and location information, etc. Then setup script for the 2d frame of the clothing display box. The script is used to control state information displaying area. In 3D Max, all clothes that need to be changed ,mainly using BB module of the HIDE and SHOW module to realize the change of the characters, should be wore on the body, exported as Nmo format. While the display of each piece of clothing, you can hide other clothing, so as to realize the smooth of task switching. After implementing the change, you can, through changing the script respectively connecting to the clothing display boxes of different clothing pictures, achieve a simultaneous change and information display. Building the changing background of 2d frame in the interface, named background display box, is used to store different background. When the mouse or fingers slide to different backgrounds it can present a different background.

7. The Conclusion

In this article, the functions of the intelligent fitting system, starting from the functional requirements of intelligent fitting system, is determined by using the method of documentary investigation, questionnaire investigation and expert interview to consumers and business research, combining with the value engineering analysis method. Through the contrast analysis, we know we need to increase three functions which include storing customer database data on RFID system function, putting forward amendments to the clothing, the evaluation function and simulating the reality in the action, which is to improve the information of users and customers.

Complete the fitting system through the implement of

3dreconstruction, 3d virtual fitting body, the online fitting implementation, as well as the realization of the concept of interface and software interface. Start from the fabric simulation technology, the module and motion simulation technology, garment simulation dynamic virtual display technology and scene simulation technology to realize the three-dimensional human body fitting function. It will be a big trend in the development of intelligent fitting system, leading technology in intelligent fitting system is the leading apparel market.

Therefore, the research is prospective in the field of clothing, pointing out the direction of the development of future intelligent fitting system and the direction for the development of future clothing retailing forms. It is both opportunity and challenge for our country in the development of garment industry. We should seize the opportunity, meet the challenge and make Chinese intelligent fitting system a large step even a leading position.

Acknowledgement

This study was funded by Science Technology Commission of Shanghai Municipality, the funded project No. 11510501600 and Municipal Education Commission of Shanghai, the funded project No.13ZS128.

Case percentage: case is the combined information of specific part of a project, the proportion of case in the total percentage is the case percentage.

Common factor: a corresponding measurement variables

References

[1] TAO Weiwei, YAO Minghai, CHEN Zhansheng. Application of RFID perception technology in clothing stores. Internet of Things Technologies, 2012, 07:22-24.

[2] JIN Hongsheng. The Research of 3D Fitting System of Clothing Network Marketing. Shandong Textile Economy, 2012, 08: 48-51.

[3] HE Fang. The Revolution of Virtual Fitting Room. Technology Review, 2012, 05:59-63.

[4] GAO Aiying, LAN tian. Application of RFID and IOT in IT Development of Japanese Garment Manufacturing Enterprises. Logistics Technology, 2011, 07:132-136.

[5] ZHANG Yufang, WANG Xiaoming. The Application of Virtual Apparel Fitting. Art and Design (Theory), 2011, 10: 104-106.

[6] ZHANG Zhi, ZENG Cheng. Design of RIA-Based Online Fitting Room System. Computer Technology and Development, 2011, 10: 143-146.

[7] PENG Shengqiong. A Comparetive Research on Virtual Clothing Display Syste. Art and Design (Theory), 2014, 06: 71-73.

[8] ZHANG Yuanyuan, ZHANG Junlong. Three guns: Competitiveness Is Tempered Analysis of Shanghai Three guns Group Competitiveness. TA Weekly, 2008, 24:18-19.

Analysis and Research on Combination Feature Extraction Method of EEG Signal

LI Jun-wei[1], Jason Gu[2], XIE Yun[1]

[1]Electronic & Information Engineering College, Henan University of Science and Technology, Luoyang Henan, China
[2]School of Biomedical Engineering, Dalhousie University, Halifax, Canada

Email address:

736587057@qq.con (LI Jun-wei), 1466416698@qq.com (Jason Gu), 756669066@qq.com (XIE Yun)

Abstract: EEG feature extraction problem is studied in this paper. EEG analysis is the core content of the Brain-computer interface technology research. How to effectively extract the reflect people's behavior intention characteristic from EEG signals, it's a hot spot in this neighborhood research. According to the characteristics of EEG signal, the single method of feature extraction can't describe the characteristics of the signal very well. So We have own designed experiment, and put forward a combination feature extraction method, which contains calculation the maximum Lyapunov exponent and use wavelet packet transform to calculate the rhythm average energy with wavelet energy entropy, then, the extract feature vector is inputted into the binary tree support vector machine (SVM) and the extreme learning machine (ELM), respectively. From the recognition result show that, when use the combination method of feature extraction to solve the problem of feature extraction and classification about this subject acquisition EEG, it's feasible and effective. At the same time, it also provides a new thought and method.

Keywords: EEG, The Maximum Lyapunov Index, Wavelet Packet Transform, ELM

1. Introduction

Today, The Brain-Computer Interface (BCI) technology has become a hot spot for researchers, it's a new control system that can help people to communicate information with the external environment, and it independent on the traditional way of using human brain to normal control the output of neural network and muscle tissue [1]. BCI neeeds to transform the behavior intention of the EEG signal into the control signals of equipment, then achieve the corresponding task of people want to. So It can make the people whose brain is normal but body is disabled through imagining movement to achieve direct communication with the outside world, such as controling computer, remote control cars, wheelchair, etc., thus it has a great value in using.

The EEG processing is the core technology content in the BCI technology, and how to correctly extracted the characteristics of people's behavior intention from the EEG is very critical in the using reasearch. Now most people use a single analysis method, such as AR model power spectrum estimation [2], wavelet transform [3], wavelet packet transform [4] and so on. Due to the complexity of EEG signal, making itself have the characteristics of randomness,

non-stationary, nonlinear, diversity, rhythmicity, et al, only use one method can't describe the signal feature complete, resulting in the poor reliability on the next study. So the combination feature extraction method has received the people's attention. Now, according to the characteristics of EEG and combine with the research methods of EEG characteristics, make a variety of methods mutual fusion to solve the problem of EEG feature extraction is a new train of thought.

This article mainly from the thought of uses the brain electrical to control the object ,such as mouse, toy car and so on, to do simple moving. Collected the four directions EEG signal of the brain imagining the object moving to the 'Upward' , 'Downward', 'Liftward', and 'Rightward'. Then, combine with the rhythm energy and chaotic characteristics of EEG to explore a method of feature extraction. Firstly, do the preprocessing and selecte the channels. Secondly, putting forward a combined feature extraction method, which contain calculation the maximum Lyapunov exponent and use wavelet packet transform to calculate the rhythm average energy with wavelet energy entropy. Finaly, use the two methods of binary tree support vector machine (SVM) and extreme learning machine (ELM) to make recognition analysis about the combination characteristics. From the identify results, the combined feature extraction method has well effectiveness and practicability in the EEG feature extraction.

2. Data Acquisition and Preprocessing

2.1. The Experimental Data Acquisition

In this paper, according to the principle and characteristics of brain electrical produces, use the medical instrument NCERP series of EEG and evoked potentiometer 16 guide EEG acquisition device from Shanghai, the sampling frequency is 128 hz, choose three voluntary health subjects to collect the EEG signal of imagine object to do the four directions movement, respectively, notes for the letter of 'U', 'D', 'L', and 'R'. Acquisition process: subjects relaxation sitting in a comfortable chair, close eyes listen to the external command and imagine the corresponding movement, each direction do 40 times, the sampling time is 6s for one time, the first 1s as prepare, before after 1s as the end of the time, so the effective time is 4s, the total sampling points are 512. Complete the four directions as a set of experiment, the total data sets of 3 volunteers are 120.

2.2. Preprocessing

In the pretreatment is mainly research on signal filtering and channel selection. Combined with the characteristics and the effective research spectrum of EEG, choosing the band-pass filter of combine FFT with IFFT, selecting the mainly useful signal spectrum of 0.5-30 hz and removing interference. Then using the imagine upward direction of C3 channel as example, seeing the effect from the Fig. 1.

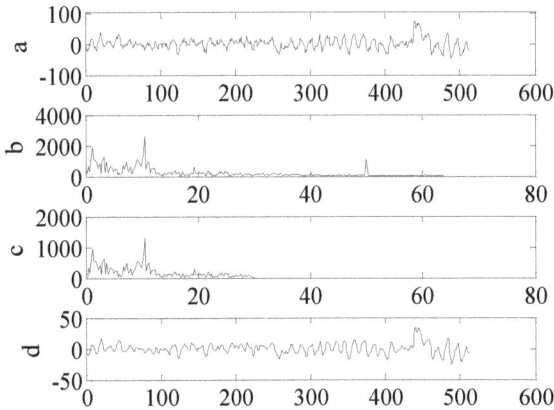

Figure 1. The instance of the imagine upward movement of C3 channel.

In the Fig.1, a is the original figure, b is the spectrum diagram, c is the spectrum diagram after bandpass filter, d is the waveform diagram after IFFT transformation. Seeing from the result that after the bandpass filter, eliminating the power frequency interference, and some low frequency and high frequency noise, improve the signal-to-noise ratio.

According to domestic and international research experience, the signal energy wave will change as soon as the brain to imagine movement, but the change is not obvious in

every channel [5], so in order to lighten the load of research, selecting the research channel. Assume $x(i, j)$ is the j-th point EEG data of the i-th test, then, the average electrical energy of the sample poin j in the total tes N is:

$$\overline{W(j)} = \frac{1}{N}\sum_{i=1}^{N} x^2(i,j) \; i = 1, 2, \cdots, N \; j = 1, 2, \cdots, 512 \qquad (1)$$

when $N = 120$, use the equation (1) to calculate the 16 electrode locations average energy for the four imagine directions, the average energy of each direction in each electrode is shown in Fig. 2.

From the Fig. 2. it is observed that average energy has significant overlaps and saltation in addition to the electrode locations of C3, C4, P3 and P4, and in this four points, the change of average energy is progressive, smooth. At the same time, from the brain anatomy, the central lobe and the parietal lobe are the feeling nervous centralis, so the four channels can be used as research object.

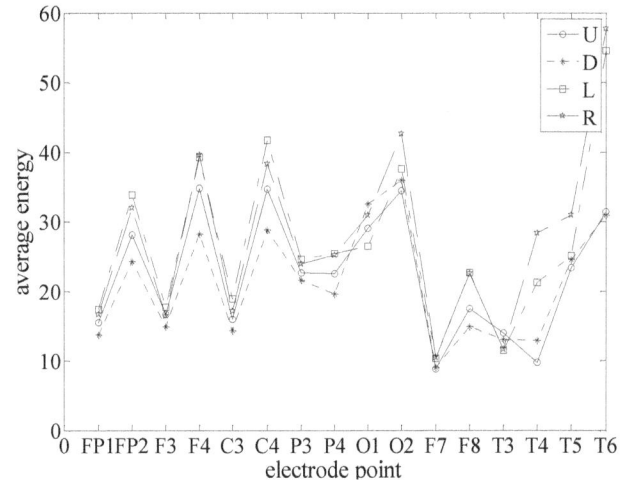

Figure 2. the average energy of the four directions on the 16 electrode locations.

3. Feature Extraction

3.1. Calculat the Maximum Lyapunov Exponents Method

Lyapunov exponent is one of the major parameters to measure a chaotic sex. It can quantitatively describe the average rate of mutual convergence or divergence over time about two tracks on approximate initial state in the system phase space. Combine with the chaotic characteristic of EEG signals, in this article, choose the maximum Lyapunov exponent to describe the characteristics of eeg changes. Under the experience of the previous studies, adopt the minimum volume method to calculate the maximum Lyapunov exponent, and use the C-C algorithm to calculate the two parameters of embedding dimension and optimal delay time [6].

Let $\{x_1, x_2, \cdots, x_N\}$ is the time chaotic sequence, with the delay coordinate method, the phase space reconstruction is given by:

$$X = \left\{ X_i \mid X_i = [x_i, x_{i+\tau}, \cdots, x_{i+(m-1)\tau}]^T \right\} \quad (2)$$

Where $i = 1, 2, \cdots, m$, τ is the delay time, m is the embedding dimension, $M = N - (m-1)\tau$ is the points of phase space. The steps of algorithm are as follow:

Step1: Use the FFT transform to calculate the time series average period T, take $\omega = T / \Delta t$ is separation interval, Δt is the sampling period of time sequence.

Step2: Use the C-C algorithm to calculate the embedding dimension m and optimal delay time τ.

Step3: Reconstruct the phase space by the embedding dimension m and optimal delay time τ.

Step4: Calculate the most near distance $X_{\hat{i}}$ of every X_i point, define the shortest distance d_i is:

$$d_i(0) = \min \|X_i - X_{\hat{i}}\|, |i - \hat{i}| > \omega \quad (3)$$

Step5: Calculate the distance of the point, which close to the each phase space point X_i, after the j-th discrete time step. The distance is:

$$d_i(j) = \|X_{i+j} - X_{\hat{i}+j}\| \quad (4)$$

Where $j = 1, 2, \cdots, \min(n-i, n-\hat{i})$.

Step6: Calculate the maximum Lyapunov exponent, it is defined as:

$$\lambda_1 = \frac{1}{k\Delta t} \frac{1}{(M-k)} \sum_{j=1}^{M-k} \frac{d_j(i+k)}{d_j(i)} \quad (5)$$

The time-varying of the maximum Lyapunov exponent about the four imaginary movement directions of C3 channel are showing in the Fig.3

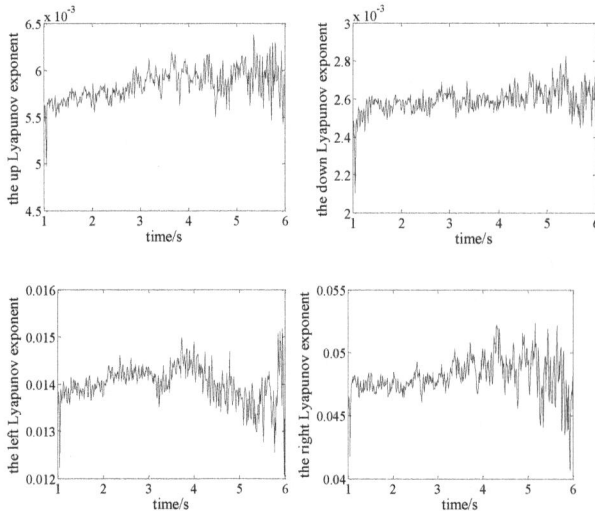

Figure 3. *The time-varying of the maximum Lyapunov exponenthere of C3 channel.*

From the Fig.3, it is noted that each Imagine movement direction has clearly change in the the maximum Lyapunov

exponent graph. After a great deal of experiments simulation, It is found that, in the C3 channel, the maximum Lyapunov exponent of upward imagine movement changes in the range of 0.0044 ~ 0.0048, the downward imagine movement changes in the range of 0.0021~0.0027, the leftward imagine movement changes in the range of 0.0121~0.0146, and the rightward imagine movement changes in the range of 0.0418~0.0501. Thus, it is also found that the chaos characteristic of the four imagine movement directions EEG signal has certain distinction in the C3 channel. Although there may be overlap part through large amounts of data test, it can still as the distinction between the four types of signals. So the the maximum Lyapunov exponent can be used to describe the feature of the four imagine movement EEG signal.

3.2. Wavelet Packet Transform Feature Extraction

The EEG signal consists of four kinds of rhythm wave, there will obviously shown a different rhythm transformation, as soon as the brain made any physiology. The wavelet packet transform is an improvement of the wavelet transform, it decomposes the signal in level and can focus on all scope, According to these reasons, it can be used to extract different rhythms of EEG signal and overcome the shortcomings of traditional spectrum analysis.

Combine with the rhythm characteristic of EEG, choose the db6 wave to decompose the EEG singnal at 4 levels [7], getting four kinds of EEG rhythm wave, there are $\delta(0.5 \sim 4\text{Hz})$, $\theta(4 \sim 8\text{Hz})$, $\alpha(8 \sim 12\text{Hz})$ and $\beta(12 \sim 30\text{Hz})$, choose one trial as example, the change of four rhythm waves on the C3 channel are showing in the Fig.4.

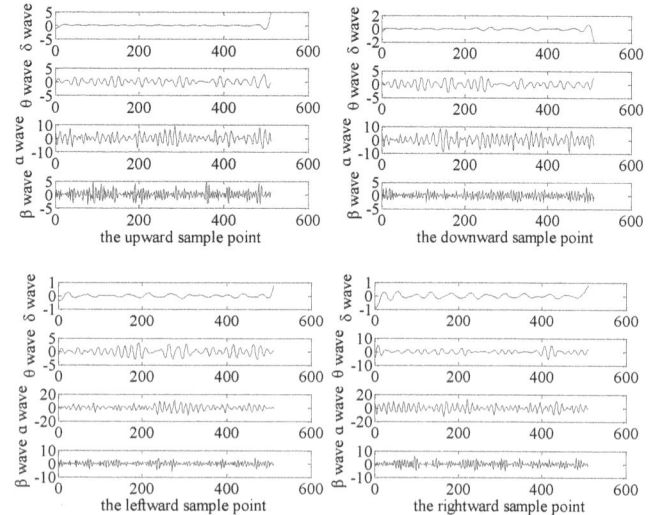

Figure 4. *The four kinds of rhythm waveform of C3 channel.*

From the Fig.4, it is observed that the four rhythm waves have clearly change in the four Imagine movement directions, so the rhythm can be used to describe the EEG characteristic.

Then, claculate the average energy as the characteristics of each direction rhythm wave, the wavelet rhythm average energy is defined as:

$$E_i = \frac{1}{N}\sum_{l=1}^{N}|d_i|^2 \ , i = 1,2,3,4 \qquad (6)$$

Where N is the total sample points, E_i is the average energy of the i-th rhythm wave. set $E = E_1 + E_2 + E_3 + E_4$, then, after normalized process, the normalized vector was become $T = [E_1 / E, E_2 / E, E_3 / E, E_4 / E]$.

Through the study of entropy, found that the wavelet packet energy entropy can also reflect the wave energy changes [8]. So in this paper, increasing the solving of wavelet packet energy entropy, it is defined as:

$$E = \sum_n E(j,n) \qquad (7)$$

$$p_n = E_n / E \qquad (8)$$

$$WE(P) = -\sum_n p_n \ln(p_n) \qquad (9)$$

Where n is the n-th subspace in the j-th wave layer, E is the total energy, P_n is the signal energy distribution probability in the subspace.

Use the equation (6), (7), (8) and (9) to calculate the overall wavelet rhythm average energy and wavelet packet energy entropy of the total 120 samples in the four directions of C3 channel, they are shown in the Table 1.

Table 1. *the four kinds of rhythm wave energy mean(mean ±standard deviation) and wavelet packet energy entropy of C3 channel.*

	U	D	L	R
δ wave	0.0058±0.0050	0.0069±0.0080	0.0034±0.0045	0.0050±0.0022
θ wave	0.1129±0.0297	0.1360±0.0321	0.0992±0.0216	0.1052±0.0443
α wave	0.7571±0.0434	0.7224±0.0695	0.7681±0.0333	0.7562±0.0442
β wave	0.1242±0.0283	0.1347±0.0440	0.1293±0.0165	0.1336±0.0215
energy entropy	0.7460	0.7106	0.7158	0.7436

From the Table 1, the energy change of α wave is conform to the EEG characteristics of close eyes, it is the largest energy value among the four rhythm waves. From the energy entropy value, the complexit of the four imagine movement directions EEG signal have few difierence. So they can be used to describe the EEG characteristic.

3.3. Combination Feature Extraction Method of EEG Signal

According to the above statement, computing the maximum Lyapunov exponent as 4 dimensionalities characteristics, combined with the wavelet packet transform method of 20 dimensionalities characteristics, consisting of 24 dimensionalities signal combination characteristics. The combination characteristics vector of the upward imagine movement EEG signal are shown in Table 2.

Table 2. *the combination characteristic value of the upward movement EEG.*

channel	Lyapunov exponent	δ wave	θ wave	α wave	β wave	energy entropy
C3	0.0041	0.0060	0.1332	0.7555	0.1409	0.7457
C4	0.0043	0.0052	0.1323	0.7415	0.1387	0.7142
P3	0.0042	0.0046	0.1311	0.7089	0.1322	0.7336
P4	0.0046	0.0042	0.1401	0.7033	0.1368	0.7425

4. Pattern Classification Verification

Based on the multiple classification problems, in consideration of the nonlinear features of EEG signal, select the support vector machine (SVM) and the extreme learning machine (ELM) to test whether the combined feature extraction method is effective for identification authentication or not, the conclusion is from the the recognition results.

The SVM is designed by the basis of the structure risk minimization combines with the statistical theory, through constructing the optimal hyperplane to realize the final minimum classification error of the unknown samples. When solve the multiple classification problems, it needs to construct multiple suitable SVM [9]. In this article, use the binary tree support vector machine, the kernel function is radial basis function. The test steps: Firstly, use the datas of 'U' and 'D' to train the SVM1; Secondly, use the combination datas of 'U' and 'D' with 'L' to train the SVM2, Thirdly, using the combination datas of 'U', 'D', and 'L' with 'R' to train the SVM3, then, complete the train and obtained a multiple SVM classification, it is confirmed effective after done simulation test. When testing, start from the SVM3 classifier and then tested samples resulto until the symbols of decision functions is positive, the signal category is comfirmed at the end.

The ELM is a typical single hidden layer feedforward neural network, which has simple operation and effective characteristic [10]. It only needs to set the number of hidden layer nodes and don't need to adjust the network weights of the input and hidden bias among the testing, and produces the unique optimal solution. Compared with the SVM classifier, in the classification problems, the ELM classifier don't need to build multiple classifier, it can be realized by setting a single neural network with multi-output votes. In this article, selecting the sigmoidal excitation function to identify the samples.

5. Results and Analysis

Selecting 120 samples of three volunteers at each imagine movement direction to do simulation test, choice 90 samples as training samples at random, the rest are testing samples. Use the two classifiers to identify the characteristic vector from the three feature extraction methods, respectively. The identify results are shown in the following table.

Table 3. The identify results of three feature extraction methods in the two classifiers(%).

classifier		U	D	L	R
SVM	Lyapunov exponent	64.3	68.5	76.6	74.3
	wavelet packet	73.4	70	74.6	73.3
	combination feature	78.4	79.2	81.7	83.9
ELM	Lyapunov exponent	68.3	70	76.8	75.2
	wavelet packet	77.4	72.6	75	73.6
	combination feature	80.4	81.6	87.6	83.3

Table 4. The average identify rate (%) and the training time (s) of the combination feature extraction method in the two classifiers.

classifier	combination feature method	
	average identify rate	training time
SVM	80.8	17.265
ELM	83.225	13.168

From the Table 3 and Table 4, compared with the two method of the maximum Lyapunov exponent and the wavelet packet transform, the combination feature method has preferable identify resuls. Meanwhile, it also proves that the single method of feature extraction can't describe the characteristics of the EEG signal very well, while the combination feature method can do it and has good results. Besides that it also shown the advantage of choosing identify in solving multiple classification problem, compared with the SVM, the ELM has good quickness and accuracy features. The method of this paper can provide a new thought the brain-computer interface for online learnig.

6. Conclusion

Along with the advancing application value of the brain-computer interface, it is very important to find a effective feature extraction way to solve the EEG signal preoblem. In this paper, gives a thought of using the maximum Lyapunov exponent combines with the wavelet packet transform to extracting the characteristics of the imagine movement directions, at the same time, uses two difierent identifies to prove the combination feature method is reliability and accuracy. In addition to do these, in order to reduce the blindness and workload of research, the paper also gives a method to selecting research channels. In general, the thought of this article can provide a new inspire in the feature extraction of EEG signal.

Acknowledgment

The author would like to thank Drs. Jason Gu for providing the theoretical support, and also thank to XIE Yun for providing sample datas.

References

[1] J. R. Wolpaw, N. Birbaumer, D. J. McFarland, G. Pfurtscheller, T. M. Vaughan, "Brain-Computer Interfaces for Communication and Control", Clinical Neurophisiology, vol. 113, pp. 767-791, 2002.

[2] A. Subasi, M. K. Kiymikl, A. Alkan, E. Koklukaya, "Neural network classification of EEG signals by using AR with MLE preprocessing for epileptic seizure detection", Math Comput Appl, vol. 10(1), pp. 57-70, April 2005.

[3] D. Cvetkovic, E. D. Ubeyli, I. Cosic, "Wavelet transform feature extraction from human PPG, ECG, and EEG signal responses to ELF PEMF exposures: A pilot study", Digital Signal Process Rev J, vol. 18(5), pp. 861-874, September 2008.

[4] Yang Banghua, Liu Li, Zan Peng, Lu Wenyu, "Wavelet packet-based feature extraction for brain-computer interfaces", Lect. Notes Comput. Sci., vol. 6330 LNBI(PART 3) , pp.19-26, 2010.

[5] Saha Anuradha, Konar Amita, Ralescu Anca, Nagar Atulya K., "EEG analysis for olfactory perceptual-ability measurement using a recurrent neural classifier", IEEE Trans. Human Mach. Syst., vol. 44(6), pp. 717-730, December, 2014.

[6] F. Shayegh, S. Sadri, R. Amirfattahi, K. Ansari-Asl, "A model-based method for computation of correlation dimension, Lyapunov exponents and synchronization from depth-EEG signals", Comput. Methods Programs Biomed., vol. 113(1), pp. 323-337, 2014.

[7] Yang Renhuan, Song Aiguo, Xu Baoguo, "Analysis of EEG basic rhythms based on discrete harmonic wavelet packet transform", Dongnan Daxue Xuebao, vol. 38(6), pp. 996-999, November 2008.

[8] Sun Yuge, Ye Ning, Xu Xinhe, "The feature extraction and recognition of EEG based on wavelet entropy and distance", Chinese Contr. Decis. Conf., CCDC, 2008, pp. 4294-4298

[9] Chen Shanshan, Meng Qingfang, Zhou Weidong, Yang Xinghai, "Seizure detection in clinical EEG based on multi-feature integration and SVM", Lect. Notes Comput. Sci., vol. 7996 LNAI, pp. 418-426, 2013.

[10] Yuan Qi, Zhou Weidong, Li Shufang, Cai Dongmei, "Approach of EEG detection based on ELM and approximate entropy", Yi Qi Yi Biao Xue Bao, vol. 33(3), pp. 514-519, March 2012.

3

Integrated Supply Chain Network Model for Allocating LPG in a Closed Distribution System

Amelia Santoso*, **Dina Natalia Prayogo, Joniarto Parung**

Industrial Engineering Department, University of Surabaya, Surabaya, Indonesia

Email address:

amelia@staff.ubaya.ac.id (A. Santoso), dnprayogo@staff.ubaya.ac.id (D. N. Prayogo), jparung@staff.ubaya.ac.id (J. Parung)

Abstract: This paper proposes a model of integrated supply chain network for allocating subsidized Liquefied Petroleum Gas (LPG) in a closed distribution system. Subsidized LPG is selected as a case study due to its specific product in Indonesia. Since 2007, the Indonesian government makes policy, namely energy conversion from kerosene to LPG. The main purpose of converting kerosene to LPG is to reduce subsidies on fuel oil. The distribution system consists of several filling stations, distributors and retailers. Currently, the distribution of subsidized LPG, does not flow smoothly because there will be a shortage or excess tubes in retailers mainly because it uses a closed distribution system. A closed distribution means that people who are eligible to buy subsidized LPG will be given a card for identifying them as a legal receiver of the LPG. The model is developed using mathematical approach with reference to previous transshipment study. Based on the developed model and by using a numerical example as a case study, the allocation of LPG from filling station to the distributors and from the distributor to the retailers with minimum distribution costs can be determined. LPG in some specific retailers is supplied by only one distributor which is authorized to distribute subsidized LPG on the retailers. However, this model has limitations to arrange the route filling and distribution route.

Keywords: Supply Chain Network, Subsidized LPG, Closed Distribution System

1. Introduction

Since 2007, the Indonesian government makes policy, namely energy conversion from kerosene to LPG. The main purpose of converting kerosene to LPG is to reduce subsidies on fuel oil. During this time, kerosene, which has a high production cost is consumed by the majority of low-income communities which are concentrated in rural areas. Therefore government provides subsidies to ease the burden of their energy costs.

LPG starter pack in the form of a gas stove, a tube with its accessories has been distributed in total of more than 56 million packs in 29 provinces in Indonesia. The problem faced now is when and where we doing the refill.

Smoothing material flow is one of the goals in the concept of supply chain, which consists of several echelons [1]. Likewise in the distribution system, if the flow of LPG distribution does not go smoothly, there will be a shortage or excess in retailers mainly because it uses a closed distribution system. A closed distribution means that people who are eligible to buy subsidized LPG will be

given a card for identifying them as a legal receiver of the LPG [2].

The lack of proper LPG's distribution can be caused by faulty allocation of LPG's distribution under the authority and responsibility of filling depots or distributors. Hence, in order to overcome the problems, it is necessary to redesign distribution network system. This distribution network design serves as an input to the government in making policy on LPG distribution system based on the real conditions of the field. In other words, we need the concept of distribution network design to manage the allocation of multi LPG filling depot to distributor and from distributor to retailers.

Figure 1 below shows the current down-stream LPG supply chain (distribution). Government applies closed distribution system for distributing subsidized LPG tubes or canisters in order to ensure the subsidized LPG would reach the proper targets.

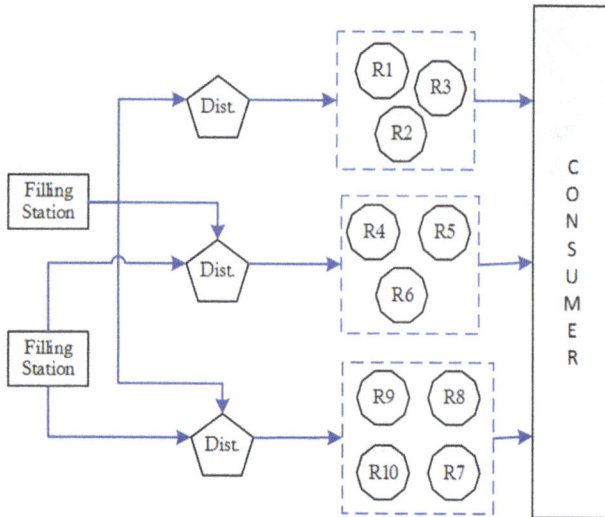

Figure 1. *LPG supply chain (distribution).*

2. Related Work

There are many distribution network designs in the literatures which concern with interaction among member of supply chain. Most of the interaction treats each member of the supply chain as a separate system. As a result, many of the problems solved with minimum integrated [3]. Here, we present previous study which is associated with the main objective of the research.

The main objective of this closed distribution network design is to minimize the total distribution cost. The total distribution cost per year consists of total distribution cost from filling stations to distributors and from distributors to retailers. Generally, network design covers supply allocation, and selecting location of supply chain members in the public and private economic sectors. Distribution network design relates to real situations where an organization needs to get the most effective and efficient distribution facilities [4]. According to Meng, Huang, and Cheu, the integration of location decisions with other relevant decisions is a basic feature that distribution design has to capture in order to support decision-making involvement in strategic supply chain planning [5].

According to Melo *et al*, [6] a company's distribution network must meet service goals at the lowest possible cost. In some instances, a company may be able to save millions of dollars in logistics costs and simultaneously improve service levels by redesigning its distribution network. To achieve this, an ideal network must have the optimum number, size, and location of facilities.

As already presented in the introduction, that because of the LPG distribution system does not run smoothly; it is necessary to redesign the distribution network of LPG. Distribution Networks is needed to be redesigned for the purpose of allocations from filling station to the distributor and from the distributor to the retailer. One of the main models that can be used is the transshipment models. Transshipment problem which was first introduced by Orden [7] refers to a development of the transportation problem by

considering the possibility of transshipment. The point is that any shipping or receiving point is permitted as an intermediate point. At the transshipment problem, an origin or destination can transport subsidized LPG to another origin or destination [8].

Development models will take into consideration the concept of transshipment [9] the model uses the concept of fixed and variable costs that proposed by Chopra and Meindl. The design of this network distribution aims to produce low distribution costs as proposed by Watson et al. [10]. The model developed is composed of 1). LPG allocations from filling station to the distributor and from the distributor to the retailer. 2) the size of the vehicle and the number of orders by distributors to the filling station.

The next part will present the development model based on this transshipment problem.

3. Research Methodology

This research using analytical based methodology to answer the questions: how to allocate, and what are the number of allocation of filled tubes from multi filling station to certain multi distributor and from each distributor to certain retailers in order to minimize total distribution cost per year.

Method of building model is as follows: Firstly, previous related work namely transshipment is analyzed and then developed with mathematical approach to create a new mathematical model. Secondly, the new distribution network design is tested using numerical example with real data as single case study problem. Based on this approach, the research can make a conclusion about model and giving several suggestions for future research.

4. Development Model

The integrated supply chain network model is developed for distributing subsidized 3-kg LPG tubes from filling stations to distributors and from distributors to retailers. This model determines number of allocation from multi filling station to certain multi distributor and from each distributor to certain retailers.

In this model, a filling station supplies multi distributor and a distributor can be supplied by more than one filling station. A distributor supplies multi retailer but only a certain distributor can supply a retailer.

Each distributor has a number of trucks with a number of empty tubes in the truck that will be filled by a filling station according to quota. After all empty tubes in a truck are filled, the truck directly distributes the LPG tubes to multi certain retailers of the distributor.

4.1. Mathematical Notations

The mathematical notations are used in developing model as follows:

Indices

s: Filling station s=1..S

a: Distributor a=1..A

p: Retailer $p=1..P$

Decision Variables

Q_{sa}^{fd}: Number of LPG tubes that are supplied by filling station s to distributor a

Q_{ap}^{dr}: Number of LPG tubes that are supplied by distributor a to retailer p

Z_{sa}^{fd}: 1 if filling station s supplies distributor a, and 0 otherwise

Z_{ap}^{dr}: 1 if distributor a supplies retailer p, and 0 otherwise

JHT_{sa}: Number of day-trucks of subsidized LPG that are supplied from filling station s to distributor a

Variables/Parameters

K_s: Capacity of filling station s

JT_a: Number of trucks owned by distributor a

D_p: Monthly LPG demand of retailer p

FC_{sa}^{fd}: Fixed cost of distributing LPG from filling station s to distributor a

VC_{sa}^{fd}: Variable cost of distributing LPG per tubes from filling station s to distributor a

FC_{ap}^{dr}: fixed cost of distributing LPG from distributor a to retailer p

VC_{ap}^{dr}: Variable cost of distributing LPG per tubes from distributor a to retailer p

LO: the order size contract of subsidized LPG per day between certain distributor and specific filling station

$JMLH$: number of days per month

$MINJHT$: minimum number of day-trucks that are used from filling station to distributor

4.2. Mathematic Formulation

The objective function is to minimize total cost of LPG supply chain. The total cost consists of fixed cost and variable cost at fulfilling station, distributor and retailer.

$$\min \quad \begin{aligned} TC = {} & \sum_s \sum_a FC_{sa}^{fd} Z_{sa}^{fd} + \sum_s \sum_a VC_{sa}^{fd} Q_{sa}^{fd} \\ & + \sum_a \sum_p FC_{ap}^{dr} Z_{ap}^{dr} + \sum_a \sum_p VC_{ap}^{dr} Q_{ap}^{dr} \end{aligned} \quad (1)$$

This model was developed by considering some constraints to ensure the model according to the condition of the distribution of subsidized LPG.

$$\sum_a Q_{sa}^{fd} \leq K_s \quad ; \forall s \quad (2)$$

$$Q_{sa}^{fd} \leq LO\, JMLH\, JT_a Z_{sa}^{fd} \quad ; \forall s, a \quad (3)$$

$$Q_{sa}^{fd} = LO\, JHT_{sa} \quad ; \forall s, a \quad (4)$$

$$JHT_{sa} \geq MINJHT\, Z_{sa}^{fd} \quad ; \forall s, a \quad (5)$$

$$JHT_{sa} \leq JMLH\, JT_a \quad ; \forall s, a \quad (6)$$

$$\sum_s Q_{sa}^{fd} \geq \sum_p Q_{ap}^{dr} \quad ; \forall a \quad (7)$$

$$Q_{ap}^{dr} \leq LO\, JMLH\, Z_{ap}^{dr} \quad ; \forall a, p \quad (8)$$

$$\sum_a Z_{ap}^{dr} = 1 \quad ; \forall p \quad (9)$$

$$Q_{ap}^{dr} = D_p Z_{ap}^{dr} \quad \forall a, p \quad (10)$$

$$\sum_p Q_{ap}^{dr} \leq LO\, JMLH\, JT_a \quad ; \forall a \quad (11)$$

$$Z_{sa}^{fd} \in \{0,1\} \quad \forall s, a \quad (12)$$

$$Z_{ap}^{dr} \in \{0,1\} \quad \forall a, p \quad (13)$$

$$Q_{sa}^{fd} \geq 0\ \&\ integer \quad ; \forall s, a \quad (14)$$

$$Q_{ap}^{dr} \geq 0\ \&\ integer \quad ; \forall a, p \quad (15)$$

$$JHT_{sa} \geq 0\ \&\ integer \quad ; \forall s, a \quad (16)$$

Constraint (2) ensures each filling station never distributes subsidized LPG tubes to their distributor more than its capacity. Constraint (3) ensures there is never a supply from the filling station to the distributor exceeds the capacity of all truck owned by the distributor. The next constraint (4) guarantees number of LPG tubes are filled and supplied from a filling station to a distributor must be a multiple of the vehicle capacity in accordance with their contracts (LO). Constraint (5) deals with the contract between a filling station and a distributor has to be equal to or greater than minimum day-trucks are used. Number of day-trucks is guaranteed not to be greater than total trucks per month that owned by each distributor (constraint 6). Constraint (7) guarantees the supply balance so that the distributor has no inventory. Constraint (8) ensures a retailer can only be supplied by a distributor that has been decided as suppliers while constraint (9) guarantees each retailer is only supplied by one certain distributor. Constraint (10) and (11) ensure the amount of allocation from a distributor to a retailer is equal to demand of the retailer and total of all allocation from a distributor to all retailer do not greater than its all vehicle capacity. Constraint (12) and (13) guarantee two decision variables are binary while constraint (14), (15) and (16) guarantee the last three decision variables have to be integer and always greater than zero.

5. Numerical Example

Supply chain structure of 3-kg subsidized LPG consists of two filling station, four distributors and 77 retailers. The capacity of filling stations and distributors can be seen in the following Table.

Table 1. Capacity of filling.

filling station	capacity	distributor	number of trucks	capacity of truck
F1	84,000	D1	2	28,000
F2	67,200	D2	3	42,000
		D3	2	28,000
		D4	3	42,000

Table 2. Demand of each retailer.

retailer	demand	retailer	demand	retailer	demand
P1	2995	P26	2643	P51	2163
P2	2085	P27	3199	P52	827
P3	2228	P28	2776	P53	1514

retailer	demand	retailer	demand	retailer	demand
P4	2497	P29	2758	P54	1020
P5	885	P30	761	P55	1610
P6	1931	P31	3093	P56	1441
P7	2683	P32	3093	P57	2378
P8	100	P33	2115	P58	763
P9	891	P34	3343	P59	1372
P10	943	P35	979	P60	2364
P11	113	P36	3040	P61	2450
P12	2189	P37	179	P62	2032
P13	1690	P38	1680	P63	1793
P14	2097	P39	850	P64	2623
P15	1442	P40	246	P65	2772
P16	2123	P41	425	P66	648
P17	1861	P42	1988	P67	2388
P18	2655	P43	1777	P68	1778
P19	1396	P44	2887	P69	3245
P20	2485	P45	334	P70	2254
P21	3061	P46	702	P71	2643
P22	1254	P47	2875	P72	1036
P23	1640	P48	2841	P73	189
P24	1346	P49	3113	P74	917
P25	2523	P50	100	P75	1223
				P76	1734
				P77	1517

Table 3. *Fixed cost of distribution from filling station to distributor.*

	D1	D2	D3	D4
F1	144,000,000	142,000,000	129,000,000	143,000,000
F2	145,000,000	141,000,000	143,000,000	127,000,000

Table 4. *Fixed cost of distribution from filling station to distributor.*

	D1	D2	D3	D4
F1	78,000	77,000	57,000	52,000
F2	51,000	58,000	59,000	69,000

With fixed and variable cost from distributor to retailer, the following result is obtained.

Table 5. *Distribution allocation from filling station to distributor.*

	D1	D2	D3	D4
F1	0	11200	28000	42000
F2	28000	30800	0	0
TOTAL	28000	42000	28000	42000

Table 6. *Number of day-trucks of subsidized LPG that are supplied from filling station to distributor.*

JHT	Distributor			
	1	2	3	4
F1	0	20	50	75
F2	50	55	0	0
TOTAL	50	75	50	75

Table 7. *Detail distribution allocation from filling station to distributor and distributor to retailer.*

Filling station	QSA	Distributor	QAP	retailer
F1	42000	Dist 4	2189	P12
			1442	P15
			2123	P16
			1861	P17
			3093	P31
			3093	P32
			3040	P36
			179	P37
			1777	P43
			334	P45
			2841	P48
			100	P50
			1514	P53
			1020	P54
			1441	P56
			2032	P62
			1793	P63
			2772	P65
			3245	P69
			2254	P70
			917	P74
			1223	P75
			1517	P77
		TOTAL	41800	23
F 1	28000	Dist 3	2085	P2
			100	P8
			891	P9
			113	P11
			2097	P14
			1396	P19
			2776	P28
			761	P30
			3343	P34
			425	P41
			2887	P44
			3113	P49
			1372	P59
			2364	P60
			648	P66
			1778	P68
			1734	P76
		TOTAL	27883	17
F 1	11200	dist 2	2497	P4
			2683	P7
			943	P10
			1690	P13
			3061	P21
F 2	30800	Dist 2	1254	P22
			2643	P26
			3199	P27
			2115	P33
			979	P35
			850	P39
			246	P40
			702	P46
			2875	P47
			2163	P51
			827	P52
			2378	P57
			763	P58
			2450	P61
			2623	P64
			2388	P67

Filling station	QSA	Distributor	QAP	retailer
			2643	P71
	42000	TOTAL	41972	22
F 2	28000	Dist 1	2995	P1
			2228	P3
			885	P5
			1931	P6
			2655	P18
			2485	P20
			1640	P23
			1346	P24
			2523	P25
			2758	P29
			1680	P38
			1988	P42
			1610	P55
			1036	P72
			189	P73
		TOTAL	27949	15

6. Conclusion

Based on the developed model, and by using a numerical example as a case study, the allocation of LPG from filling station to the distributor and from the distributor to the retailer with minimum distribution costs can be determined. Every retailer can be supplied by only one distributor which is authorized to distribute subsidized LPG on the retailer. It means retailers cannot be supplied by other distributors. Distributors can only fill an empty tube on the filling station that is authorized to supply distributor.

The developed model has been able to establish the allocation of filling stations that will supply a particular distributor. The model has also been able to establish which distributor that will supply a particular retailer. Based on the developed model, and by using a numerical example as a case study, the allocation of LPG from filling station to the distributor and from the distributor to the retailer with minimum distribution costs can be determined. LPG in some specific retailers is supplied by only one distributor which is authorized to distribute subsidized LPG on the retailers.

The model has been able to establish the allocation of filling stations that will supply a particular distributor. The model has also been able to establish which distributor that will supply particular retailers. However, this model has limitations to arrange the route filling and distribution route. This initial model will be developed in further research to establish the fleet distributors' route.

Acknowledgements

We would like to thank the Ministry of Research, Technology and Higher Education, which provides financial grants so that our research can be done.

References

[1] Chopra, S., & Meindl, P.," Supply Chain Management: Strategy, Planning and Operation", 5th Edition, 2013, Prentice-Hall.

[2] R. D. Fadillah, "LPG Stock 'Safe' Despite Scarcity in Market," Retrieve June 29, 2012, from The Jakarta Post: http://m.thejakartapost.com/news/2012/06/04/lpg-stock-safe-despite-scarcity-market.html

[3] Cohen M.A, Hau.l., "Strategic Analysis of Integrated Production-Distribution Systems: Models and Methods", 1988, Operations Research, Vol. 36, No. 2.

[4] M. Hlyal, A. Ait Bassou, A. Soulhi, J. El Alami, N. El Alami, "Designing A Distribution Network Using A Two Level Capacity Location Allocation Problem: Formulation And Efficient Genetic Algorithm Resolution With An Application To A Moroccan Retail Company", 2015 Journal of Theoretical and Applied Information Technology. Vol.72 No.2.

[5] Q. Meng, Y. Huang, and R. L. Cheu, "Competitive facility location on decentralized supply chains," 2009, Eur. J. Oper. Res., vol. 196.

[6] Melo M.T., S. Nickel and F. Saldanha-da-Gama, "Network Design Decisions in Supply Chain Planning, 2001, Fraunhofer-Institut fur Techno- und Wirtschaftsmathematik ITWM Fraunhofer-Platz 1.

[7] Orden A (1956) "Transshipment problem", 1956. Management Science, Vol.2, Issue.3.

[8] Here Y.T., M. Tzur and EY, Cesan, "The multilocation transshipment problem", 2006, IIE Transactions Vol.38.

[9] Benham Malakooti, "Operation and Production systems with multiple objectives, Chapter 9 supply chain and transportation", 2014. John Wiley and Sons, Inc., New Jersey.

[10] Watson M., S. Lewis, P. Cacioppi and J. Jayaraman., "Supply Chain Network Design: Applying Optimization and Analytics to the Global Supply Chain", 2013. FT Press, New Jersey.

4

The Estimation of Thin Film Properties by Neural Network

Chi-Yen Shen[1], Yu-Ju Chen[2], Shuming T. Wang[1], Chuo-Yean Chang[3], Rey-Chue Hwang[1, *]

[1]Electrical Engineering Department, I-Shou University, Kaohsiung City, Taiwan

[2]Information Management Department, Cheng-Shiu University, Kaohsiung City, Taiwan

[3]Electrical Engineering Department, Cheng-Shiu University, Kaohsiung City, Taiwan

Email address:

cyshen@isu.edu.tw (Chi-Yen Shen), yjchen@isu.edu.tw (Yu-Ju Chen), smwang@isu.edu.tw (S. T. Wang),
cychang@csu.edu.tw (Chuo-Yean Chang), rchwang@isu.edu.tw (Rey-Chue Hwang)

[*]Corresponding author

Abstract: This paper presents a method based on neural network (NN) for estimating the properties of semiconductor thin film. Through the effective learning process, NN is able to catch the relationship between input and output pairs bypassing the complicated statistical steps such as model hypothesis, identification, estimation of model parameters, and verification. Such an estimator then can be developed to be a smart mechanism which can help the technician to set the relevant control parameters in the manufacturing process of thin film. In this research, the thickness and refractive index (RI) of thin film were estimated by the well learned NN model. From the studied results shown, the properties of thin film indeed could be estimated in advance according to the relevant control parameters in the manufacturing process. That also means the estimator we developed could be built and fulfilled its function.

Keywords: Neural Network, Thin Film, Manufacturing Process

1. Introduction

In recent twenty years, the high-tech skills have been developed widely and vitally in various electronic appliances, such as photoelectric, semiconductor and biomedical chips. The electronic industry has become the focus of economic development for many countries. Besides, due to the fast improvement of manufacturing technique, more and more electronic products are requested to be small and exquisite. Their function is requested to be more powerful either.

It is well-known that thin film is an important and indispensable part for many electronic products. Taking the wafer manufacturing as an example, the filming process plays an important and key role in the wafer front-end manufacturing step. Figure 1 shows the flowchart of wafer manufacturing process [1]. Thus, if the relevant manufacturing parameters of filming process could be set quickly and accurately, not only the efficiency of working machine can be greatly improved, but also the time and frequency of machine test can be reduced effectively.

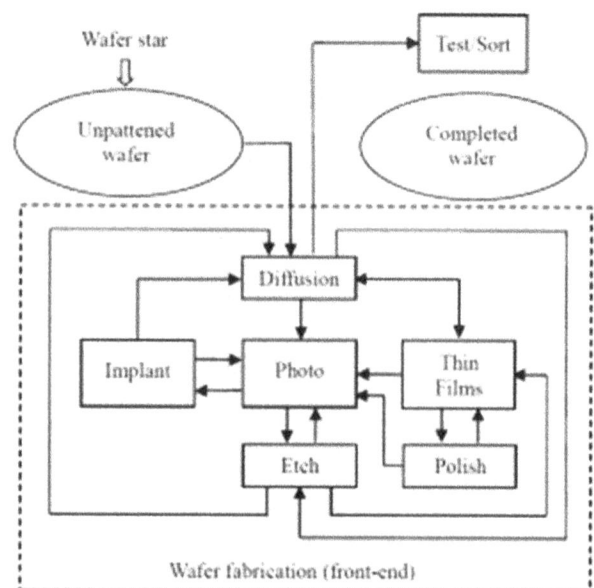

Figure 1. The manufacturing process of wafer.

Generally, Chemical Vapor Deposition (CVD) is the popular method used in the field of thin film manufacturing process. In CVD process, many complicated and nonlinear chemical and physical reactions are hardly analyzed. The phenomena of particle drift and variation are happened very often in the filming process either. Undoubtedly, these uncontrolled factors will affect the quality of filming process very seriously. Thus, how to set the adequate manufacturing parameters for improving the yield rate and reducing the times of machine test has become the most important work in the thin film manufacturing process. Unfortunately, trial-and-error is still the common method taken by the technicians in many companies. As we know, the number of failed manufacturing process and the defective product could be possibly raised, if the manufacturing parameters were determined by the technician based on personal experience only. Thus, several studies about the optimal thin film manufacturing have been proposed [2-5].

Recently, due to the fast development of artificial intelligent techniques, some studies about the optimal control of film manufacturing process were reported. For instance, the genetic algorithm (GA) had been used for searching the optimal parameters of physical vapor deposition manufacturing control process [4]. Hsieh, Tong, Cui and Hwang et al. proposed the film's control and estimation by using NN techniques [5-14].

Since the powerful learning and modeling capabilities, NN has been widely used in different applications, such as the signal processing and control [15-19]. Basically, through the well learning, NN could generate an efficient mapping between input and output pairs bypassing the complicated statistical steps. The well-trained NN model then can be used for the specific work.

In this research, an artificial intelligent (AI) system based on NN model for the estimation of film properties is studied. Its aim is that according to the estimation information provided, the junior technician with no full experience is able to make a good setting work for the manufacturing parameters in the filming process. Thus, such an AI system can not only help the technician to do the work of film manufacturing very efficiently and easily, but also reduce the rate of defective products and then save the production cost.

2. Neural Network

NN technique is the main tool used for constructing the estimator of film properties. As previous descriptions, the relationship between input and output pairs is expected to be obtained through the well learning of NN.

The NN structure commonly known as multi-layered feed-forward network is used in this study. The supervised NN with error back-propagation (BP) learning algorithm is taken for NN's training [15-17]. An example of a three-layered feed-forward NN architecture as shown in Figure 2 is the model of selected topology. Each layer is connected to a layer above it in a feed-forward manner, which means no feed-back from the same layer or a layer

above. All connections have a multiplying weight associated with them. Training is equivalent to find the proper weights for all connections such that a desired output is generated for a given input set. Once the neural network is well trained, the proper input information could be inferred in accordance with an expected output. In other words, the useful information can be found for helping the technician to do the well control in the manufacturing process.

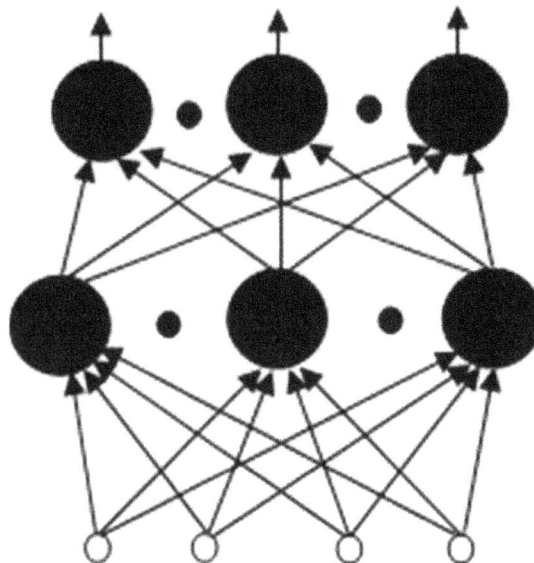

Figure 2. A three-layered feed-forward NN architecture.

In this study, the error back-propagation (BP) learning algorithm is used for NN's training. The major steps of BP learning rule algorithm is briefly summarized as follows [15-17].

1st step: Initialize all weights (ω_{ij}) to the small random values firstly.

2nd step: Present an input pattern with the corresponding desired outputs and then calculate the outputs.

3rd step: Find the error term for all nodes.

4th step: Adjust weights by

$$\omega_{ij}(n+1) = \omega_{ij}(n) + \alpha\delta_j X_i + \zeta(\omega_{ij}(n) - \omega_{ij}(n-1)) \qquad (1)$$

where n+1, n and n-1 are the next, present, and previous iteration numbers, respectively. δ_j is the error of node j and X_i is the ith input of node j. α is the learning rate, the step size in the gradient search algorithm. ζ is the momentum and its value is between 0 and 1.

5th step: Present another input pattern and go back to 2nd step.

3. Experiments

In our study, the thin film data manufactured by using two recipes, LDRXX and TEOSXX, were collected and simulated. Table 1 and Table 2 present the examples of two collected data sets. The numbers of LDRXX and TEOSXX data sets are 102 and 89, respectively.

Table 1. The examples of data manufactured by using recipe LDRXX.

Manufacturing Parameters							
DT	cSHDist	TEOS factor	TEOS (mgm)	HE (sccm)	O_2 (sccm)	HFRF (pre)	LFRF (pre)
25.5	420	0.9736	800	9000	8000	196	50.4
25.5	420	0.9736	800	9000	8000	196	51.4
26.5	420	0.9736	800	9000	8000	196	50.4
26.5	420	0.9736	800	9000	8000	196	51.4
26.5	420	1.0453	800	9000	8000	194	51.2
26.5	420	1.0453	800	9000	8000	193	50.4
25.5	420	1.0253	800	9000	8000	194	31.2
25.5	420	1.0253	800	9000	8000	164	51.2
27.5	420	0.9936	800	9000	8000	196	50.4
27.5	420	0.9936	800	9000	8000	216	50.4

Thin Film Properties	
Thickness	RI
443	1.4590
433	1.4590
455	1.4619
443	1.4627
457	1.4609
429	1.4624
419	1.4816
416	1.4821
436	1.4820
443	1.4804

Table 2. The examples of data manufactured by using recipe TEOSXX.

Manufacturing Parameters							
DT	cSHDist	TEOS factor	TEOS (mgm)	HE (sccm)	O_2 (sccm)	HFRF (pre)	LFRF (pre)
38.0	310	1.0386	5250	4000	4200	797	209
38.0	310	1.0386	5250	4000	4200	798	209
38.0	310	1.0236	5250	4000	4200	793	209
38.0	310	1.0236	5250	4000	4200	795	210
38.0	310	1.0036	5250	4000	4200	763	199
38.0	310	1.0036	5250	4000	4200	765	200
38.0	310	1.0236	5250	4000	4200	765	200
38.0	310	1.0236	5250	4000	4200	795	200
38.0	310	1.0236	5250	4000	4200	780	200
38.0	310	1.0236	5250	4000	4200	765	200

Thin Film Properties	
Thickness	RI
4983	1.4584
4988	1.4587
4814	1.4628
4830	1.4628
4885	1.4634
4903	1.4633
4832	1.4636
4836	1.4629
4839	1.4632
4682	1.4629

The values of thickness and RI of thin film are expected to be estimated by the well-trained NN model. In order to fairly demonstrate the effectiveness of NN model in the estimation of thin film properties, three same size data sets, i.e. LDR-1a, LDR-1b, and LDR-1c, are randomly reorganized from data LDRXX. Similarly, three same size data sets, TEOS-1a, TEOS-1b and TEOS-1c, are randomly reorganized from data TEOSXX. In the simulations of data LDR-1a, LDR-1b and LDR-1c, 70 sets were used for NN's training and 32 sets were used for testing. For data TEOS-1a, TEOS-1b and TEOS-1c, 59 sets were used for NN's training and 30 sets were used for testing.

For all data sets, the size of NN in thickness estimation is 4-10-1. The inputs are DT, TEOS factor, HFRF, LFRF. In RI estimation, the size of NN is 5-10-1. The inputs are DT, TEOS factor, HFRF, LFRF and thickness. The mean absolute error (MAE) and mean absolute percentage error (MAPE) are used as the estimated measurements.

$$\text{MAE} = \frac{\sum_{i=1}^{N} | y_i - \hat{y}_i |}{N} \tag{2}$$

$$\text{MAPE} = \frac{\sum_{i=1}^{N} | {}^{y_i - \hat{y}_i}\!/\!{}_{y_i} |}{N} \times 100\% \tag{3}$$

Where, y_i and $\hat{y}_i$ are actual and estimated values. N is the total number of estimation data.

In order to observe the distribution behaviors of all training and test data sets, the simple statistical analysis was done. For example, Table 3 lists the information of mean value, variance and standard deviation for the whole LDR-1a data, LDR-1a 70 training data and 32 LDR-1a test data. Similarly, Table 4 and Table 5 list the statistics of data LDR-1b and LDR-1c, respectively. Table 6 lists the estimation errors of LDRXX data series performed by NN. For TEOSXX data, Table 7, Table 8 and Table 9 list the statistics of data TEOS-1a, TEOS-1b and TEOS-1c, respectively. Table 10 presents the estimation errors of TEOSXX data series performed by NN.

Table 3. The distribution behaviors of data LDR-1a.

Total Data (LDR-1a)		
Statistics	Thickness	RI
Mean	483.39215	1.47009
Var.	862.65673	0.000068761
Std.	29.37102	0.00829221
Training Data (LDR-1a)		
Statistics	Thickness	RI
Mean	477.87142	1.47021
Var.	1090.7511	0.000073625
Std.	33.02652	0.00858049
Test Data (LDR-1a)		
Statistics	Thickness	RI
Mean	495.46875	1.4698187
Var.	163.41835	0.000060046
Std.	12.78352	0.0077489

(Var.: Variance, Std.: Standard Deviation)

Table 4. *The distribution behaviors of data LDR-1b.*

Total Data (LDR-1b)		
Statistics	Thickness	RI
Mean	483.39215	1.47009
Var.	862.65673	0.000068761
Std.	29.37102	0.00829221
Training Data (LDR-1b)		
Statistics	Thickness	RI
Mean	479.58572	1.4696674
Var.	1126.9996	0.000074579
Std.	33.570816	0.00863592
Test Data (LDR-1b)		
Statistics	Thickness	RI
Mean	491.71875	1.4710001
Var.	197.82158	0.000056770
Std.	14.064906	0.00753455

(Var.: Variance, Std.: Standard Deviation)

Table 5. *The distribution behaviors of data LDR-1c.*

Total Data (LDR-1c)		
Statistics	Thickness	RI
Mean	483.39215	1.47009
Var.	862.65673	0.000068761
Std.	29.37102	0.00829221
Training Data (LDR-1c)		
Statistics	Thickness	RI
Mean	481.62857	1.4718357
Var.	800.9036	0.0000731506
Std.	28.300241	0.00855281
Test Data (LDR-1c)		
Statistics	Thickness	RI
Mean	487.25	1.4662563
Var.	1005.5484	0.0000391548
Std.	31.710382	0.00625738

(Var.: Variance, Std.: Standard Deviation)

Table 6. *The statistics of estimations for data LDRXX.*

	LDR-1a			
Statistics	**Thickness**		**RI**	
	Training	**Test**	**Training**	**Test**
MAE	6.3079	7.3324	0.001173	0.003112
MAPE	1.3182%	1.4888%	0.0799%	0.2112%
	LDR-1b			
Statistics	Thickness		RI	
	Training	Test	Training	Test
MAE	6.2145	6.8923	0.001089	0.003557
MAPE	1.3011%	1.4027%	0.0751%	0.2328%
	LDR-1c			
Statistics	Thickness		RI	
	Training	Test	Training	Test
MAE	6.8258	7.8785	0.001208	0.003345
MAPE	1.4255%	1.5989%	0.0832%	0.2275%

Table 7. *The distribution behaviors of data TEOS-1a.*

Total Data (TEOS-1a)		
Statistics	Thickness	RI
Mean	4967.382	1.4596821
Var.	6752.8295	0.00000177311
Std	82.175605	0.00133158
Training Data (TEOS-1a)		
Statistics	Thickness	RI
Mean	4942.0	1.4597865
Var.	8020.3105	0.00000241364
Std	89.55618	0.00155359
Test Data (TEOS-1a)		
Statistics	Thickness	RI
Mean	5017.3	1.4594768
Var.	562.2862	0.00000048738
Std	23.71257	0.000698123

(Var.: Variance, Std.: Standard Deviation)

Table 8. *The distribution behaviors of data TEOS-1b.*

Total Data (TEOS-1b)		
Statistics	Thickness	RI
Mean	4967.382	1.4596821
Var.	6752.8295	0.00000177311
Std	82.175605	0.00133158
Training Data (TEOS-1b)		
Statistics	Thickness	RI
Mean	4945.085	1.45999
Var.	7853.424	0.00000210681
Std	88.619544	0.00145149
Test Data (TEOS-1b)		
Statistics	Thickness	RI
Mean	5011.2333	1.4590766
Var.	1783.7709	0.000000594971
Std	42.23471	0.00077134

(Var.: Variance, Std.: Standard Deviation)

Table 9. *The distribution behaviors of data TEOS-1c.*

Total Data (TEOS-1c)		
Statistics	Thickness	RI
Mean	4967.382	1.4596821
Var.	6752.8295	0.00000177311
Std	82.175605	0.00133158
Training Data (TEOS-1c)		
Statistics	Thickness	RI
Mean	4983.356	1.4593307
Var.	4049.8537	0.00000110908
Std	63.638462	0.00105313
Test Data (TEOS-1c)		
Statistics	Thickness	RI
Mean	4935.967	1.4603734
Var.	10851.551	0.00000241653
Std	104.17078	0.00155452

(Var.: Variance, Std.: Standard Deviation)

Table 10. The statistics of estimations for data TEOS-xx.

Statistics	TEOS-1a			
	Thickness		RI	
	Training	Test	Training	Test
MAE	39.8665	16.2448	0.00059	0.00032
MAPE	0.812%	0.3243%	0.040%	0.022%
	TEOS-1b			
Statistics	Thickness		RI	
	Training	Test	Training	Test
MAE	41.2543	17.5334	0.00052	0.00027
MAPE	0.852%	0.3394%	0.0387%	0.0251%
	TEOS-1c			
Statistics	Thickness		RI	
	Training	Test	Training	Test
MAE	32.5743	19.225	0.00064	0.00032
MAPE	0.823%	0.0351%	0.0414%	0.0257%

The examples of superposition plot for thickness estimation are shown in Figure 3 and Figure 4. Figure 3 shows the example of NN's training result and Figure 4 shows the example of NN's testing results. In figures, the solid line stands the actual thickness values and the dotted line stands NN's estimated values.

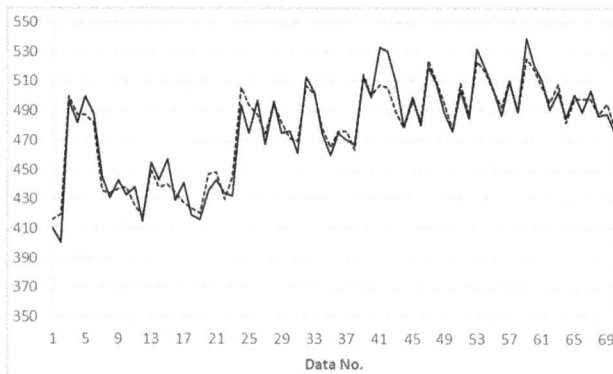

Figure 3. The superposition plot of thickness for NN's training.

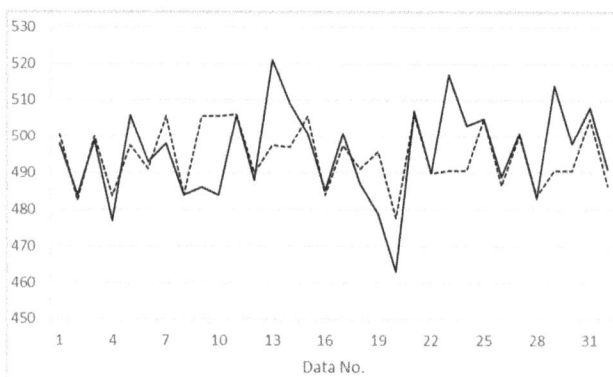

Figure 4. The superposition plot of thickness for NN's test.

Similarly, the examples of superposition plot for RI estimation are shown in Figure 5 and Figure 6. Figure 5 shows the example of NN's training result and Figure 6 shows the example of NN's testing result. Same as above figures, the solid line stands the actual thickness values and the dotted line stands NN's estimated values.

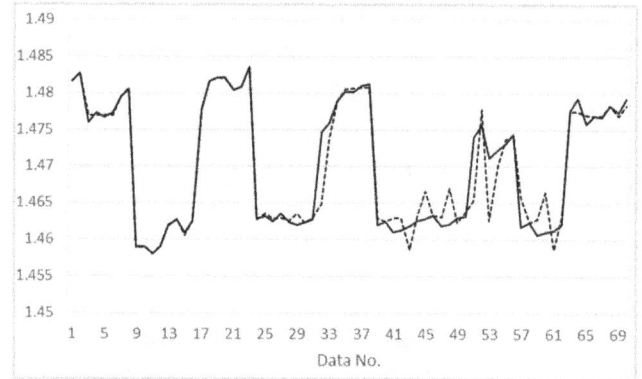

Figure 5. The superposition plot of RI for NN's training.

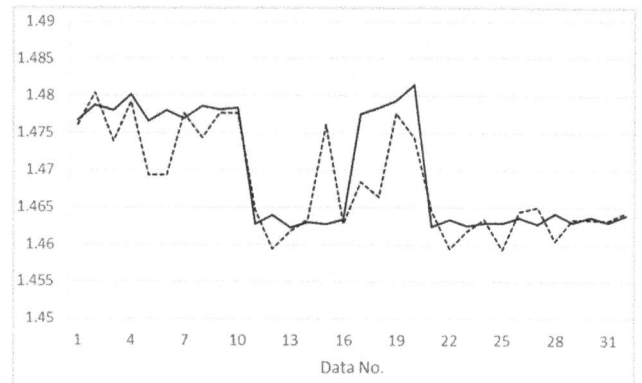

Figure 6. The superposition plot of RI for NN's test.

4. Results & Discussion

It is known that many unknown factors will affect the properties of thin film in its real manufacturing process. These unknown factors are usually uncontrolled and can be treated as the disturbances. Besides, different manufacturing machines have different physical characteristics. All these conditions might make the films have different properties even they are filmed under the same manufacturing parameters. In our study, we tried to use NN technique to catch the relationships among the film's properties and the relevant manufacturing parameters so that the film's properties could be estimated in advance. From the simulation results shown, the relationships between the film's properties and the manufacturing parameters indeed can be obtained by a well-trained NN model. For both thickness and RI estimations, the plots show that the trends of film's properties still can be estimated by NN.

5. Conclusion

In this research, the estimation for the properties of semiconductor's thin film based on NN technique was studied. From the study results shown, we conclude that NN model indeed has the ability to estimate the properties of thin film if NN was well-trained. In other words, these well-trained NN estimators could provide the important information to the technician for setting the proper manufacturing parameters in the filming process. The technician is able to make the whole filming process more effective and successful. However, in this research, only a few manufacturing parameters were collected. We do believe that the estimation accuracy for the film's property could be improved greatly if more relevant manufacturing parameters can be considered and collected.

Acknowledgements

This research was supported by the Ministry of Science and Technology, Taiwan, ROC under the contract No. MOST-103-2221-E-214-050.

References

[1] Q. Michael, & S. Julian, Semiconductor Manufacturing Technology, Prentice Hall, 2000.

[2] C. H. Lai, "Fault diagnosis on CVD equipment via neural network approach," Master Thesis, National Cheng Kung University, 2007, (In Chinese).

[3] E. Ritter, "Deposition of Oxide film by reactive evaporation", Journal of Vacuum Science and Technology, vol. 3, issue 4, pp. 225, 1966.

[4] C. H. Li, "Application of genetic algorithm optimization on the physical parameters of the meteorological forecast deposition of thin film semiconductor manufacturing process," Master Thesis, Southern Taiwan University of Science and Technology, 2009, (In Chinese).

[5] K. L. Hsieh, and L. L. Tong, "Optimization of multiple quality responses involving qualitative and quantitative characteristics in IC manufacturing using neural networks", Computers in Industry, vol. pp. 46, 1-12, 2001.

[6] W. Z. Cui, C. C. Zhu, and H. P. Zhau, "Prediction of thin film thickness of field emission using wavelet neural networks", Thin Solid Films, vol. 473, pp. 224-229, 2005.

[7] W. Z. Cui, C. C. Zhu, H. P. Zhao, "Prediction of thin film thickness of field emission using wavelet neural networks," Thin Solid Films, vol. 473, no. 2, pp. 224-229, 2005.

[8] B. Kim, H. Lee, D. Kim, "Modeling of thin film process data using a genetic algorithm-optimized initial weight of backpropagation neural network," Applied Artificial Intelligence, vol. 23, no. 2, pp. 168-178, 2009.

[9] C. C. Huang, H. C. Huang, Y. J. Chen, R. C. Hwang, "An AI system for the decision to control parameters of TP film printing", Expert Systems With Applications, vol. 36, no. 5, pp. 9580-9583, 2009.07.

[10] P. H. Weng, Y. J. Chen, S. M. T. Wang, R. C. Hwang, "The predictions of optoelectronic attributes of LED by neural network", Expert Systems With Applications, vol. 37, no. 9, pp. 6282–6286, 2010.09.

[11] Y. J. Chen, J. C. Chien, C. Y. Chang, S. M. T. Wang, Y. C. Chiang, R. C. Hwang, "AI transmittance estimator for multi-layer coating TP film", Applied Mechanics and Materials, vol. 312, pp. 996-1000, 2013.03.

[12] Y. D. Ko, P. Moon, C. E. Kim, M. H. Ham, M. K. Jeong, G. D. Alberto, J. M. Myoung, I. Yun, "Predictive modeling and analysis of HfO2 thin film process based on Bayesian information criterion using PCA-based neural networks," Surface and Interface Analysis, vol. 45, no. 9, pp. 1334-1339, 2013.

[13] N. M. Sabri, N. D. Md Sin, M. Puteh, M. Rusop Mahmood, "Prediction of nanostructured ZnO thin film properties based on neural network," Advanced Materials Research, vol. 832, pp 266-269, 2014.

[14] L. Cai, Y. G. Tan, Q. Wei, "On-line thickness measurement of thin film based on neural network," Applied Mechanics and Materials, vol. 484-485, pp. 307-310, 2014.

[15] A. Khotanzad, R. C. Hwang, A. Abaye, D. Maratukulam, "An adaptive modular artificial neural network: Hourly load forecaster and its implementation at electric utilities," IEEE Transactions on Power Systems, vol. 10, pp. 1716-1722, 1995.

[16] C. Y. Shen, C. L. Hsu, R. C. Hwang, J. S. Jeng, "The interference of humidity on a shear horizontal surface acoustic wave ammonia sensor", Sensors & Actuators: B. Chemical, vol. 122, pp. 457-460, 2007.

[17] P. H. Weng, Y. J. Chen, H. C. Huang, R. C. Hwang, "Power load forecasting by neural models", Engineering Intelligent Systems for Electrical Engineering and Communications, vol. 15, pp. 33-39, 2007.

[18] S. Malinov, W. Sha, J. J. McKeown, "Modelling the correlation between processing parameters and properties in Titanium alloys using artificial neural network," Computational Materials Science, vol. 21 pp. 375–394, 2001.

[19] M. Toparli, S. Sahin, E. Ozkaya, S. Sasaki, "Residual thermal stress analysis in cylindrical steel bars using finite element method and artificial neural networks", Computers and Structures, vol. 80, No. 23, pp. 1763-1770, 2002.

5

Using IEC 61850 Protocol in Automation Systems of High Voltage

Asghar Ehsani Fard[1], Masoud Masomei[2], Mehdi Hedayeti[2], Hamid Chegini[3]

[1]Telecommunication, of Non-profit Institution of Higher Education, ABA, Abyek, Qazvin, Iran

[2]Non-profit Institution of Higher Education, ABA, Abyek, Qazvin, Iran

[3]Research Assistant of Non-profit Institution of Higher Education, ABA, Abyek, Qazvin, Iran

Email address:

Ali.Ehsani_8520@gmail.com (A. E. Fard), massoud_masoumi@yahoo.com (M. Masomei), Mahdi-hedayati@yahoo.com (M. Hedayeti), Hamidchegini26@gmail.com (H. Chegini)

Abstract: Growing Development of new metric technology, communications on one side, and variety in IT and incremental information on the other side in designing and manufacturing automation station systems make the pioneers of this industry change the system with the aim of creating "a world, a technology, a standard" in this industry. This evolution led to create a new technique for designing automation systems which is called IEC 61850 standard. Considering the agreement on developing and using this standard, applying this in station automation systems was taken into account since 2005. The new technology needs new interfaces to connect the high voltage switchgears such as modern current and voltage measurement transformers to station automation control systems. By introducing the new technology and using IEC 61850 protocols, designing automation systems with decentralized structure is possible and has a lot of advantages, however great care is needed in designing such system so that system reliability will not be impaired. In this thesis, for better understanding, we first explain briefly the old station structures and their transition and then the IEC 61850 standard structure will be investigated and this investigation includes the way of data transfer. Finally we refer to the effect of this standard on the structure of high voltage station automation system. In power distribution systems, IP and IEC-61850 are introduced as a security ring with more importance of communication, that in the meantime, IEC-61850 as the new international standard electrical station, is capable of providing the Integration, protection, control, measurement and monitoring functions within the stations with high-speed. So not only can prove the importance and the need to replace the IEC-61850 standard as the protocol with high speed data transfer capability ,but also we will also see reducing the time and power transmission losses in the automation network

Keywords: Station of Automation System, Protocol, IED, IEC 61850, Standard

1. Introduction

The new industries have complex activities and many functions inside themselves. They need a lot of control devices and intelligent electronic devices (IED) to have a safe activity. To operate automatically, these devices should work very harmoniously together. The IEC 61850 standard is a valid international one for building station communications and has created a new opportunity for evolution in system protection and control. This standard denotes a big step in making multi-objective intelligent electronic devices integrated based on development and implementation of advanced distributed protection and control procedures. To fulfill this evolution, it is necessary to know how to make use of it.

The advances in designing intelligent electronic devices and cooperative communication protocols like IEC61850 led to new generation of intelligent electronic devices. The new electronic intelligent devices, which are used to protect and control, can accept different voltage and current levels as their input and are capable of analyzing the data very quickly. The main advantage of using such devices which contain microprocessors is;

i. Ease of connect to each other,

ii. Cost reduction of components,

iii. Maximization of system reliability

iiii. And capability of saving data extensively.

One effective method for using these intelligent electronic devices and reducing connections between them is using Goose messaging method in IEC61850 communication protocol which has a high speed and it is in a cooperative form. To understand the way of establishing connection between these intelligent electronic devices, automation design and station control, it is essential to be completely familiar with IEC 61850.

Communication protocol collection used in station level is one of important subjects in designing station automation system. This protocol needs to offer all required services which enable the optimal fulfillment of different station functions: The following are the most important necessities:

1- Appropriate definition of functional and applicable necessities,

2- Sufficient recognition of station communication protocols.

IEC61850 standard and its applicable programs are of important subjects in electricity due to being comprehensive and taking all control requirements and station protection into account.

2. Theoretical Foundations of Research

2.1. The New Communicative Technologies

A relatively new standard is IEC61850 in communicative networks of high voltage stations in which core element contains the following:

- A software module for every function in high voltage stations which is known as logical node and describes all available data.

- A definition of characteristics and way of communication between intelligent electronic devices in the station automation control system. Intelligent electronic devices refer to all devices that are available in the station computer network and exchange information with network which include numerical protection relays, BCU, station server computers and so on.

- A definition of programming language and configuration which is a tool for system engineering in automation stations and enables communication between tools and engineering software in systems. It can be used, for example, for system engineering in connection with dispatching center. It also facilitates definition of functions and special programs if necessary.

To draw the benefits of a standard, it is of high importance to satisfy present and future requirements and IEC61850 standard has this merit.

The main subject of this standard is designing a communicative system which provides interoperability of communication between functions in devices for different devices embedded in station produced by different

manufacturers and meets all the needs well.

To achieve this, the functions are divided into sub functions which are known as logical nodes.

The logical nodes are 4 and core elements of information modules that should be exchanged between different devices in an automation system. In other words, an automation system includes exchange of logical nodes data with each other which is shown in figure 1.

As you see in Fig. 1, logical nodes including PDIS, TCTR, TVTR and XCBR are related to high voltage key.

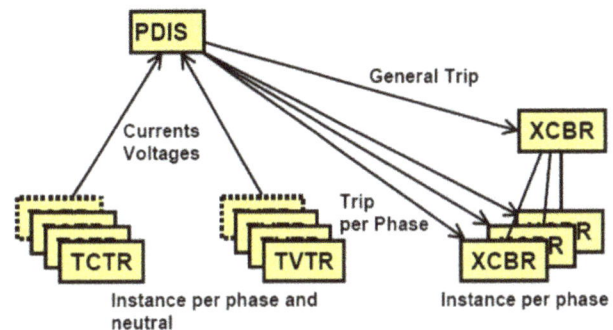

Fig. 1. *Samples of logical needs exchange in a protection function.*

As it is shown in Fig. 1, although different functions can be in separate physical devices depending on their respective manufacturer, all logical nodes can exchange information with each other through network if they are designed according to IEC61850 and so they can be produced by different manufacturers. This standard also defines logical and physical devices.

Logical devices are a collection of logical nodes that are collected in a joint group like protection group. Of course these logical devices are physically in a real device, e.g. protection relay and a protective relay has three logical devices, namely protection, control and measurement and the logical nodes, according to their functions, are in one of these logical devices. In figure 2, there are some examples of this model.

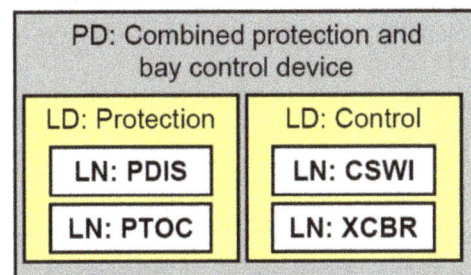

Fig. 2. *An example of transformation between devices and logical nodes with physical devices.*

The examined standard and structure of this and programming language are designed in such a way that they can support possible future functions. Also, it is necessary to separate the discussion of fast technology development from low speed of adding functions in station industry.

Every changes and optimization that occur in

communications industry, including speed increasing that functions on protocol basis, will influence on 61850 IEC.

This separation of functions is done by an interface called ACSI2 that really maps the applied data by SCSM.3 The approach is shown by fig. 3.

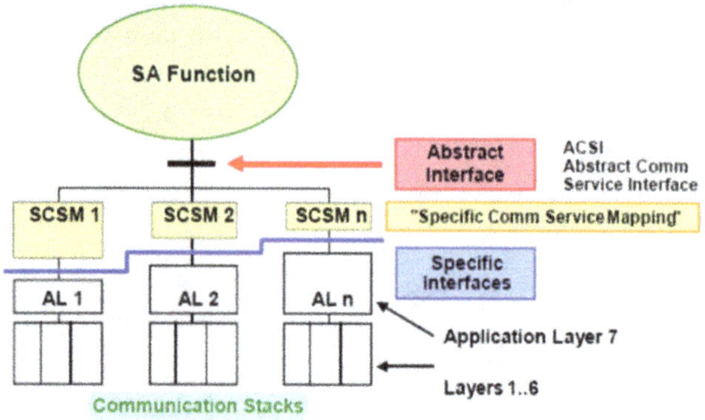

Fig. 3. A stack of information medium and the way of sending information to different parts of structure

In fact the IEC 61850 carries out two separate mappings; one based on IEC 61850 for station passage (bus) that establishes a connection between Bay level intelligent electronic devices with station level or control room. This mapping is used for transforming sampled quantities which pass the seven-layers of ISO standard. And it does the second mapping that uses just two layers to increase speed.

This mapping is used to communicate between data layers and ISO standard physical layer in processing bus network which is between substation switchgear process level and bay level in modern automation systems.

2.2. Automation Systems Levels with IEC 6180

The modern automation designs of high voltage station are divided into three levels: processing level, bay level, and standard level. The station level communication generally refers to the control room devices and includes the network main servers (with bay level) and protection relays and bay controllers. The connection between them is done by station bus network in IEC-161850-8. In Figure 4, the three levels are shown.

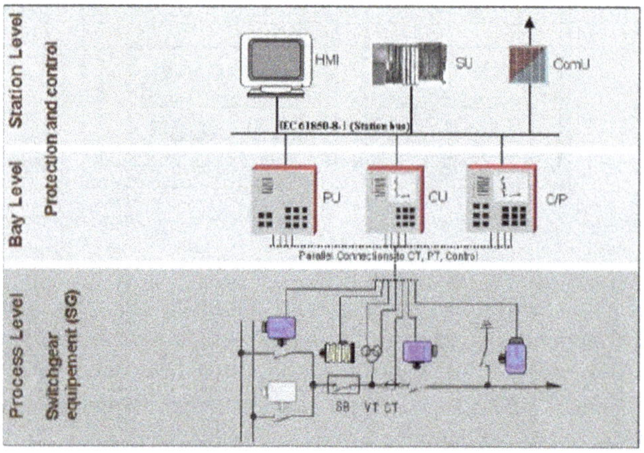

Fig. 4. The topology of station devices in three different levels.

The processing level devices generally include intelligent sensors and distant input outputs which has network and communication level called process bus network.

In other words, in stations which have modern automatic system with intelligent switchgear and have communication with the other devices, have two separated networks called station bus and process bus. As it is shown in Fig. 4, according to the other computer networks that function based on Ethernet protocols, the switchers of the two networks and information transform bus, have the role of Filter and responsibility of sending data to different parts of LAN network. If the station level node is likely to transform data with processing level, it will send its message through the network switches in bay level. Then the mentioned node will fulfill the message and send the result through the switch or connected unit, the responsibility of connected unit is mixing the information of 3-phase current and 3-phase voltage. This can also be the role of switch in bus network to transfer the information.

Many electronic devices that have different functions in the station are in the bay level and communicate with intelligent switchgear through the process bus network. In automation stations which are currently in our country, due to lack of modern switchgear the bay level communication is done by copper wires and cables.

Perhaps you consider that due to the presence of modern devices and their outcome, the communication network in processing level may decrease the quantity of consuming wiring and cables.

Basically, communication in processing bus network is similar to station bus communication, however there are two additional services for protecting bus network.

The first problem is quick and confident exchange of protection trip order between protection devices and switchgear, and the second one is quick transform of instance deals of electronic transformers for protection relays. These two services need quick execution on communication stacks and there should not be any delays.

So, the main selection is generally applying the quick networks in basic technology (processing bus) that must be executed in optical fiber communication.

2.3. The New Technology in Primary Devices

The new measurement transformers for current and intelligent voltage (optical current and voltage Trans) and also the electronic modern switchgears need new interfaces for automation systems. The new used technology in power key drivers, like new mechanism functions, which are controlled by serving motors, gives us this opportunity to have a direct communication with driver and breaker electronic mechanism.

The new serial interfaces with IEC61850 standard will provide communication. Using these keys with intelligent electronic abilities in switchgear with new interfaces will create a better situation for identifying diagnostic condition for observing breaker. This situation can be very useful for better exploitation and real time scheduling.

3. The Future Structure / Station Automation System

The above mentioned technology, will direct the automation system toward decentralized structure and communication topology and this may lead to omission of cables and copper wires.

3.1. Research Methodology

Automation systems and numeric relays introduction with old switchgears, during the 15 years in our country, led us to a structure similar to figure 5.

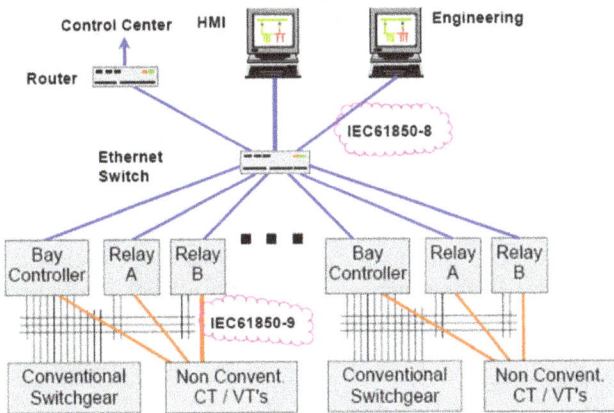

Fig. 5. Bus level and traditional wiring to controllers.

In the first step by entering the new transformers, the modern measurement of serial communication and point by point of these trances with relays will be possible.

Protection 61850-IEC and bay controllers will be created as shown in fig. 6.

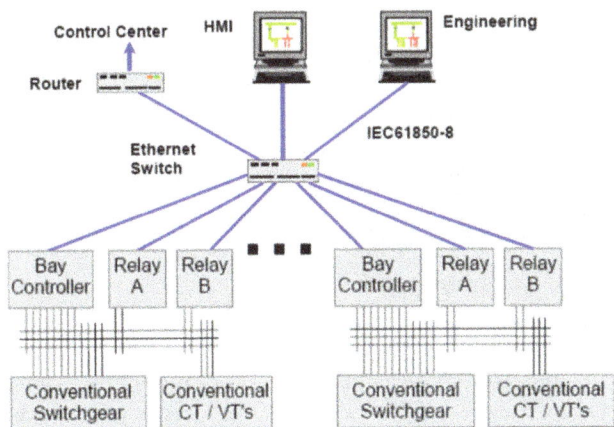

Fig. 6. Station bus and accurate transformers communication.

In the next step, the modern switchgears and keys with electronic drivers and mechanism, which have communication interfaces, will enter the market. They will have a structure as shown in fig. 7, the new bounty for automation systems which are called E C P or equipment close process. Processing to switchgears and high voltage devices are suitable for E C P.

E C P will create a communication network between switchgears and bay level and this will cause the omission of cables and copper wires and also the simplification of this part topology.

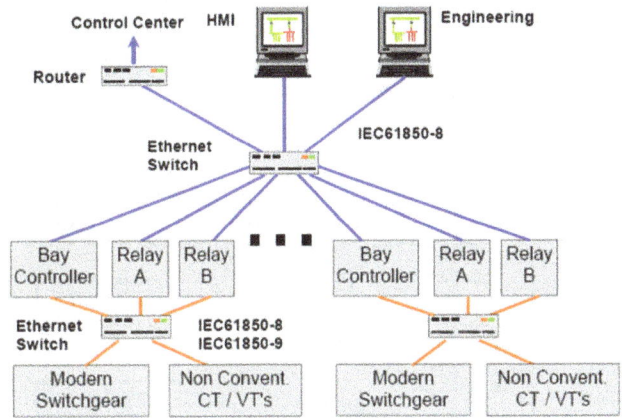

Fig. 7. The system structure by using the keys with mechanism and electronic drivers.

As 61850-8 IEC uses the equal communication topology for station and processing bus, the final structure will be as fig. 8. The system will bring an akin structure for accessory to information.

In the system and also make it possible to use the structure and Ring Topology.

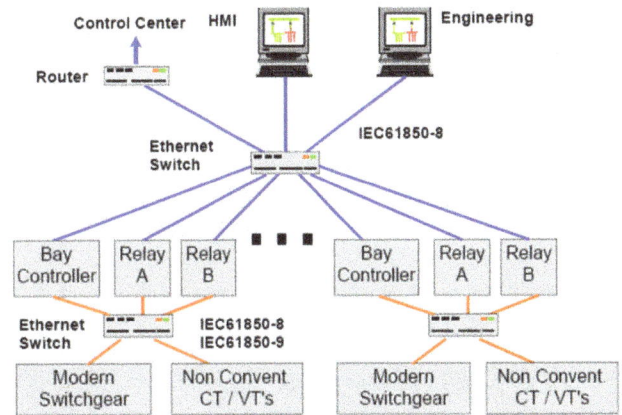

Fig. 8. The integrated communication network by using IEC 61850-9 and IEC 61850-8.

3.2. The Details of the Close Processing Structure

As discussed previously, there is a merging unit in the processing level that joins the current trances and voltage to the control and protection devices.

The output of this integration is according to 61850-9 IEC standard and current and voltage trances that are connected to the integration unit can be both conventional and traditional or the new intelligent one or a combination of the two kinds.

The output analog sample quantities of integration unit should be time sanction and have time tag.

This can be simultaneous with sampling which the total system dose for the total analog quantity or can have time tag for sampling itself. The processing structure of close device can be based on sending signals sent from switchgear and its integration method can be in the cable or copper wires or every modern system based on IEC 61850-9. For example, in Fig. 10, which is usual in our country, the integration of switchgear with bay level is in the form of copper wires. In addition to the presence of logical node connected to PDIS and PTOC, there is a controller device which includes logical nodes for controlling (CSWI, GCBR).

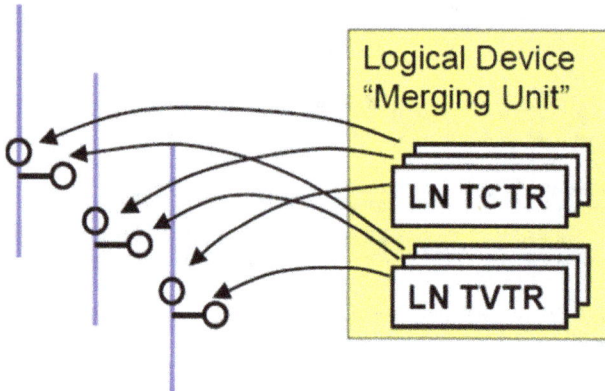

Fig. 9. *The method of integration unit connected to line voltage and current trances.*

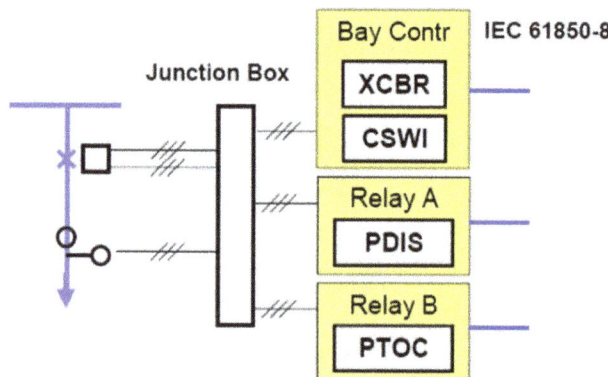

Fig. 10. *The conventional connection of switchgears to lines.*

In this structure there are a lot of cables and copper wires for each bay that need to be tested and wired. In the first step by entering the new modern measurement transformers which are connected to protection relays and buy controllers through combined units and connected to 61850-9 according to figure 11.

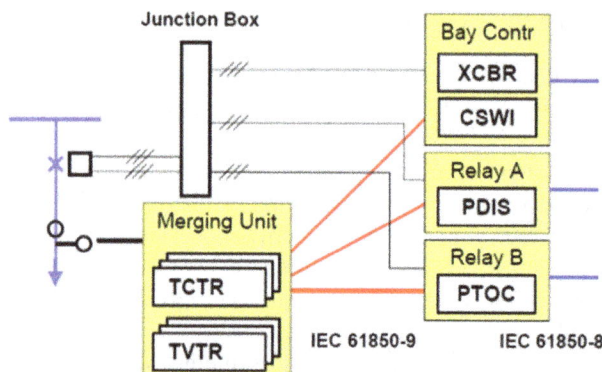

Fig. 11. *The switcher connection (junction) to lines by using the modern tools transformers.*

In the next step, addition to merging unit of modern sensors, there is a monitoring unit for electronic drivers of high voltage keys and the other devices, a switcher is added and made a new structure similar to fig. 12.

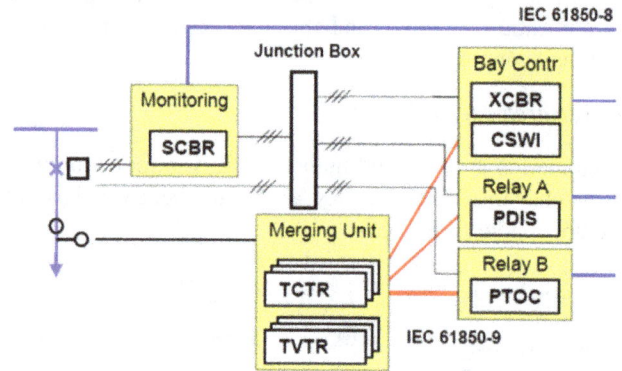

Fig. 12. *Junction of monitoring unit for electronic control of high voltage keys*

The monitoring unit has a suitable communication with electronic sensor of breaker and also a wire and cable communication with control devices and protection relays that is used for sending information to station level according to 61850-8 IEC. The commands of trip and alarm signals between breaker and protection relays are in copper cables and software form.

Finally by entering the fully intelligent switchgears, which is shown in fig. 13, monitoring and merging units are in network form and based on 61850-9 IEC and connected to bay devices, that is, by controllers and protection relays that are in copper cable and software form.

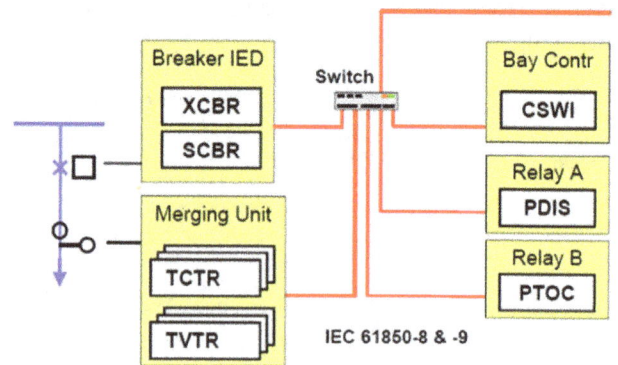

Fig. 13. *Monitoring unit connection to merging unit in modern network.*

In other words, processing bus network is used for junction between bay level and processing level and there is not any wire or cable junction between switchgear and protection relays and control devices.

3.3. Function Integration

Function integration in a set is an appropriate solution to decrease the number of devices and cost reduction. By decreasing the devices one can trust in system more. Today it is common to concentrate on all protection functions; control and monitoring in bay level and in separate devices in automation systems. By introducing intelligent switchgear systems and electronic sensors and merging units we can consider some integration. Look at fig. 14.

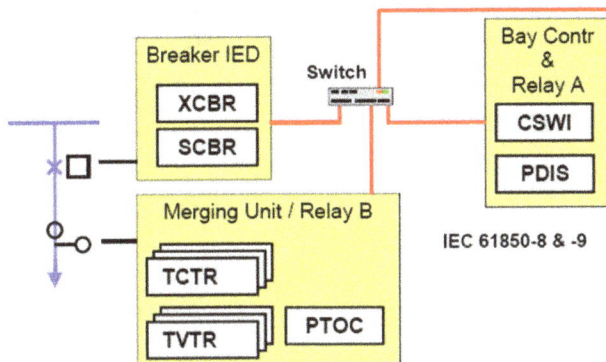

Fig. 14. *Function integration merging at bay level and processing for entering.*

Here by we should consider the following:

- Reliability of function ability of devices or system for fulfillment of functions of necessity times that directly refer to a parameter called MTBF.4 Considering the accessibility it the MTBF is larger and nearer to 100 percent it is a better one. In other words, when the distance is bigger, it means it is farther than problem.

Accessibility means getting access to the devices or system in the required time and it has opposite proportion to MTTR. That is, when the repair time is shorter, that means the situation is better.

The MTTR is dependent on power and diagnosis speed and knowledge at the problems. So the total equation and accessibility and reliability depend on identifying constant development of devices and self-examining of devices in the system and system itself. So, in modern systems, considering the numerical and intelligent devices even in processing level and omission of devices which are not visible, there is a better condition than in old systems.

In old systems there are not much support, additional devices like control devices, monitoring and protection of bases measurement. While in modern automation, there are many additional devices and hard ware for switchgear and other functions.

It seems that by introducing processing technology of modern systems which are close to switchgear level, there will be concern about reliability of function specially protection devices. This anxiety is at least limited to current systems and as mentioned above there are more function freedom. The freedom includes the following:

Design step, arrangement of place, control and protection function with self-examine ability, decreasing of devices.

So, there can be increasing of redundancy with respect to the importance at system.

In 115 key systems and similar systems we can connect the additional relays to current trances without the necessity of additional wiring. As the current transformers are connected to switchers and put their information in the network, several relays can use the information currently.

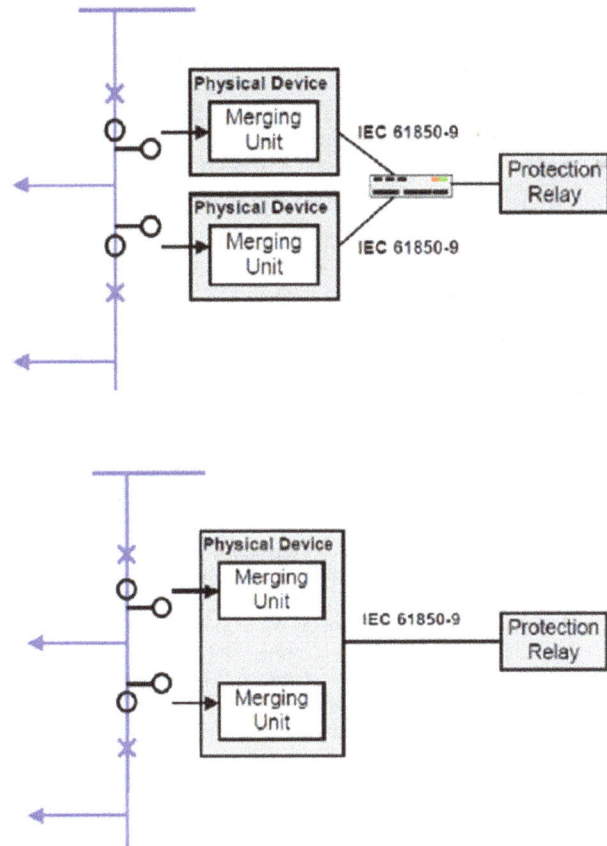

Fig. 15. *Merging units junction in 1/5 key arrangement.*

4. Conclusion

On the one hand, the results of this research reflect the replacement of the old generation with new posts that have many advantages. And the other is economically affordable. A typical life for a post, the payback time is only 10 years old. Thus, taking into account the complete life of post, alternative is economically quite affordable. The new processing technology of switchgear devices and 61850 IEC has a lot of advantages for high voltage station design. The advantages cause the omission of many copper wires and cables and consequently decreasing of work volume on wiring and testing process. The number of devices will decrease without self-examining ability and cause decreasing of system fouls and increasing of reliability.

Making support track at switchgear level for system and functions is becoming possible with acceptable cost and by integrating functions and number of devices, can lead to increase system reliability.

References

[1] Christoph Brunner, "How the IEC61850 support the migration phase" CIGRE 2002.

[2] Lars Andersson, "Discussion on cost differences between conventional and intelligent sensors as well as the different process connections for intelligent sensors", CIGRE 2002.

[3] Ch. Brunner, G Schimmel, H. Schubert, "Standardisation of serial links replacing parallel wiring to transfer process data - Approach, state and practical experience", CIGRE 2002 Paris.

[4] Lars Andersson, Klaus-Peter Brand, Wolfgang Wimmer, "The Impact of the coming Standard IEC61850 on the Life-cycle of Open Communication Systems in Substations", Distribution 2001, Transmission & Distribution, Brisbane 2001.

[5] Lars Andersson, Klaus-Peter Brand, Wolfgang Wimmer. "The communication standard IEC61850 supports flexible and optimized substation automation architectures", 2nd International conference on Protection and Control in NEW DELHI 2001.

[6] C. Brunner, A. Ostermeier, "Serial Communication between Process And Bay Level - Standards And Practical Experience –", CIGRE 2000.

[7] M. Saitoh, T. Kimura, Y. Minami, N. Yamanka, S. Maruyama, T. Nakajima, M. Koskada, "Electronic Instrument Transformers for Integrated Substation Systems", IEEE/PES T&D Conference and Exhibition 2002: Asia Pacific.

[8] Masayuki Kosakada, Hiroshi Watanabe, Tokuo Ito, Yoshito Sameda, Yuji Minami, Minoru Saito, Shiro Maruyama, "Integrated substation systems – harmonizing primary equipment with control and protection systems", IEEE/PES T&D Conference and Exhibition 2002: Asia Pacific.

[9] IEC 61850 Communication Networks and systems in substations, IEC standard in ten main parts, first parts published in 2002.

[10] J. Rosenberg, H. Schulzrinne, G. Camarillo, A. Johnston, J. Peterson, R. Sparks, M. Handley, E. Schooler, "SIP: Session Initiation Protocol", RFC 3261, June 2002.

[11] J. Rosenberg, "A Framework for Conferencing with the Session Initiation Protocol (SIP)", RFC 4353, February 2006.

[12] R. Even, N. Ismail, "Conferencing Scenarios", RFC 4597, August 2006.

[13] http://www.iptel.org/sems.

[14] C. Partridge, "Isochronous Applications Do Not Require Jitter-Controlled Networks", RFC 1257, September 1991.

[15] H. Schulzrinne, S. Casner, R. Frederick, V. Jacobson, "RTP: A Transport Protocol for Real-Time Applications", RFC 3550, July 2003.

[16] A. H. Ashouri, F. Samsami, A. Akbari, "E-Learning Media Server Evaluation and its architecture modeling with signaling load tests," ICeLT, IUST, Tehran, Iran, Dec 2009.

[17] A. H. Ashouri. "Media Server Evaluation and Real-Time Tests" Iran University of Science and Technology, B. Sc Thesis, p46-61, Sep 2009.

6

An Overview of Application of Artificial Immune System in Swarm Robotic Systems

Johar Daudi

Department of Aerospace Engineering, School of Engineering, University of Glasgow, Glasgow, UK

Email address:

daudij@yahoo.com

Abstract: The Artificial Immune System (AIS) is a biologically inspired computation system based on vertebrate immune system. AIS applications in last one decade have been developed to address the complex computational and engineering problems related to classification, optimization and anomaly detection. Many investigations have been conducted to understand the principles of immune system to translate the knowledge into AIS applications. However, a clear understanding of principles and responses of immune system is still required for application of AIS to Swarm Robotics. This paper after a review of AIS models and algorithms proposes an integration of AIS and Swarm Robotics by developing a very clear understanding of immune system structures and associated functions.

Keywords: Immune System (IS), Artificial Immune Systems (AIS), AIS Algorithms, Neutrophils, Swarm Robotics (SR)

1. Introduction

Man has survived through millions of year and the credit goes to the natural defense system our bodies are blessed with, the "Immune System (IS)". The immune system provides protection to a living body against number for foreign molecules (referred to as antigens) e.g. viruses, bacteria, fungi and other parasites. The immune system achieve this objective by observing, analyzing and identifying foreign molecules that enter our bodies, then, it prompts its response against them by creating and releasing antibodies that attack these, thus eliminating them from our bodies and freeing us from infections. To eliminate the threat, IS has to distinguish between foreign molecules and the molecules/tissues that constitute itself to avoid auto-immune responses.

Immune System (IS) possesses excellent ability to recognize the foreign molecules, when they are encountered for the very first time, retain their memory and identify them when encountered at a later stage. IS uses genetic characteristics for biological functioning and thus provide the base for computational modelling of adaptive or in-borne responses. Following human central nervous system IS the most complex biological system due to wider and variable responses. Based on its complex and multiple behaviours, its understanding for computational adoption still is inadequate.

2. Types of Immune System

Immune system is sub-divided into two types of systems i.e. adaptive and innate immunity.

Adaptive immunity is directed against specific disease causing foreign agents and is modified by exposure to such organisms or antigens and keeps strong immunological memory. Adaptive immunity targets the specific pathogens, either previously encountered or not and gets modified by exposure to such pathogens. Adaptive immunity system consists of lymphocytes (white blood cells, more specifically B and T type) which function to recognize and destroy specific substances, and are antigen-specific (Castro and Zuben, 1999). Adaptive immune system has Immunological memory which gives the system, capability of more effective immune response against an antigen after its first encounter, leaving the body in a better position to be able to resist in the future against same pathogens. On primary response, the immune system identifies an antigen and responds with release of large number of antibodies to fight with infection and to eliminate the antigen from the body.

Innate immunity is aimed to target any invaders or disease causing agents or pathogen in the body and is non-specific. It is not modified by repeated exposure and thus plays an important role in the initiation and regulation of immune responses. Innate immunity involves number of specialized white blood cells involved to recognize and bind to common molecular patterns of disease causing microorganisms. It does not provide complete protection and is primarily static in nature and does not modify (Castro and Zuben, 1999). The innate immune system is considered as the first line of defense (Grasso et al.2002) which comprised of cells and mechanisms that defend the host from infection by disease causing organisms. Unlike the adaptive immune system (which is found only in vertebrates), it does not confer long-lasting or

protective immunity to the host (Alberts et al. 2002). Innate immune system recruits the immune cells to sites of infection, through the production of specialized chemical mediators, called cytokines. This causes the identification and removal of foreign substances infecting organs, tissues, blood and lymph vessels by specialized white blood cells. Generally, innate and adaptive immunity systems have following characteristics indicated as Figure-1:

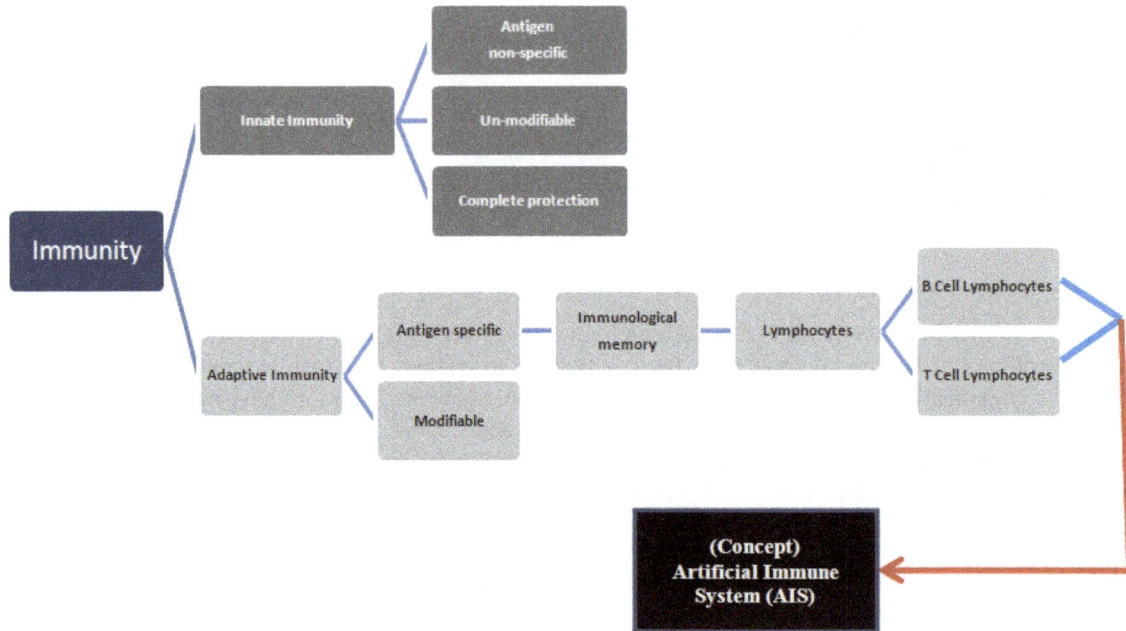

Figure 1. Classification of Human Immune System.

All white blood cells (WBC) are known as leukocytes. Leukocytes are not associated with a particular organ or tissue; thus, they function like a unicellular organism. Leukocytes are able to move as singular cell or population of cells and respond against foreign agents. The innate leukocytes or white blood cells include: Natural killer cells, mast cells, eosinophils, basophils, phagocytic cells including macrophages, neutrophils, and dendritic cells.

Artificial Immune System (AIS) is actually based upon the capabilities of immune system such as robustness, de-centralization, error tolerance and adaptiveness which enabled the researchers for system computation as "Artificial Immune System (AIS)" (Kephart, 1994; Dasgupta, 1996 and Forrest et al.1994). Following the patterns and behaviours of IS three basic computational areas are developed such as immune modelling, theoretical AIS and applied AIS. Immune modelling is focused on mathematical models and simulations of natural and artificial immune systems. Theoretical AIS is concerned with the theoretical aspects including mathematical modelling of algorithms, convergence analysis, and performance and complexity analysis of such algorithms. Applied AISs includes working on immune-inspired algorithms, building immune-inspired computer systems, to apply AISs to diverse real world applications (Dasgupta et al. 2011).

AIS follows the IS characteristics for computational application development such as feature extraction, pattern recognition, memory, learning, classification, adaption for utilization in computer security, fraud detection, machine learning, data analysis, optimization algorithms (Dasgupta, 2006).

3. AIS Algorithms

A critical review of literature has concluded that four major AIS algorithms are focused to develop various AIS applications (Figure-2):

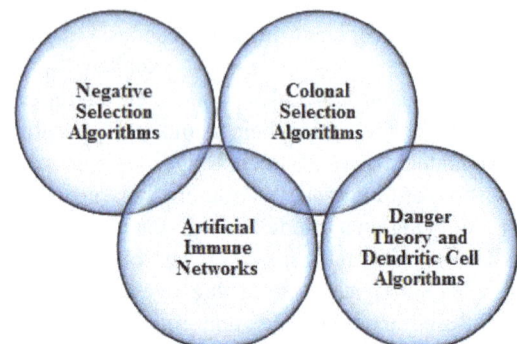

Figure 2. AIS Algorithms.

3.1. Negative Selection Based Algorithms

Negative Selection Algorithms are based on the principle function that protects the body against self-reactive lymphocytes. IS identifies foreign antigens without reacting with the 'self cells'. Receptors are produced during a pseudo-random genetic re-arrangement process for the generation of T-cells. These receptors undergo negative selection mechanisms in thymus. Those T cells which react against "Self Cells" or against self-proteins are discarded and remaining which bind to self-proteins are permitted to leave the thymus. These are called as matured T-cells which are allowed free movement throughout the body to contribute in immunological process against foreign antigens (Somayaji et al., 1997). This biological phenomenon has given an inspiration for the developments of most of the existing Artificial Immune systems.

3.2. Clonal Selection Based Algorithms

This Clonal Selection Based Algorithms is formulated on the principle of mechanism of antigen-antibody recognition, binding, cell propagation and separation into memory cell (Burnet, 1959). This is called as Colonal Selection Theory which has resulted in development of several artificial immune algorithms named as clonal selection algorithm (CSA) by Castro and Zuben (2000). Based on clonal selection and affinity maturation principles they named this as CLONALG. CLONALG in 2002.

3.3. Artificial Immune Networks (AINs)

Artificial immune networks (AINs) are other successful models in AISs. Framer's et al proposed their immune network model (Farmer et al.1986) which became the fundamental for various AINs algorithm and hence based on this the first immune network algorithm was proposed by Ishida (1990). Later, Timmis et al. (2008) re-defined these immune networks, which were formally named as AINE (Artificial Immune Network). According to Knight and Timmis (2001) AINs uses Artificial Recognition Ball (ARB) to represent identical B-cells. Two B-Cells are joined together, if the affinity between two ARBs is below a network affinity threshold (NAT). A Resource Limited Artificial Immune System (RLAIS) based on AINE is developed in 2001 by Timmis and Neal. This up-gradation of models included the knowledge of the fixed total number of B-cells presented in ARBs with centralized control but having the specific role of each ARB in obtaining resources from the mainstream. Those ARBs not able to obtain resources are removed from the network. The cloning, mutation and interactions of B-Cells all take place at the ARB level.

3.4. Danger Theory and Dendritic Cell Algorithms

Dendritic cells within an innate immune system are cells which respond to some specific danger signals. The three main types of dendritic cells such as: 1) Immature Dendritic Cells, which collect parts of the antigen and the signals, 2) Semi-mature Dendritic cells, are immature cells which have decision power against local signals and represent safe and present the antigen to T-cells resulting in tolerance, and 3) Mature Dendritic cells, that are capable to react strongly identifying that the local signals represent danger and deliver the antigen to T-cells for reaction against pathogens. The Dendritic Cell Algorithm is biologically inspired development taking inspiration from the Danger Theory of the mammalian immune system with specific function of dendritic cells. Matzinger, first proposed this Danger Theory stating that the roles of the acquired immune system is to respond to signals of danger, rather than discriminating self from non-self (Matzinger,1994 and 2002). The theory states that the helper T-cells activate an alarm signal providing the co-stimulation of antigen-specific cells to respond. The Dendritic Cell Algorithm (DCA) is inspired by the function of the dendritic cells of the human immune system.

An abstract model of dendritic cell (DC) behaviour was also developed and used to form an algorithm, the artificial DCA by Greensmith et al, 2008. For this purpose, population based DCA algorithm was applied to numerous intrusion detection problems in computer security including the detection of port scans and botnets, where it has produced impressive results with relatively low rates of false positives (Greensmith et al, 2008).

There are several other immunology areas reported in the literature to inspire the development of algorithms and computational tools, for example, humoral immune response (Dasgupta et al. 2003), Danger Theory (Aickelin and Cayzer. 2002) dendritic cell functions (Greensmith et al. 2005), and Pattern Recognition Receptor Model (Yua and Dasgupta. 2008).

4. Review of AIS Application Areas

Dasgupta in 1999 under the title "Artificial Immune Systems and Their applications" compiled several useful literature resources including AIS textbooks and application papers. These comprehensive literature resources also addressed the computational models of the immune system and their applications till 1998. Various modeling techniques based on ordinary differential equations, delay differential equations, partial differential equations, agent-based models, stochastic differential equations as well as associated algorithm and simulation frameworks are in use to simulate IS (Kim. 2009).

This diversity of AIS knowledge has helped the researchers to apply AIS to solve several bench-mark problems of the field such as; computer security, numerical function optimization, combinatorial optimization, learning, bioinformatics, image processing, robotics, adaptive control system, data mining, and anomaly or error diagnosis (Hart and Timmis, 2008). As a more advance innovation, the applications of AIS in controlling a robotic arm is evident from studies (Lee et al. 2002), remote-sensing classification of satellite images (Zhong et al.2006), compensating exposure in images with back-lighting (Su et al. 2006), web-mining applications

(Nasraoui et al. 2006) and application in industrial manufacturing process (Bailey, 1984).

Karaboga; Lee et al. 2002; Castro & Timmis, 2002; Castro & Zuben, 2000; Castro & Zuben, 2002) has concluded the application of AIS in following three major categories as shown below. However, the newly identified and explored areas may fall beyond these categories (Figure-3).

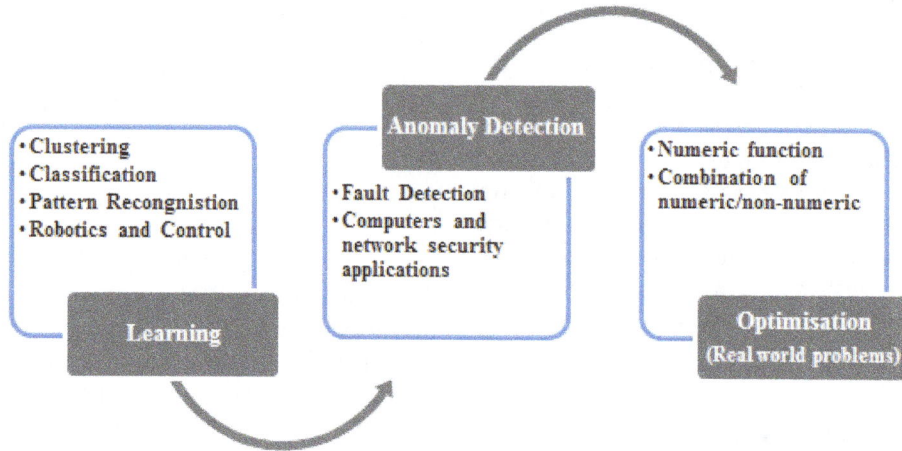

Figure 3. Categorical Areas for AIS Applications.

5. AIS Application in Robotics

AIS can provide a baseline for robots to learn new skills, adapt to new environment throughout its lifetime. With the progress in mechatronics, MEMS (Micro-Electro Mechanical Systems) and nanotechnology the sizes and cost of electronic components (e.g. sensors, actuators and electronic boards) are decreasing, thus robots built from such components if produced at larger size will be very cheap. It would be possible to deploy a large number of such small robots and should be rightly called swarm of robots to achieve the desired task. AIS plays its role by bringing intelligence into this swarm of robots to achieve/accomplish the desired task. This idea has been specifically taken from Immune cells of adaptive or innate immunity. In this regard, the behaviour of neutrophils is particularly the interest of author with the potential towards AIS application and Swarm Robotics.

Neutrophils are the most abundant white blood cells as constitute about 40–70% of the white blood cells in the blood stream (Beers et al. 2006). As a first defense line, they respond early to threats against the hosts by detecting changes in the vascular endothelium induced by tissue damage and/or infection (Janeway et al. 2005). The behaviour of Neutrophils along with other immune cells is quite complex and through the microvascular system initially get access to the source of antigen. This causes changes in the vascular endothelium, giving impetus to neutrophils to exit the microvasculature and to move through the tissue by sensing molecules, or chemokines, produced by damaged or infected tissue (Ariel and Serhan. 2007; Janeway et al. 2005).

Thus, at the initial phase of inflammation caused by bacterial infection, environmental exposure or by some cancers, the first defensive response is received from the neutrophils which migrate towards the site of inflammation. Neutrophils first migrate through the blood vessels and then through the interstitial tissue, following chemical signals (Chemokines produced by damaged or infected tissues) such as Interleukin-8 (IL-8), C5a, fMLP and Leukotriene B4 in a process called *"Chemotaxis"* (Aerial and Serhan, 2007; Janeway et al, 2005; Kadirkamanathan et al, 2012) as indicated in Figure-4.

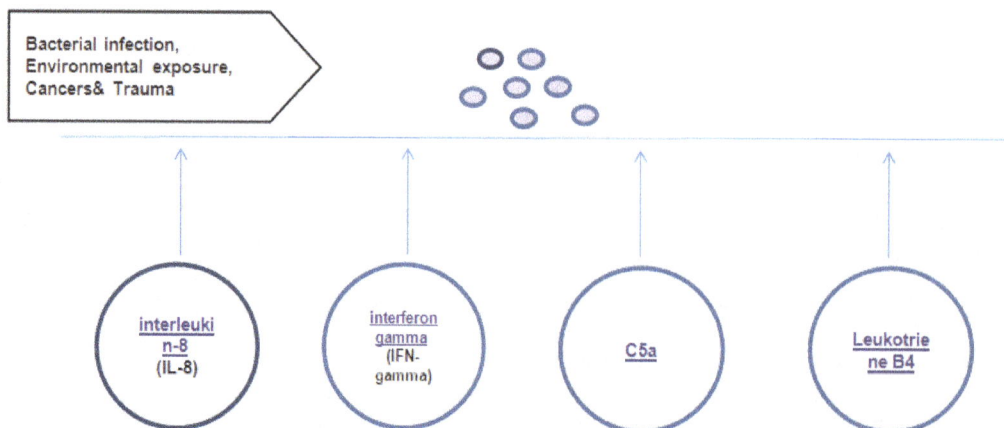

Figure 4. Concept of Chemotaxis.

This process, termed chemotaxis, plays a significant role in immune cell motion computation models (Nathan. 2006). These fundamental behaviours of neutrophils are mathematically well explained through the convection–diffusion and reaction–diffusion equations (Holmes, et al. 2012; Sua et al. 2009). The immune system swarms like neutrophils swarms initially develop from the primary arrest of a small count of neutrophils followed minutes later by massive numbers. This directed migration or dynamic swarming is probably caused by intercellular communication via signaling molecules. These signaling molecules are produced by, and are attracted to, neutrophils. Streaming is another dynamic behavior of neutrophils. It is assumed that neutrophils generate signals to induce swarming and once a swarm reaches a certain size, a large enough signaling center exists to overcome the competing signals of nearby smaller swarms (Chtanova et al., 2008).

6. AIS Applications in Swarm Robotics

Sahin (2005) has defined swarm robotics as "swarm robotics is the study of how large number of relatively simple physically embodied agents can be designed such that a desired collective behaviour emerges from the local interactions among agents and between the agents and the environment". Millonas (1994) has also proposed that a swarm system must encapsulate the principles of proximity, quality, diverse response, stability and the adaptability.

Swarm robotics is an innovative approach to the coordination of large numbers of robots. Basic motivation comes from the observation of birds/insects as to how these individual entities in itself can cooperate together to carry out complex tasks/goals that cannot be accomplished individually. This sort of coordination capabilities are still beyond the reach of current multi-agent robotic systems. The main characteristics of swarm robotics are depicted below as Figure-5:

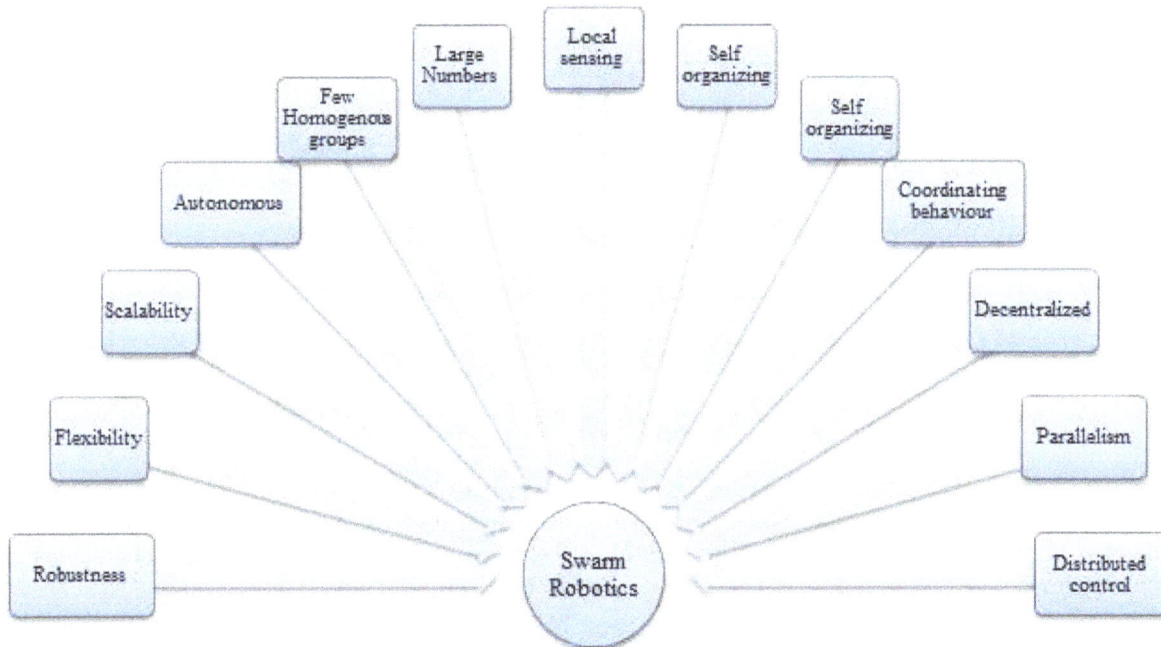

Figure 5. *Characteristics of Swarm Robotics.*

The main feature of swarm behavior is that each individual (robot) follows simple rules and there is no centralized control dictating their behavior. Each robot is capable of observing and responding to its environment and directing its activity towards achieving common goal. Robots will collaborate through interaction among themselves exhibiting simple behaviors like self-assembly, self-repair, co-operation, monitoring and responding.

Following this, it is assumed that mathematical modeling, design and methodology of emergence as well as testing of swarm systems are key tasks to apply neutrophils behavior in swarm robotics.

7. Research Needs in AIS and Swarm Robotics

Based on above facts, it is concluded that immune system is a very complex system and still there is a lack of reliable data about many of its constituent cells and molecules, and thus any simulation that intends to model the immune system at full scale will require huge computational resources. Thus, it allows the interconnection of predictive models, defined at multiple scales (molecules, cells, tissues and organs, body-wide systems, the whole organism, and collection of organisms) for conversion into in systemic networks. These networks can address and validate the systemic hypotheses by combining clinical observations, experimentation and predictive modelling.

AIS based on neutrophils behavior, will play potential role by bringing intelligence into swarm of robots. An extensive literature review has concluded that Swarm Robotics has high prospects and utilization for a variety of challenging applications as shown in the Figure-6 below:

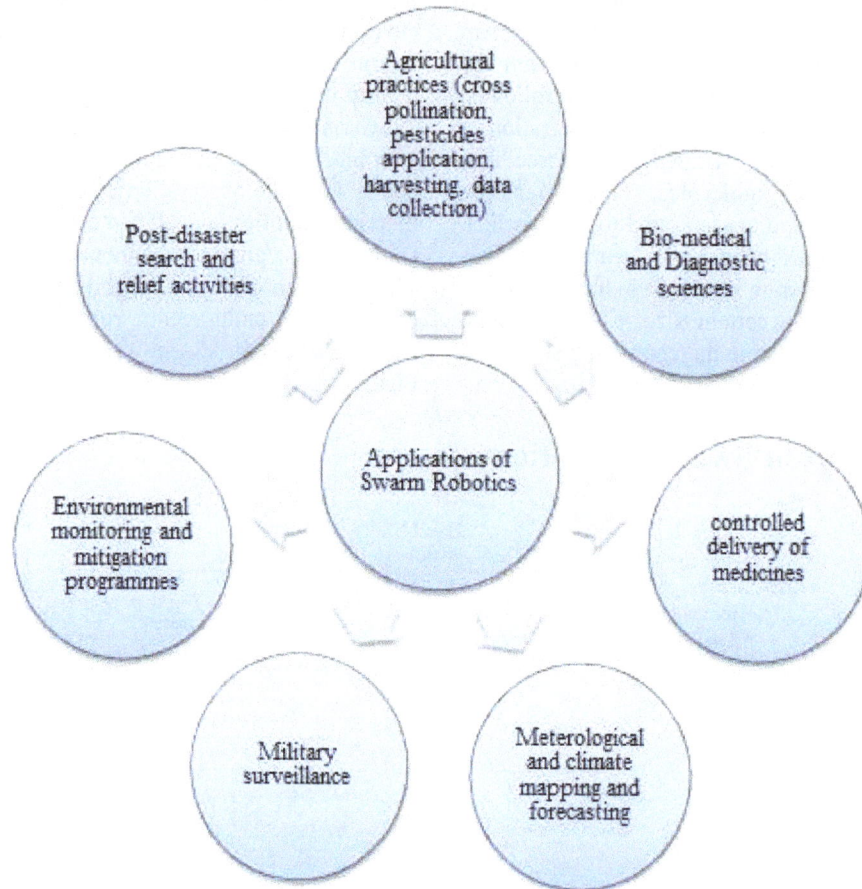

Figure 6. Application of Swarm Robotics in real world.

Application of swarm robotics in wide range of real life fields cannot be ignored for serious consideration for further developments. These applications include biomedical applications for developing micro-nano systems by applying high-resolution monitoring, fast prototyping of micro-environments such as by application of micro-nano robots may be for on-site drug delivery, development of unconventional robots (for DNA devices, nano-particles, synthetic bacteria, bio-bots for advance treatments), synthetic biology applications based on fabrication and manipulation at the micro-nano scale, development of energy-based robot control (magnetic, light-based self-organizing biological and robotic systems) and finally the control of swarm robotic systems with large-scale simulations will be applicable to all such applications. This can be further explored for the development of tools to explore swarm behaviours (bio-inspiration, crowdsourcing, machine learning).

Swarm robots can disperse and perform multiple tasks at difficult and inaccessible sites such as in forests, lakes, hilly areas etc. Swarm of robots because of the robustness of the swarm, can prove highly useful for dangerous tasks including monitoring and mitigating the environmental hazards, like a

leakage of a chemical substance, clearing off the environment from such hazards, mining detection and cleaning off the associated mining wastes. Environmental pollution is one of the biggest concerns of planet earth in context to survival of all living organisms and thus application of AIS and SI in environmental monitoring and mitigation may prove most beneficial, if applied precisely. Under such situation, if one robot fails and the mine explodes, the rest of the swarm continues working and this is not accomplishable in case of working of a single robot.

Dramatic incidences of aquatics pollution due to oil spills, organic pollutants and suspended matters in surface water bodies such as oceans, rivers, lakes etc. have highlighted the potential to address the man-made environmental damage. Mitigation of this unwanted situation requires resources in terms of man power, machinery, labour cost and time. Hence, to provide alternative to all such requirements and constraints, research based on application of advance computational approaches is required to address this real world problem of environmental pollution.

Another environmental application may be based on developing the computational simulation for in-pipe inspection swarm robots capable to overcome the issues of human factor in labour and time intensive monitoring and also to act in inaccessible environment inside the water or gas pipelines of an area within shortest possible time for long-distance inspection.

Future research is required to investigate deployment possibility of swarm of robots to move on the water's surface autonomously and gather organic pollutants and suspended matters to preserve water resources from quality deterioration. The robots may work together to cover a large area of the water and communicate with one another and with land-bound researchers. This will also help in developing an insight so as to how the problems such as large scale simulation, control of mirco/nano swarm robots can be dealt with by using AIS.

References

[1]   Ariel and N.C. Serhan. 2007. Resolvins and protectins in the termination program of acute inflammation, Trends Biochem. Sci. 28(4) (2007), pp. 176–183.

[2]   Alberts, B., Alexander, J., Julia, L., Martin, R., Keith, R and Peter, W. 2002. *Molecular Biology of the Cell; Fourth Edition.* New York and London: Garland Science. ISBN 0-8153-3218-1.

[3]   Aickelin, U and Cayzer, S. 2002. The danger theory and its application to artificial immune systems, in: The 1st International Conference on Artificial Immune Systems (ICARIS 2002), Canterbury, England.

[4]   Burnet F.M., 1959. The Clonal Selection Theory of Acquired Immunity. Cambridge University Press

[5]   Bailey, S. 1984. From desktop to plant floor, a CRT is the control operators window on the process.Control Engineering, 31 (6) (1984), pp. 86–90

[6]   Beers, M.H., Porter, R.S and Jones, T.V.2006. The Merck Manual, 18th edition, Merck & Co., Inc.

[7]   Chtanova T, Schaeffer M, Han S, van Dooren G, Nollmann M, Herzmark P, Chan S, Satija H, Camfield K and Aaron H. 2008. Dynamics of Neutrophil Migration in Lymph Nodes during Infection. Immunity.

[8]   Campelo, F.,Guimarães, F.G., Igarashi, H.,Ramírez, J.,Noguchi, S 2006. A modified immune network algorithm for multimodal electromagnetic problems. IEEE Transactions on Magnetics, 42 (4).

[9]   Castro, L.N.D andTimmis, J. 2002. Artificial immune systems: A novel paradigm to pattern recognition. J.M. Corchado, L. Alonso, C. Fyfe (Eds.), Artificial neural networks in pattern recognition, University of Paisley, Paisley (UK) (2002), pp. 67–84

[10] Castro, L.N.D., Zuben, F.J.V. 2000. The clonal selection algorithm with engineering applications, Genetic and Evolutionary Computation Conference (GECCO'00) – Workshop Proceedings, Las Vegas, Nevada, USA,

[11] Castro, L.N.D and Zuben, F.J.V. 2002. Learning and optimization using the clonal selection principle, IEEE Transactions on Evolutionary Computation, vol. 6, 2002, pp. 239–251.

[12] Castro L. N. de and Zuben, F. J. V. 1999. Artificial Immune Systems: Part I -Basic Theory and Applications", School of Computing and Electrical Engineering, State University of Campinas, Brazil, No. DCA-RT 01/99.

[13] Dasguptaa, D., Yua, S and Nino, F. 2011. Recent Advances in Artificial Immune Systems: Models and Applications. Applied Soft Computing 11 (2011) 1574–1587.

[14] Dasgupta, D. 2006. Advances in artificial immune systems. Computational Intelligence Magazine, IEEE 1(4) (2006) 40–49.

[15] Dasgupta, D., Yu, S and Majumdar, N.S. 2003. MILA – multilevel immune learning algorithm, in: Genetic and Evolutionary Computation Conference (GECCO 2003), Chicago, IL, USA.

[16] Dasgupta, D.1996. Using immunological principles in anomaly detection. In: Proceedings of the Artificial Neural Networks in Engineering (ANNIE96), St Louis, USA.

[17] Dasgupta, D 1999. Artificial Immune Systems and their Applications. Springer, Nerlag (1999) ISBN 3-540- 64390-7

[18] Dasgupta, D. 1998. An overview of artificial immune systems, in: D. Dasgupta (Ed.), Artificial Immune Systems and Their Applications, Springer-Verlag, 1998, pp. 3–19

[19] Forrest, S., Perelson, A.S., Allen, L., Cherukuri, R. 1994. Self-nonself discrimination in a computer. In: SP 1994: Proceedings of the 1994 IEEE Symposium on Security and Privacy, pp. 202–212. IEEE Computer Society, Washington, DC .

[20] Geoffrey R. Holmes, Giles Dixon, Sean R. Anderson, et al. 2012. "Drift-Diffusion Analysis of Neutrophil Migration during Inflammation Resolution in a Zebrafish Model," Advances in Hematology, vol. 2012, Article ID 792163, 8 pages, 2012. doi:10.1155/2012/792163

[21] Greensmith, J., Aickelin, J and Cayzer, S. 2008. Detecting Danger: The Dendritic Cell Algorithm. HP Laboratories HPL-2008-200. Copyright Robust Intelligent systems, IGI Publishing.

[22] Greensmith, J., Aickelin, U and Cayzer, S. 2005. Introducing dendritic cells as a novel immune-inspired algorithm for anomaly detection, in: 4th International Conference on Artificial Immune Systems (ICARIS 2005), Banff, Alberta, Canada.

[23] Grasso, P.; Gangolli, S and Gaunt, I. 2002. *Essentials of Pathology for Toxicologists.* CRC Press. ISBN 978-0-415-25795-4.

[24] Hart, E. and Timmis, J. 2008. Application areas of AIS: the past, present and future. Journal of Applied Soft Computing, 8.

[25] Ishida, Y. 1990. Fully distributed diagnosis by PDP learning algorithm: towards immune network PDP model, IEEE International Joint Conference on Neural Networks, San Diego, USA.

[26] Kalini, N. Karaboga. 2005. Artificial immune algorithm for iirfilter design. Eng. Appl. Artif. Intell., 18 (8) (2005 December), pp. 919–929

[27] Kim, H. 2009. Asymptotic problems for stochastic processes and related differential equations. PhD Dissertation. Department of Mathematics. University of Maryland, USA.

[28] Janeway C.A. et al. 2005.(eds), Immunobiology: The Immune System in Health and Disease, Garland Science Publishing, New York.

[29] Kadirkamanathan, V., Anderson, S.R., Billings, S.A., Zhang, X., Holmes, G.R., Reyes-Aldasoro, C.C., Elks, P.M., and Renshaw, S.A. 2012. The Neutrophil's Eye-View: Inference and Visualisation of the Chemoattractant Field Driving Cell Chemotaxis In Vivo. PLoS One 7(4): e35182.

[30] Knight, T., Timmis, J. 2001. AINE: an immunological approach to data mining, in: IEEE International Conference on Data Mining (ICDM 2001), San Jose, CA, USA.

[31] Kalinli, N. Karaboga. 2005. Artificial immune algorithm for IIR filter design. Engineering Applications of Artificial Intelligence, 18 (2005), pp. 919–929.

[32] Kephart, J.O., 1994. A biologically inspired immune system for computers artificial life IV. In: Brooks, R.A., MaesP., (Eds.), Proceedings of 4th International Workshop on the Synthesis and Simulation of Living Systems, MIT Press, pp. 130–139.

[33] Kephart, J.O. 1994. A biologically inspired immune system for computers. In: Brooks, R.A., Maes, P. (eds.) Artificial Life IV: Proc. of the 4th Int. Workshop on the Synthesis and Simulation of Living Systems, pp. 130–139. MIT Press.

[34] Lee, Z.-J., Lee, C.-Y., Su, S. F. 2002. An immunity based ant colony optimization algorithm for solving weapon-target assignment problem. Appl. Soft Comput., 2 (August (1)) (2002), pp. 39–47

[35] Matzinger,P. 2002.The danger model: are newed sense of self. *Science* 296, 301–305.

[36] Matzinger, P.1994.Tolerance, danger, and theextendedfamily. *Annu.Rev. Immunol.* 12, 991–1045.

[37] Nasraoui, O., Rojas, C and Cardona, C.2006. A framework for mining evolving trends in web data streams using dynamic learning and retrospective validation Comput. Networks, 50 (July (10)) (2006), pp. 1425–1429.

[38] Nathan, C. 2006. Neutrophils and immunity: Challenges and opportunities, Nat. Rev. Immunol. 6 (2006), pp. 173–182.

[39] Millonas MM. 1994. Swarms, phase transitions, and collective intelligence. In: Artificial life III. Addison-Wesley, Reading, pp 417–445.

[40] Sua, B..Zhoua, W, Dormanb, K.S and Jones, D.E. 2009. Mathematical modelling of immune response in tissues. Computational and Mathematical Methods in Medicine. Vol. 10, No. 1, March 2009, 9–38.

[41] Şahin, E. 2005 "Swarm robotics: from sources of inspiration to domains of application," in Swarm Robotics Workshop: State-of-the-Art Survey, E Şahin and W. Spears, Eds., Lecture Notes in Computer Science, no. 3342, pp. 10–20, Berlin, Germany, 2005.View at Scopus

[42] Somayaji A., Hofmeyr S., and Forrest S., 1997. Principles of a Computer Immune System. In: Proceedings of the Second New Security Paradigms Workshop, pp. 75–82.

[43] Timmis, J., Hone, A. Stibor, T and Clark, E. 2008. Theoretical advances in artificial immune systems, Theoretical Computer Science 403 (2008) 11–32.

[44] Timmis, J., Neal, M and Hunt, J. 2000. An artificial immune system for data analysis, Biosystems 55 (2000) 143–150. (10).

[45] Yus, S and Dasgupta, D. 2008. Conserved Self Pattern Recognition Algorithm, in: 7th International Conference on Artificial Immune Systems, Phuket, Thailand.

[46] Zhong, Y., Zhang, L and Huang, B andLi, P.2006. An unsupervised artificial immune classifier for multi/hyperspectral remote sensing imagery. IEEE Trans. Geosci. Remote Sens., 44 (February (2)) (2006), pp. 420–431

7

Workflow Management in Intelligent ERP at Aseman Airlines Company

Hojatollah Rashidi Alashty[1], Filimonov Aleksandr Borisovich[2]

[1]Cybernetics Department of the MATI University – Russian State Technological University named after K. E. Tsiolkovsky and member of research group of the Islamic Azad University, Qaemshahr, Mazandaran, Islamic Republic of Iran (Department of management)
[2]Cybernetics Department of the MATI University– Russian State Technological University named after K. E. Tsiolkovsky

Email address:
rsh_hojat@yahoo.com (H. R. Alashty), filimon_ab@mail.ru (F. A. Borisovich)

Abstract: The real goal of this paper is to study a framework for intelligent ERP in repair & maintenance at Aseman Airlines Company based upon workflow management. For this purpose, a workflow management system was designed at Aseman Airlines Company by presenting an intelligent model and applying different concepts like genetic algorithm, repair & maintenance at depot level, social capital and long-life management system. All managerial & technological aspects were considered in the design of this system in order for upgrading- efficiency and quality of aviation industry services through applying intelligent systems.

Keywords: Workflow Management, Intelligent ERP, Aviation Industry

1. Introduction

Managers of organizations and departments involved in the maintenance, repair and overhaul of airplanes are obliged to seek new methods and programs for reducing costs and increasing the efficiency because of two reasons. The first reason is related to high levels of man power and consuming parts and total primary investments which are expensive and the second reason is related to extraordinary importance of performance time and other factors like dollar dependency, technical and somehow applicable dependencies from the other. Upon daily-increase entrance of computer into the world of industry and management, all related parts of industries were introduced to this factor accordingly. Of course, modern and complex industries like aviation industry have benefited from these facilities more ethan other industrial groups even beginning from the stage of primary designing of flying objects up to stage of manufacturing the autopilot systems.

Recently, it is common to use computer in various fields like repair, maintenance, overhaul, natural testing systems, programming for periodic checks, life-saved parts replace, controlling of repairs without any need to issue of paper notes, light digital tasks instead of thick books, warehousing and relevant fields, wide range of controlling networks, educational & repairing simulations and most of other common fields. Today repairs and maintenance need something more than wrench and usual physical tools. Rather than searching for new methods, here we intend to introduce relevant repairs and maintenance of them as the under-experimented or tested guidelines.

2. Workflow Management Model in Intelligent ERP

2.1. Applicable Model of Workflow in ERP

Introduced model of this research is mainly oriented on workflow collaborative environment. Figure 1 illustrates introduced model where tasks and roles are defined for collaboration among members.

Such collaborative environment includes three key concepts as follows: Members, roles and duties. Members are key factors in a collaborative environment including all important members in aviation industry in the fields of control, handling, maintenance, guidance, flight safety, flight engineers, servants and generally all active parts in aviation industry. Information is produced by members and while performing their assignments. For instance, flight safety engineers control and report relevant information of safety system of airplane in each flight and supply flight safety in collaboration with other technical units. Members in *Aseman Airlines Company* may have one or more official roles. For instance, we have Airplane mechanic engineer, airplane maintenance engineer, flight safety engineer, mechanical designer and so on. Furthermore, there are various up-down relations among them. They will provide a hierarchy of duties named as "Duties Tree". Duties tree illustrates the architecture of collaborative team of *Aseman Airlines Company*. In addition to members drawing, duties tree has also duties model drawing. Regarding any relations among members we should point out to the important role of social capital of members in upgrading of services level at this airlines company. For explanation of this conception, which can influence on workflow, assume there are infrastructure capitals in aviation industry like enough hardware and

software facilities. Also we may assume that all members of company passed enough education and in fact company human capital has a suitable situation. But only mentioned infrastructures and human capital are not enough in improvement of services, but it is necessary to have suitable cooperation among organizational members. They should find suitable cooperation and interaction with each other for organizational goals achievement. Such a cooperation among members is considered as required social capital for achieving better effects of workflow system. Therefore any attention to social capital plays a great role. But benefiting from advanced software systems will assist to more interaction among members and reduce organizational costs and increase quality of services at *Aseman Airlines Company*.

Diagram 2 illustrates the roles of all members.

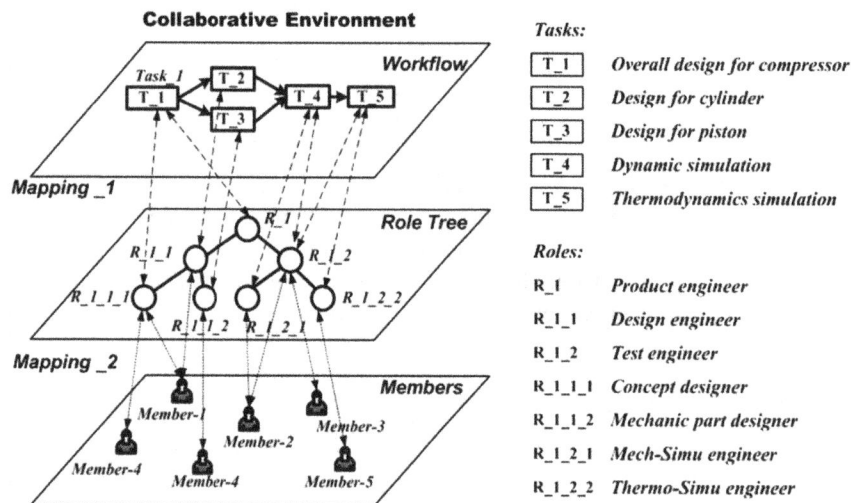

Figure 1. Workflow model based on collaborative environment with definition of tasks and roles.

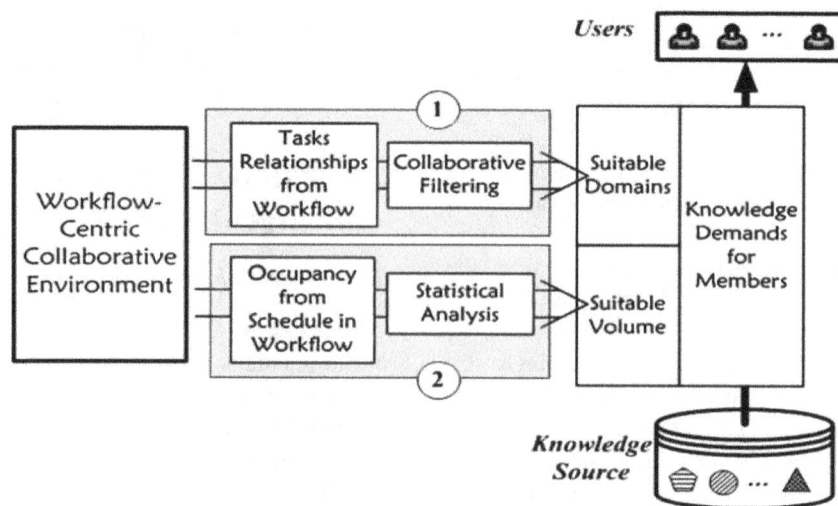

Diagram 1. Workflow & Members of Airlines Company.

Diagram 1 illustrates major framework of instructing system based upon workflow. There are two kernel modules as specified in this model.

There is a new collaborative filtering for the first problem based upon duty relations (or roles) of work flow in order to

obtain communicative table of "Member fields of information". It may reflect any requests of members to the suitable information fields. Workflow will understand any cooperation among team members through a final and logical process and benefit the same for generalizing distributed and non-harmonized duties into an independent process. There are lots of information in workflow. For instance, we have logical sequential relations among any duties (activities) of team members and resource information of member – roles – duties.

It illustrates which member plays what role or carry out what duties. It is possible to combine above-mentioned information with a cooperative filtering in order to find out any requests of members for more information of similar colleagues. It may guarantee that suitable modules of information resources could be recommended to suitable members in a cooperative team [4-6].

There is a fixed analysis method for second problem and finding out involvement rate in proposed workflow for recognition of suitable volume of recommended information [3-5].

Following two parts present partial information about above-mentioned key problems as well.

In second module statistical analysis has a fundamental role for guidance, control and safety of flight . Here we may present a modern analysis of this module:

One of the most common and oldest methods of Optimization Approaches is analytical solving and applying mathematical methods. This is applicable by the use of Objective Function derivation and also current limitations in different issues. Then it is possible to determine maximum or minimum point by the use of second derivative function [2-3].

Regarding any issues in which target function is a little more complex and with hardly derivation, we may use Numerical Approaches. Although it may reduce work accuracy and/or cause non-fixed condition in problem solving, but it may assist to simplify and solve the most engineering problems. Both methods, in complex situations, both mentioned methods need difficult, long-term and bothering calculations with lots of possible mistakes. In addition, these methods may be trapped with local optimizations and do not specify general optimization in alternative functions or Local Optimums and General Optimum.

The most important point is the lack of a special function for defining of problem. Most of mentioned problems are resulted from various functions even with special forms in different conditions of variants. Therefore in such a condition, it is impossible to apply analytical & numeric methods for any derivation.

For removing such a problem we may apply Genetic Algorithm (GA) as a modern method and a powerful tool in solving complex problems of optimization [7, 8].

Genetic Algorithm is a searching algorithm with a guided search which tries to find a good answer for an assumed optimized function with examination of it. But what makes it different from other methods is parallel and random search. This means that genetic algorithm will test a collection of search spaces in each generation and benefits from transfer and random selection in next generation appointments. Since it may benefit from answer test, it is not necessary to have an derivable function. It will find its optimization in any way. In

addition and by random search, algorithm genetic has broken some of complexity bounds. In a clear form and from theoretical viewpoint, if calculation complexity of problem solving need great number of instructions out of the range of facilities in current computer. Then it is possible to solve them by the help of genetic algorithm and quantitative instructions [7, 8].

The other point is that functions which are considered in different issues may be discontinuous and non-derivable and non-convex. The mentioned conditions will make it difficult problem or impossible to solve function optimally,. As a result, it is necessary to have a wide range of search in any problems with great solving space. But exact methods have usually high calculation costs. As a result, in spite that random search methods are included in the group of random algorithms, but they are used for solving of geometrical problems. Usually there are acceptable close solutions for practical optimization, because it is preferred to find a close optimized answer for a logical period of time in order find complete optimized solution in a long-term and expensive time [7, 8].

2.2. Encoding in Intelligent Systems Based Upon Genetic Algorithm

Encoding is one of the major issues in genetic algorithm as mentioned in following encoding methods:

A) Binary Encoding

It is a common method in which we have genetic algorithm. Variables would be replaced with suitable strings of 0's and 1's in this method. The number of bits for encoding of variables depends upon required precise for answers and change limitation of variables. Encoded variables are placed in a sequential string. Bits may be encoded in binary encoding in integers or real numbers [7, 8].

Oliveira and Loucks specified that Real-Value encoding is more effective than binary one. Coding components of possible solutions into a chromosome is the first part of a GA formulation. Each chromosome is a potential solution and is comprised of a series of substrings or genes, representing components or variables that either form or can be used to evaluate the objective function of the problem. [9]

Each position in a chromosome is a real value. Real-value vectors are especially useful for solving real-value optimization problems. Permutations area popular representation for some combinatorial optimization problems [11].

As a result, the strings include all encoded variants in concerned issue for illustrating a point in solution space known as a chromosome. Followings are encoded chromosomes:

Chromosome A: 1011000111101010000 1110101
Chromosome B: 1111100000010101000 01101000

There are two types of encoding such as continuous and discrete variables encoding. If the upper and lower bounds of a variable like x are respectively x_u and x_u and ε is the accuracy rate of it for binary encoding of continuous variables, the following formula is used for determining a minimum number of required bits for variable [7, 8].

$$2^m \geq \frac{x_u - x_l}{\varepsilon} + 1 \qquad (1)$$

where m is the minimum number of bits required for the variable.

For binary encoding of discrete variables the length of m in discrete variables depends upon the number of these variables. For instance, if there are 16 discrete variables we should consider four bits for the string length of each variable. Table 1 illustrates any relation between discrete variables and binary string [7, 8].

Table 1. *Relation of discrete variables & binary string.*

Value of discrete variants	Variable number	Binary string similar to variable
D_1	0	0000
D_2	1	0001
D_3	2	0010
.	.	.
.	.	.
.	.	.
D_{15}	15	1111

Excessive-Distribution method is recommended if the number of discrete variables are non-definable by 2^m. There are more than one binary digit would be allocated for every variable in this method [7, 8].

B) Permutation encoding

Permutation encoding can be used in ordering problems. In permutation encoding, every chromosome is a string of numbers, which represents number in a sequence as follows:

Chromosome A: 153264798

Chromosome B: 856723149

It is necessary to make some modifications after some combinations and mutations in permutation encoding. This is for ensuring about real sequence in a string [7, 8].

The mutation depends on the encoding as well as the crossover. For example when we are encoding permutations, mutation could be exchanging two genes [10].

C) Value Encoding

Direct value encoding can be used in problems, where some complicated value, such as real numbers, are used. Use of binary encoding for this type of problems would be very difficult. In value encoding, every chromosome is a string of some values. Values can be anything connected to problem, form numbers, real numbers or chars to some complicated objects.

Value encoding is very good and useful for some special problems. On the other hand, for this encoding is often necessary to develop some new crossover and mutation specific for the problem [7, 8].

D) Tree Encoding

Tree encoding is used mainly for evolving programs or expressions, for genetic programming. In tree encoding every chromosome is a tree of some objects, such as functions or commands in programming language.. Tree encoding is good and useful for evolving programs. Programing language LISP is often used for this, because programs in it are represented in this form and can be easily parsed as a tree, so the crossover and mutation can be done relatively easily , like a tree as illustrated in following figure, we may assume input and output values. The target is finding a function with the best outputs for the inputs [7, 8].

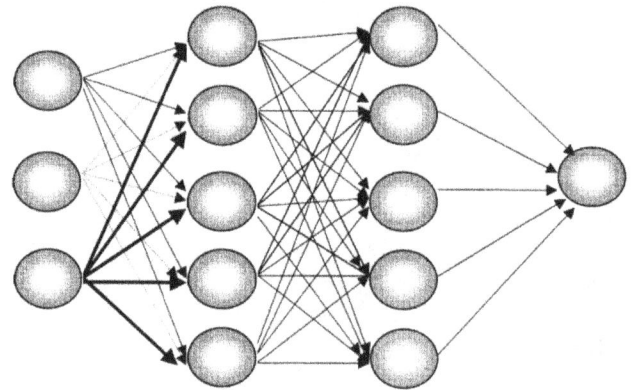

Figure 2. *Tree encoding in genetic algorithm.*

Finally it is possible to apply genetic algorithms in designing of intelligent ERP systems at Airlines Company for statistical analysis and designing of intelligent control systems as mentioned in this research [7, 8].

Furthermore following models are proposed in repair and maintenance of airplanes:

A- Repair & Maintenance at depot level

The mentioned Repair & Maintenance IT could provide required information of equipment according to serial number. After providing of this technology, the system would be enabled for program repair and maintenance at parts level (depot level). Then it is possible to estimate required repairs and maintenance for benefiting from flying through specified times for this purpose. This system activates automatic link of recorded information at flight for prevention from any interfere for persons in charge of repair and maintenance due to very emphatic information. Also it makes it possible to view and evaluate the fleet with more than 750 aircraft and 2000 motors. Also it may browse general view and conditions for approving repair and maintenance management [1, 4].

Repair and maintenance information are reserved in a way to enable controlled system works as softness as possible. The mentioned information would be sent to central data base and make it possible to estimate required spare parts (if a network connection is available) in future at fleet and making requests from resources through internet. Also it is used for sending repair & maintenance modifications from central department to the headquarters for daily updates. Now this system is able to send any recorded information about repairs and maintenance of landing airplane from one headquarter to another. Flight technicians prefer this system more because of its friendly and easy usage. It enables all technicians to allocate more times for real repairs and maintenance. Also they should spend little time for preparing of complex reports and filling of forms [1, 4].

B: Management system for long-life

The real goal of any cargo, passenger or military plane is facility of repair, maintenance and its permanence through the long-life. With creation of clear and suitable responsibilities for reaching more confidence and efficiency from ownership time up to the final step with economy, management system for long-life may improve support process as well. Various modifications like end-to-end are originated from non-military airplane facilities. Airplane manufacturers supply a program for supporting of parts, updating, regular documentation, and relevant information of repair and maintenance and information packs as a part of total value of collection [4, 5].

3. Conclusion & Future Work

This research is about workflow management system and its application in intelligent ERP systems at aviation industry. The mentioned innovation in this research is introducing and applying advanced genetic algorithm in designing the system and also statistical analysis as a fundament module in workflow system. This results in upgrading repair and maintenance of ERP systems at *Aseman Airlines Company*. Therefore, it is proposed to apply genetic algorithm and pay more attention to social capital of this process. In addition, this research presents any repair and maintenance at depot level along with long-life management system based upon workflow management at aviation industry. Upon applying of intelligent systems, all managerial and technological aspects of this system have been considered for upgrading of efficiency and quality of services at aviation industry

accordingly.

References

[1] R.H. Sprague, Jr., "A Framework for the Development of Decision Support Systems," *MIS Quarterly*, 4, no. 4 (December 1980), pp. 1-26.

[2] P. Gray and H.J. Watson, *Decision Support in the Data Warehouse*, Upper Saddle River, NJ, Prentice-Hall, 1998.

[3] C. White, "Now is the Time for Real-Time BI," *DM Review*, (September 2004), pp. 47-54.

[4] Richard Hackathorn, "The BI Watch: Real-Time to Real Value," *DM Review*, 14, no. 1 (January 2004).

[5] Additional information about the real-time BI best practices at Continental is described in R. Anderson-Lehman, H.J. Watson, B.H. Wixom, and J.A. Hoffer, "Continental Airlines Flies High with Real-Time Business Intelligence," *MIS Quarterly Executive*, 3, no. 4 (December 2004), pp. 163-176.

[6] The story of Continental's turnaround can be found in G. Brenneman, "Right Away and All at Once: How We Saved Continental," *Harvard Business Review*, 76, no. 5 (September/October 1998), pp. 162-74.

[7] Syswerda, G. (1989): "Uniform Crossover in Genetic Algorithms, In: Proceedings of the Third International Conference on Genetic Algorithms", Schaffer, J. (Ed), Morgan Kauman Publishers, Los Altos, CA, PP. 2–9.

[8] Mitchell, M. (1995): "Genetic Algorithms: An Overview", Cambridge, Mass Complexity, 1(1), PP. 31-39.

[9] Oliveira, R., and Loucks, D. P. (1997). "Operating rules for multireservoir systems." Water Resour. Res., 33(4), PP. 839–852.

[10] ZHANG. J, Chung. H and Lo. W. L, "Clustering-Based Adaptive Crossover and Mutation Probabilities for Genetic Algorithms", IEEE Transactions on Evolutionary Computation vol.11, no.3, pp. 326–335, 2007.

[11] J. D. Schaffer, "A study of control parameters affecting on-line performance of genetic algorithms for function optimization," in *Proceedings of the 3rd International Conference on Genetic Algorithms*, pp.675–682, 1989.

8

The Sectored Antenna Array Indoor Positioning System with Neural Networks

Chih-Yung Chen[1], Yu-Ju Chen[2], Ya-Chen Weng[1], Shen-Whan Chen[3], Rey-Chue Hwang[4, *]

[1]Department of Computer and Communication, Shu-Te University, Kaohsiung City, Taiwan

[2]Department of Information Management, Cheng Shiu University, Kaohsiung City, Taiwan

[3]Department of Communication Engineering, I-Shou University, Kaohsiung City, Taiwan

[4]Department of Electrical Engineering, I-Shou University, Kaohsiung City, Taiwan

Email address:

mikechen@stu.edu.tw (Chih-Yung Chen), yjchen@csu.edu.tw (Yu-Ju Chen), s97115209@stu.edu.tw (Ya-Chen Weng),
jasonchen@isu.edu.tw (Shen-Whan Chen), rchwang@isu.edu.tw (Rey-Chue Hwang)
[*]Corresponding author

Abstract: This paper presents a sectored antenna array indoor positioning system (IPS) with neural network (NN) technique. The hexagonal positioning station is composed of six printed-circuit board Yagi-Uda antennas and Zigbee modules. The values of received signal strength (RSS) sensed by wireless sensors were used to be the information for object's position estimation. Two NN models, including NN with back-propagation (BP) learning algorithm and probabilistic NN (PNN), were applied to perform the positioning work for a comparison. In the experiments, an 8x8 square meters indoor scene was performed and 288 points and 440 points were experimented in this area. The positioning results show that both NN models have the average error less than 0.7 meter. In other words, the proposed positioning system not only has the high positioning accuracy, but also has the potential in real application.

Keywords: Sectored Antenna, Indoor Positioning System (IPS), Neural Network (NN), Received Signal Strength (RSS)

1. Introduction

It is well known that IPS has become more and more popular in the object searching due to the rapid developments of wireless communication technique and personal network [1-3]. IPS is used to provide the location information of person and device. It is a system of location based service (LBS) through the integration of the wireless communication and information services. The object's accurate position could be determined by using such a service system. It has been used widely in the various applications, such as cargo management, patient monitoring, public guiding system, etc. Undoubtedly, IPS will certainly play a significant role for the smart life of human beings in the future.

Generally, the structure of IPS could be divided into two parts, i.e. the positioning algorithm and the sensing infrastructure. The positioning algorithm is the method of determining the object's location. So far, three algorithms, including triangulation, scene analysis and proximity, are mainly used for the object's position estimation [4-9]. The sensing infrastructure is related to the wireless communication technology used for IPS. Nowadays, the various wireless communication technologies such as wireless local area network (WLAN) [10-13], wireless sensor network (WSN) [14-15], radio frequency identification (RFID) [16-18], Bluetooth [19-20], Zigbee [21-22], etc. have been wildly used in the sensing technique of IPS. Each IPS has its advantage and limitation in accordance with the developed element's function. In which, many positioning computational algorithms used the values of RSS sensed from the known reference nodes to calculate the object's coordinate [23-24]. But, since the influences of external factors such as the obstacle of hindrance, the noise disturbance and the diffraction of electromagnetic wave, the positioning computational method is still a challenging topic in the research of IPS application.

In recent years, due to the powerful learning and adaptive capabilities, NN technique has been employed into the positioning applications [25-31]. It is used to catch the nonlinear mapping between the coordinate of object and RSS signals. Through a training process, the well-trained NN model then can be used to estimate the object's location based on RSS measurements. In this research, two NN models, including NN with BP learning algorithm and PNN, were applied to perform the positioning work for a comparison. The detailed NN models will be described in the following section.

In order to improve the positioning accuracy, Cidronali et al [32] designed a new switched beam array antenna for wireless indoor positioning application. The antenna is intended to augment a wireless device operating as the coordinator or base station, and its design is suitable for installation on the ceiling of any large indoor space [33-34]. Similar to [32], this paper presents a novel indoor positioning scheme which is composed of array antennas and Zigbee modules. The information of signal angle and RSS are used to estimate the object's location. The whole paper is organized as follows. The proposed indoor positioning system is presented in Section 2. Section 3 describes the NN models for the positioning estimation. Section 4 presents the relevant experiments and results. At last, a conclusion is given in Section 5.

2. The Proposed Indoor Positioning System

In this research, Figure 1 shows the developed IPS module which consists of two parts. One part is the indoor positioning station and the other part is the embedded positioning device. The indoor positioning station is composed of a sectored antenna array, a microcontroller and six Zigbee modules. The sectored antenna array has six printed-circuit board (PCB) Yagi-Uda antennas with hexagonal arrangement which can provide 360 degrees coverage.

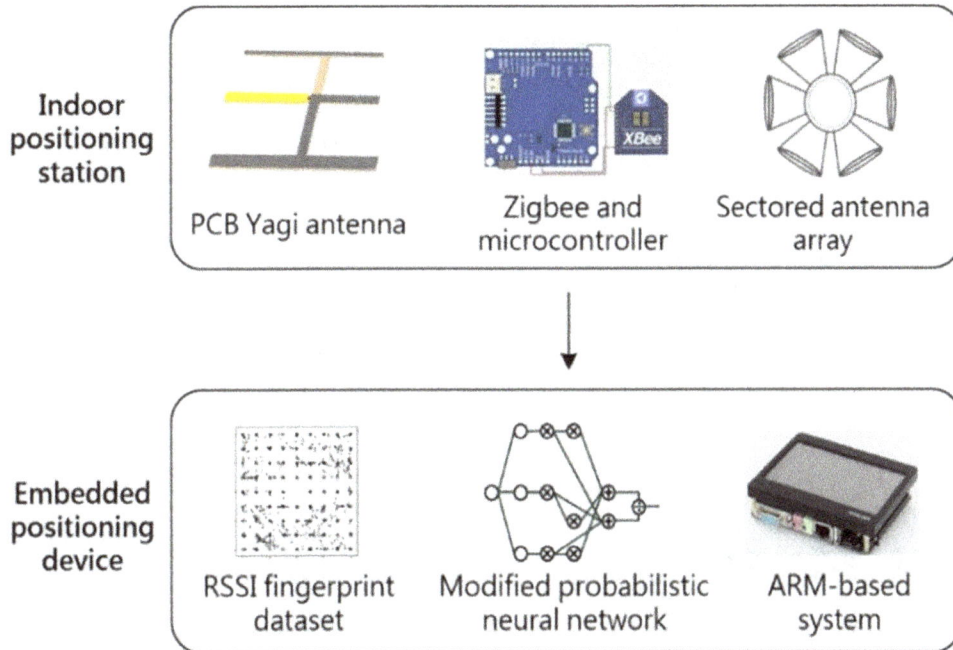

Figure 1. The proposed indoor positioning system.

Yagi-Uda antenna is one of the most successful radio frequency directional antenna designs. It has been used in a wide variety of applications that the antenna design needs gain and directivity [35]. Figure 2 shows the figures of Yagi-Uda antenna and its radiation pattern.

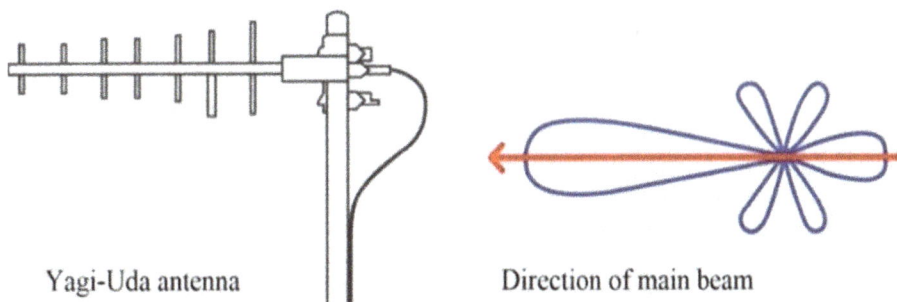

Figure 2. Yagi-Uda antenna and radiation pattern.

The embedded positioning device is developed to perform the indoor positioning function. Through a training process, NN model could calculate the object's position by using RSS fingerprint dataset obtained from the station. Then, the positioning result could be displayed on the screen of ARM-based system.

The RSS based positioning technique estimates the position from the signals of RSS vector which can be obtained by wireless sensors. The radio propagation model with positioning algorithm is a common way to determine the distance between an object and the station. Its equation is generally expressed as

$$RSS = -1 \cdot \left(10 \cdot n_p \cdot \log_{10} d + A\right) \tag{1}$$

Where n_p is the signal propagation constant, d is the distance between object and sensor and A is the object's RSS when the distance is 1 meter.

3. NN and Modified PNN

In our studies, traditional NN model with BP learning algorithm and modified PNN model were employed into the positioning works for a comparison. Two NN models are briefly described as follows.

3.1. NN Model

A three-layered feed-forward fully connected NN is used in the studies. The size of NN is 6-13-2 which means NN has 6 input nodes, 13 hidden nodes and 2 output nodes. It structure is shown in Figure 3. Two output nodes estimate the values of x and y axes of object's coordinate, respectively.

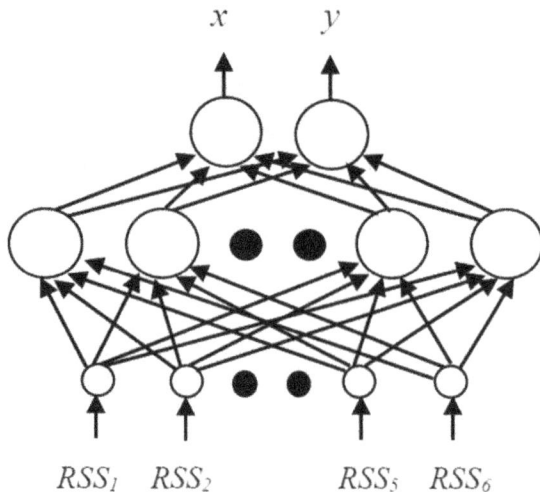

Figure 3. *The NN structure.*

The error back-propagation (BP) learning algorithm is adopted by NN model. The learning process is all training inputs are presented cyclically until all weights of NN are stabilized. The major steps of BP learning algorithm are summarized as follows [36-37].

Step 1: Initialize all weights, $\omega_{i,j}$ to small random values (typically between -0.5 to 0.5).

Step 2: Present an input pattern and specify the desired output. Calculate output using the presents $\omega_{i,j}$.

Step 3: Find the error term, δ for all nodes. If D_j, O_j and H_j denote the desired value of jth output node, the computed value of jth output node, and the computed value of jth hidden node, then the error terms of all nodes could be calculated by using the following equations.

The error of output layer node j:

$$\delta_j = (D_j - O_j)O_j(1 - O_j) \tag{2}$$

The error of hidden layer node j:

$$\delta_j = H_j(1 - H_j)\sum_k \omega_{jk}\delta_k \tag{3}$$

where k is over all nodes in the layer above node j.

Step 4: Adjust weights by

$$\omega_{i,j}(n + 1) = \omega_{i,j}(n) + \eta\delta_j u_i + \zeta(\omega_{i,j}(n) - \omega_{i,j}(n - 1)) \tag{4}$$

Where $(n+1)$, (n) and $(n-1)$ indicate the next, present, and previous iteration numbers, respectively. u_i is the ith input connected with node j. η is the learning rate. ζ is a momentum that effectively filters out high-frequency variations of the error surface.

Step 5: Present next input pattern and go back to step 2.

3.2. Modified PNN

The modified PNN was initialized by Zaknich [25]. In this research, PNN is applied to estimate the coordinate of object. The architecture of modified PNN is shown in Figure 4. It consists of one input layer, one pattern layer, one summing layer and one output layer. The algorithm of modified PNN is described as follows.

Let **C** be a set of class vectors i.e. IPS training data, which is given by

$$\mathbf{C} = \left\{(c_1, y_1),(c_2, y_2)\ldots,(c_m, y_m)\right\} \tag{5}$$

where m is the number of class vectors. c_i contains six RSS signals sensed by antenna and y_i is the scalar output related to c_i. Here, the probability density function (PDF) of modified PNN is defined as

$$\Phi(x, c_i, \sigma) = \exp\left(-\frac{(x - c_i)^T (x - c_i)}{2\sigma^2}\right) \tag{6}$$

where σ is the smoothing parameter of Gaussian function, x is the training vector for class i in the input space. Thus, the output $\hat{y}$ i.e. the coordinate of object can be obtained by

$$\hat{y}(x) = \frac{\sum_{i=1}^{m} z_i y_i \, \Phi(x, c_i, \sigma)}{\sum_{i=1}^{m} z_i \, \Phi(x, c_i, \sigma)} \qquad (7)$$

where z_i is the number of x_i associated with c_i.

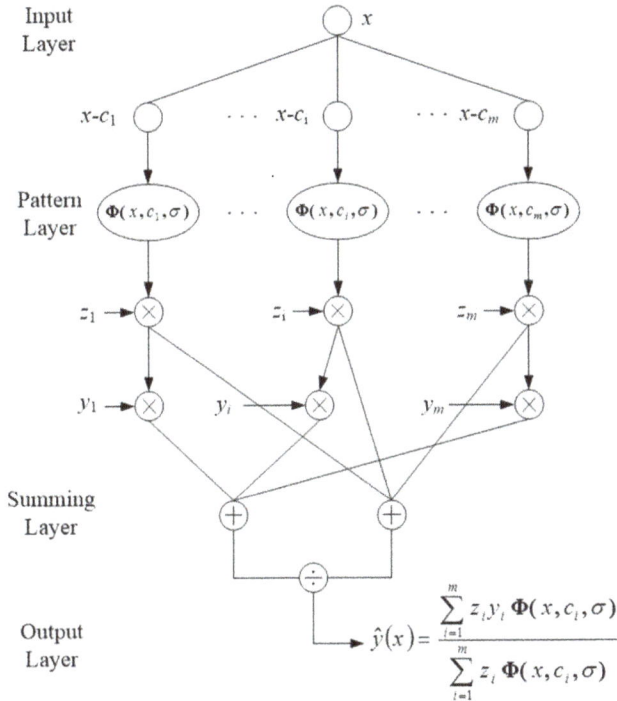

Figure 4. The architecture of PNN.

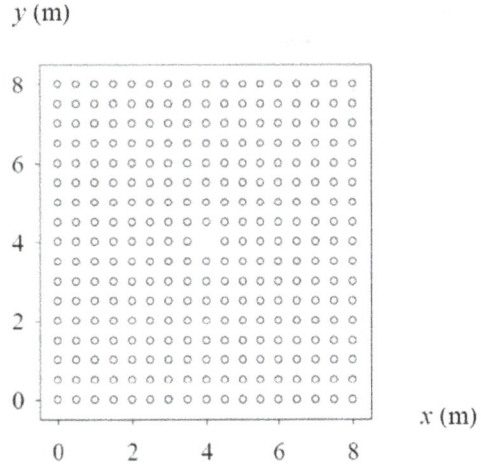

Figure 6. The illustrated figure for 288 positions.

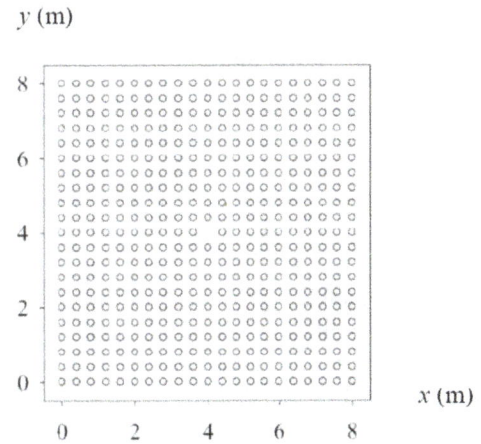

Figure 7. The illustrated figure for 440 positions.

4. Experiments and Results

In this research, an 8x8 square meters indoor field as shown in Figure 5 is used for the experiments. In order to test the indoor positioning system developed, 288 and 440 positions (features) within the intervals of 0.5 meter and 0.4 meter were measured, respectively. The features collected by IPS station are RSS signals measured by wireless signal receiver. Figure 6 and Figure 7 present the illustrated figures for 288 and 440 test positions, respectively.

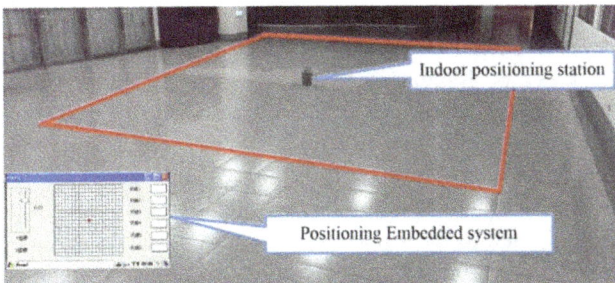

Figure 5. The indoor experimental field.

4.1. The Experiments by NN with BP Learning Algorithm

Firstly, NN model with BP learning algorithm was used to do the object's position estimation. For 288 points experiment, two data groups were collected randomly, each group has 288 data sets and each set includes the information of point's coordinate (x, y) and six RSS sensed values (RSS_1, RSS_2, ..., RSS_6). First data group was used for NN's training and second data group was used for test. Table 1 lists the mean absolute errors (MAEs) of positional estimations by using 6-13-2 NN model with 10 different learning rates. From the results shown, the best performance is taken by NN model with 0.1 learning rate. The training MAE and test MAE are 48.5 cm and 52.09 cm, respectively. Similarly, two data groups for 440 points were collected either. The same experiment was redone by 6-13-2 NN model. The positioning results is also shown in Table 1. Again, the best performance is taken by NN model with 0.1 learning rate. The training MAE and test MAE are 62.11 cm and 63.80 cm, respectively.

In this part of experiments, NN is viewed as the nonlinear regression model for performing a nonlinear input-output mapping. NN generates an approximate function to the training data.

Table 1. The statistic errors of 288 and 440 positional estimations by NN.

Learning rate	288 points		440 points	
	Training MAE	Test MAE	Training MAE	Test MAE
0.1	48.50	52.09	62.11	63.80
0.2	55.04	55.26	62.73	68.02
0.3	59.26	56.88	65.83	69.63
0.4	50.90	57.71	60.56	65.95
0.5	57.22	58.41	63.55	70.36
0.6	59.20	58.84	62.48	69.06
0.7	60.64	59.60	64.58	71.39
0.8	56.57	59.86	67.88	69.41
0.9	58.19	61.32	63.89	67.92
1.0	60.99	64.99	66.22	70.61
Avg.	56.65	58.50	63.98	68.62

4.2. The Experiments by Modified PNN

In the experiment by using PNN model, the same data groups were performed by modified PNN with 10 different σ values. Table 2 presents the estimated positioning errors. From the results shown, PNN model with $\sigma = 0.01$ has the best estimation. The estimated MAEs of 288 points and 440 points could reach to 1.96 cm and 1.72 cm. Figure 8 and Figure 9 show the plots of 288 and 440 positional estimations. The symbols of circle and dot are actual and estimated coordinates of the experimental data.

Unlike previous NN's experiments, in this part of research, the modified PNN is viewed a classifier which is used to estimate the object's position in accordance with the features (RSS signals) sensed. Compare the results of Table 1 with Table 2, it is clearly found that the positioning accuracy performed by modified PNN is much better than NN with BP learning rule.

Table 2. The statistic errors of 288 and 440 positional estimations by modified PNN.

σ	288 points	440 points
	MAE	MAE
$\sigma=0.1$	109.88	108.97
$\sigma=0.09$	95.77	95.97
$\sigma=0.08$	82.00	83.40
$\sigma=0.07$	68.82	71.15
$\sigma=0.06$	55.82	58.70
$\sigma=0.05$	42.33	45.20
$\sigma=0.04$	28.08	29.77
$\sigma=0.03$	13.76	14.45
$\sigma=0.02$	4.26	4.49
$\sigma=0.01$	1.96	1.72
Avg.	50.27	51.38

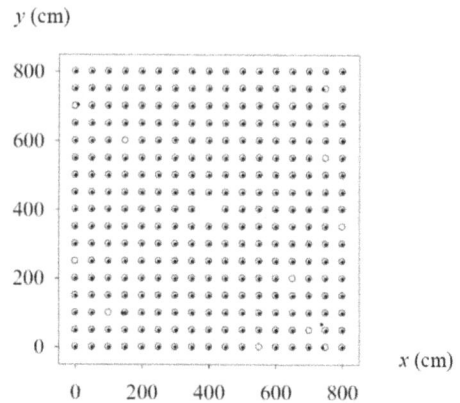

Figure 8. The plot of 288 positional estimations.

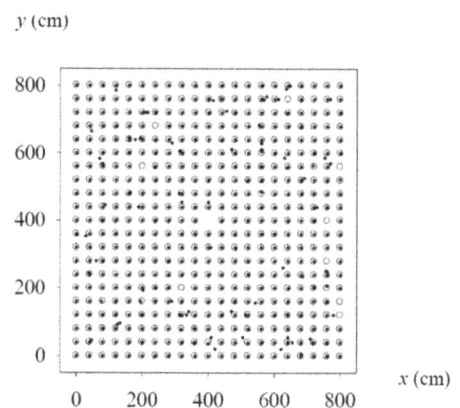

Figure 9. The plot of 440 positional estimations.

5. Conclusion

This paper presents a sectored antenna array indoor positioning system which structure includes the positioning algorithm and the sensing infrastructure. The infrastructure is composed of six printed-circuit board Yagi-Uda antennas and Zigbee modules which are used to generate and sense the RSS signals. Two NN models are the positioning methods to perform the object's position estimation according to the signals of RSS sensed. In our studies, the positioning accuracy performed by modified PNN model is much better than traditional NN. That means the classifier by modified PNN has the outperformance than the regression function generated by traditional NN in indoor positioning application. However, such a conclusion is given by the environment of RSS based positioning system must be stable and the signals of RSS have no serious problem caused by the effects of interference, diffraction or reflection. From the results shown in Table 1 and Table 2. It is able to be found that the estimated accuracies performed by modified PNN model are highly related to the values of σ. On the contrary, compare with the modified PNN model, the NN model with BP learning is more stable in performing the estimations. The variances of estimations are small. Thus, how to establish a more excellent and stable indoor positioning system is still the future work we will continue.

Acknowledgements

This research was supported by the Ministry of Science and Technology, Taiwan, ROC under Contracts No. MOST-104-2221-E-366-004, No. MOST-104-2632-E-366-001 and No. MOST-103-222-E-214-050.

References

[1] Y. Y. Gu, A. Lo, I. Niemegeers, "A survey of indoor positioning systems for wireless personal networks," IEEE Communications Surveys & Tutorials, vol. 11, no. 1, pp. 13-32, 2009.

[2] H. Liu, H. Darabi, P. Banerjee, J. Liu, "Survey of wireless indoor positioning techniques and systems," IEEE Trans. on Systems, Man, and Cybernetics, vol. 37, no. 6, pp. 1067-1077, 2007.

[3] G. W. Shi, Y. Ming, "Survey of indoor positioning systems based on ultra-wideband (UWB) technology," Lecture Notes in Electrical Engineering, Wireless Communications, Networking and Applications, Proceedings of WCNA 2014, Vol. 348, pp. 1269-1278, 2016.

[4] B. Kim, W. Bong, Y. C. Kim, "Indoor localization for Wi-Fi devices by cross-monitoring AP and weighted triangulation," In the Proceedings of IEEE Consumer Communications and Networking Conference (CCNC), NV, U.S.A., pp. 933-936, 2011.

[5] Y. Mo, Z. Z. Zhang, Y. Lu, G. Agha, "A novel technique for human traffic based radio map updating in Wi-Fi indoor positioning systems," KSII Transactions on Internet and Information Systems, vol. 9, no. 5, pp. 1881-1903, 2015.

[6] X. F. Jiang, C. J. Mike Liang, K. F. Chen, B. Zhang, J. Hsu, J. Liu, B. Cao, F. Zhao, "Design and evaluation of a wireless magnetic-based proximity detection platform for indoor applications," In the Proceedings of 11th ACM/IEEE International Conference on Information Processing in Sensor Networks (IPSN/SPOTS), Beijing, China, pp. 221-231, 2012.

[7] J. Hightower, G. Borriello, "Location sensing techniques," Technical Report UW CSE 2001-07-30, Department of Computer Science and Engineering, University of Washington, 2001.

[8] K. Kaemarungsi, P. Krishnamurthy, "Properties of indoor received signal strength for WLAN location fingerprinting," In the Proceedings of 1st Annual International Conference on Mobile and Ubiquitous Systems: Networking and Services (MobiQuitous '04), MA, USA, pp. 14-23, 2004.

[9] D. Focken, R. Stiefelhagen, "Towards vision-based 3-D people tracking in a smart room," In the Proceedings of 4th IEEE Intl Conference on Multimodal Interfaces, PA, USA, pp. 400-405, 2002.

[10] A. Kotanen, M. Hannikainen, H. Leppakoski, T. D. Hamalainen, "Positioning with IEEE 802.11b wireless LAN," In the Proceedings of 14th IEEE Proceedings on Personal, Indoor and Mobile Radio Communications, Beijing, China, pp. 2218–2222, 2003.

[11] Y. B. Xu, M. Zhou, L. Ma, "Hybrid FCM/ANN indoor location method in WLAN environment," In the Proceedings of IEEE Youth Conference on Information, Computing and Telecommunications, Beijing, China, pp. 475–478, 2009.

[12] V. Honkavirta, T. Perala, S. Ali-Loytty, R. Piche, "A comparative survey of WLAN location fingerprinting methods," In the Proceedings of 6th Workshop on Positioning, Navigation and Communication (WPNC'09), Hannover, Germany, pp. 243–251, 2009.

[13] M. Y. Umair, K. V. Ramana, D. K. Yang, "An enhanced K-Nearest Neighbor algorithm for indoor positioning systems in a WLAN," 2014 IEEE Computers, Communications and Its Applications, pp. 19-23, January 20, 2014.

[14] K. F. S. Wong, I. W. Tsang, V. Cheung, S. H. G. Chan, J. T. Kwok, "Position estimation for wireless sensor networks," In the Proceedings of IEEE Global Telecommunications Conference, MO, USA, pp. 2772–2776, 2005.

[15] S. Aomumpai, K. Kondee, C. Prommak, K. Kaemarungsi, "Optimal placement of reference nodes for wireless indoor positioning systems," 11th International Conference on Electrical Engineering, Electronics, Computer, Telecommunications and Information Technology. Paper no. 6839894, 2014.

[16] P. Bahl V. N. Padmanabhan, "RADAR: An in-building RF-based user location and tracking system," In the Proceedings of INFOCOM 2000, Nineteenth Annual Joint Conference of the IEEE Computer and Communications Societies, Tel Aviv, Israel, pp. 775-784, 2000.

[17] H. D. Chon, S. Jun, H. Jung, S. W. An, "Using RFID for accurate positioning," Journal of Global Positioning Systems, vol. 3, pp. 32–39. 2004.

[18] H. L. Ding, W. W. Y. Ng, P. P. K. Chan, D. L. Wu, X. L. Chen, D. S. Yeung, "RFID indoor positioning using RBFNN with L-GEM," In the Proceedings of IEEE 2010 International Conference on Machine Learning and Cybernetics, Qingdao, China, pp. 1147–1152, 2010.

[19] A. K. M. M. Hossain, W. S. Soh, "A comprehensive study of Bluetooth signal parameters for localization," In the Proceedings of 18th Annual IEEE International Symposium on Personal, Indoor and Mobile Radio Communications (PIMRC'07), Athens, Greece, pp. 1-5, 2007.

[20] F. Subhan, H. Hasbullah, A. Rozyyev, S. T. Bakhsh, "Indoor positioning in Bluetooth networks using fingerprinting and lateration approach," In the Proceedings of 2011 International Conference on Information Science and Applications (ICISA), Jeju Island, Korea, pp. 1-9, 2001.

[21] W. P. Chen, X. F. Meng, "A cooperative localization scheme for Zigbee-based wireless sensor networks," In the Proceedings of 14th IEEE International Conference on Networks, Singapore, pp. 1-5, 2006.

[22] G. Goncalo, S. Helena, "Indoor location system using ZigBee technology," In the Proceedings of Third International Conference on Sensor Technologies and Applications, Athens/Glyfada, Greece, pp. 152-157, 2009.

[23] S. Merat, W. Almuhtadi, "Wireless network channel quality estimation inside reactor building using RSSI measurement of wireless sensor network," In the Proceedings of Canadian Conference on Electrical and Computer Engineering, Calgary, AB, Canada, pp. 339-341, 2009.

[24] H. C. Chen, Y. J. Chen, C. Y. Chen, S. M. T. Wang, J. P. Yang, R. C. Hwang, "A new indoor positioning technique based on neural network," Advanced Science Letters, vol. 19, no. 7, pp. 2029-2033, 2013.

[25] A. Zaknich, "Introduction to the modified probabilistic neural network for general signal processing applications," IEEE Transactions on Signal Processing, vol. 46, no. 7, pp. 1980-1990, 1998.

[26] R. C. Chen, Y. H. Lin, "Using ZigBee sensor network with artificial neural network for indoor location," In the Proceedings of Eighth International Conference on Natural Computation, Chongqing, China, pp. 290-294, 2012.

[27] M. Altini, D. Brunelli, E. Farella, L. Benini, "Bluetooth indoor localization with multiple neural networks," In the Proceedings of 5th IEEE International Symposium on Wireless Pervasive Computing (ISWPC), Modena, Italy, pp. 295-300, 2010.

[28] Y. S. Lin, R. C. Chen, Y. C. Lin, "An indoor location identification system based on neural network and genetic algorithm," In the Proceedings of 3rd International Conference on Awareness Science and Technology (iCAST), Dalian, China, pp. 193-198, 2011.

[29] H. Mohammad, A. F. Ozan, A. N. Ali, P. Aveh, "Neural network assisted identification of the absence of direct path in indoor localization," In the Proceedings of IEEE Global Telecommunications Conference, Washington, DC, USA, pp. 387–392, 2007.

[30] S. H. Fang, T. N. Lin, "Indoor location system based on discriminant-adaptive neural network in IEEE 802.11 environments," IEEE Transactions on Neural Networks, vol. 19, no. 11, pp. 1973–1978, 2008.

[31] C. Laoudias, D. G. Eliades, P. Kemppi, C. G. Panayiotou, M. M. Polycarpou, "Indoor localization using neural networks with location fingerprints," Lecture Notes in Computer Science - 19th International Conference on Artificial Neural Networks, Limassol, Cyprus, vol. 5769, no. 2, pp. 954–963, 2009.

[32] A. Cidronali, S. Maddio, G. Giorgetti, G. Manes, "Analysis and performance of a smart antenna for 2.45-GHz single-anchor indoor positioning," IEEE Trans. Microw. Theory Tech., vol. 58, no. 1, pp. 21-31, 2010.

[33] C. H. Lim, Y. Wan, B. P. Ng, C. M. S. See, "A real-time indoor WiFi localization system utilizing smart antennas," IEEE Transactions on Consumer Electronics, vol. 53, no. 2, pp. 618-622, 2007.

[34] X. M. Qing, Z. N. Chen, T. S. P. See, "Sectored antenna array for indoor mono-station UWB positioning applications," In the Proceedings of 3rd European Conference on Antennas and Propagation, Berlin, Germany, pp. 822-825, 2009.

[35] Antennas and Propagation. http://www.radio-electronics.com/info/antennas/yagi/yagi.php

[36] A. Khotanzad, R. C. Hwang, A. Abaye, D. Maratukulam, "An adaptive modular artificial neural network: Hourly load forecaster and its implementation at electric utilities," IEEE Transactions on Power Systems, vol. 10, pp. 1716–1722, 1995.

[37] R. C. Hwang, P. T. Hsu, J. Cheng, C. Y. Chen, C. Y. Chang, H. C. Huang, "The indoor positioning technique based on neural networks," In the Proceedings of IEEE International Conference on Signal Processing, Communications and Computing (ICSPCC 2011), Xi'an, China, pp. 225-228, 2011.

9

E-Business Logistics Based on the Storage of Robot Picking Mode Study

Binbin Fu, Juntao Li, Yiming Wei

School of Information, Beijing Wuzi University, Beijing, China

Email address:

binbin6807@163.com (Binbin Fu), Ljtletter@126.com (Juntao Li), 879353329@qq.com (Yiming Wei)

Abstract: With the continuous development of e-business, e-business distribution center operating mode is gradually changing. Small batch and more frequency is the characteristics of the e-business. The new characteristics gives birth to new distribution center picking mode. This paper analyzes disadvantaged of the business mode of the current e-business distribution center, put forward a new type of e-business distribution center picking mode, which is based on storage e-business logistics mode order picking robot. Paper on the business model of the pattern has carried on the detailed analysis and logistics process design, and said that the mode is the trend of the development of e-business in the future.

Keywords: Warehouse Robot, E-Business Logistics, Picking Model

1. E-Business Logistics and Present Situation

E-business refers to using the Internet or other electronic tools(including telegraph, telephone, radio, television, fax, computer, computer network, mobile communication, etc.) based on computer network of a variety of business activities in the global business trade activity, including goods and services provider, advertisers, consumers, middlemen and related the sum of the conduct of parties.

Logistics as an important link in the process of e-business, for the physical distribution service between merchants and customers, e-business plays a vital role by efficient logistic system, reduce costs, enhance competitiveness in improve service. As more varieties, high frequency, small batch, the characteristics of e-business logistics gradually, it has high flexibility, high efficiency, high quality of service features[1]. This model not only led to the rapid development of modern logistics technology and equipment, but also makes the updating cycle of logistics mode more and more short. The key to improve the processing capacity of e-business logistics network nodes is to improve e-business logistics capability, warehousing, sorting, transport and distribution link to adopt intelligent information system and automation equipment, to improve the logistics speed and ability, to avoid the delivery do not in time, achieve finally "timing" "time limit for highly

flexible service". And in the aggregate, e-business logistics mainly has the following characteristics: 1. information; 2. automation; 3. network; 4. intelligent; 5. Flexibility[2]. These characteristics make the e-business logistics center operation different from other areas.

2. E-Business Operation Status and Development Analysis

The continuous development of e-business makes e-commerce industry new characteristics, such as order of many varieties, small batch, high frequency inferior, our country e-commerce is the main traditional manpower assignment on logistics distribution center in present, the high cost operation mode, low service efficiency, at the same time on the efficiency and service level can not meet the demand of this kind of e-business logistics in fast response, which makes the electricity distribution center are different from traditional logistics system, construction of electricity distribution center operations should adapt to the change of market construction more rapid response ability of the distribution center, the application of the automation equipment as well as the overall solution has to be adapted to the development of electricity industry characteristics. To the guideline of foreign Amazon which is using KIVA, put forwards a new mode of distribution center operation provides a guide to the e-business industry in our country. E-business operation need to be flexible and efficient logistics support system, in order to implement the limited resources to focus on the e-business enterprise real advantage areas, conform to the e-distribution center of the robot arises at the historic moment.

Distribution center based on warehouse robot has changed operation mode, the automatic, intelligent, integrated logistics center system can not only improve the operational efficiency and customer satisfaction of logistics, with more and more expensive human cost of the environment at the same time, it can saves logistics cost and development prospect of logistics mode in the long run. The model compared with the traditional manual operation mode are a lot of advantages, such as save labor resources, improve the space utilization, reduce the cost, etc., and have chosen as the main operation of distribution center link, its efficiency affects the overall operating efficiency directly, this article based on a new type of e-business storage robot distribution center operating mode, analyzed the picking operation mode and business process, the distribution center selection model suited to the characteristics of the electricity industry, and find out the main factors influencing the distribution center selection efficiency, improve the efficiency of distribution center selection operation.

The most typical points in storage system application is Amazon's warehouse, the application of the new logistics equipment KIVA AGV, people in the form of active bearing storage unit of the advanced warehouse machine, it can moves in the warehouse and fetching goods quickly and accurately, goods dispatch can be carried out in accordance with the orders for the first time at the same time, improves the efficiency of its goods delivery[3-5]. New warehouse operation mode has changed the traditional way of storage and pick, from passive to active, namely displacement of fixed shelves and warehouse the goods is not limited by geographical location, AGV controlled by software instruction fetching goods to the destination. This means that each staff do not need to change work position, it can be accurate to obtain the customer needs the goods in a very short time. The goods are effectively organized, greatly improve the efficiency of the goods accurately chosen, and can obtain the goods which customer needs in a short time, improves efficiency at the same time, and also improves security. Through practical application, Amazon has saved a maximum of 916 million dollars a year.

3. Storage Characteristics and Robot Parameters

Warehouse robot equipped with automatic guiding device carries along the predefined rules on the road, load shelves safety accurately, and handle in the warehouse without the need for manual handling automatically, and its power source can keep at any time, charging storage robot control the route through the upper machine, and behavior. Warehouse robot is key laboratory research and development of intelligent logistics system which can be used in the warehouse inventory,

picking, dispatch independently controlled robots such as homework, namely the logistics handling car in this paper.

Warehouse robot working mode: warehouse robot receive picking and other scheduling task, automatic handling shelves along the shortest path to the side, if the robot need to turn to stop in case and spin to continue along a prescribed path, when the shelf arrive the target, the robot stopped running after spin to the target at the bottom of the racks, spin around to make the support shaft rising way of lifting shelf to a certain height, the racks of the path as stipulated operation work needs to be done, in case of need to turn, robot spin in place to ensure that static way to body rotation, shelf, and ensure the stability of the shelves, and to prevent the collision between the shelves[6]. After robot shipped to the destination within the given waiting area waiting for the next steps, when completion of the operation, the robot carries the shelves back to its place in the same way, arrived in destination, the robot stopped, also reduce the supporting shaft height in the form of spin gradually around to unload the carrying shelves, after completed, robot itself back to wait for next time scheduling tasks scheduling area.

Warehouse robot operation model of special, new, automated not only broke the traditional mode of the layout of the electricity distribution center, it also put forward higher requirements and challenges. First, in the limited channel width, robot should spin around the operation of the way of lifting shelf, second, robot turning action also take place at the crossroads turn out the way, at the same time, in order to avoid the collision problem between the steering rack, the robot needs to do to "turn the car body shelves still" and to maintain as far as small possible turning radius to finish turning action within the channel in addition, in order to minimize the robot working space. Finally, in order to guarantee the efficiency, the robot should not too low load rate, load bearing of the robot not too low robot weight at the same time try to stay low.

The mode shelves through the above analysis as well as the basic setup of the channel and so on, set the warehouse robot basic parameter requirements are shown now in table 1:

Table 1. Warehouse robot parameters.

Warehousing project robot parameters	Parameter requirements
size	750*600*425mm
self-respect	<=200kg
bearing	300kg
No-load speed	1.0m/s
Loading speed	0.5m/s
Lifting time	5s
Lifting way	Body spin around lifting
Turned to the way	Spin, body rotation shelves still
Turning radius	Robot steering shaft axis of longest distance to the body

Parameters required by the work mode and warehouse robot, the robot structure design schematic diagram and the real figure, respectively, the following figure 1, as shown in figure 2:

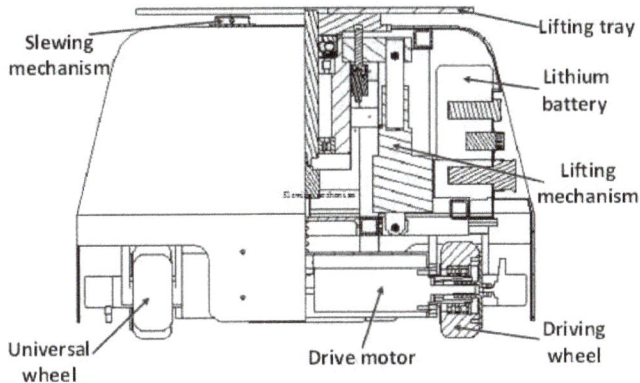

Figure 1. *Warehouse robot structure design sketch.*

Figure 2. *Warehouse robot real figure.*

By the picture above, e-business logistics distribution center system based on warehouse robot operating mode requirements this kind of submersible logistics handling robot by inertial navigation and QR code for indoor positioning, implement six institutions, four-wheel for universal wheel, the middle two rounds of driving wheels in the before and after, drive to mode for the differential steering, also can realize the storage machine work together, mainly for electricity for ins and outs of logistics distribution center of the work, such as access, chosen to run in the bottom of the racks, walking, steering and in narrow space, lifting the mode to the shelves, the top has the slewing mechanism, slewing mechanism and steering mechanism of relative movement, shelves with robot collision rotation, to ensure that narrow space around the logistics handling robot with multiple laser collision sensor.

4. E-Business Distribution Center Model Analysis

E-business enterprise type variety, each e-business enterprise must according to their own site layout, product features, order structure choice suits own operation mode, the design of electric logistics center need to choose the suitable model according to the respective business situation, for example, the detailed scheme, process, personnel, investment estimate and so on, and adjust the corresponding plan optimization and upgrade with the development of business scale[7-8].

E-business logistics picking based on warehouse robot is a kind of new mode. The model realized the automatic logistics imagination into reality, namely in the channel in the factory are the thousands of robot. Goods storage and distribution model for the system innovation of goods stored in removable shelves. When the goods need to be obtaining background control system automatically place, robot path to reach goods shelves, robot along the ground according to the instructions of the encoded network route to the corresponding shelf touchpad, lifting shelf again to choose the operator station[3]. Robot will be transported back to the original shelves after workers taken the goods again by storing. In this advanced warehousing logistics center stores and distribution of goods show no workers busy scene, across and carrying shelves replaced by a group of robots under computer control in an orderly way. Compared with the traditional warehousing and distribution system, the system can be one or two times higher order processing efficiency, this advantage for highly dependent on warehousing and distribution of large-scale retail and make e-business site strong attraction. The following are based on different selection of warehouse robot mode analysis and design, the main research according to the picking orders has chosen and according to the order of two kinds of operation mode, and comparing the two modes work.

4.1. List Picking Operation Mode Analysis

To the list picking operation mode, first is order of distribution center by the system analysis and processing, in order with the same product or similar products or order, assigned to each port with a batch of order processing to become picking list as the picking task. With picking orders generated, the system will release instruction in order shelves distribution selection box, and workers correspond to the order number and order code label. After receive the task from the cache, robot start from warehousing zone to labeling shelves completed, then carrying shelves to choose the appropriate position of the mouth, after entering the picking area by the system automatically, robot recognized chosen shelves and sending it to the specified port. Workers chosen plenty of goods according to the light and screen display specified in the corresponding box until the shelf cases chosen completely in the task[9]. Finally, goods which have chosen will be back to its place by warehousing robots, operation of storage shelves and then will complete the order which has chosen shelves to second place, the chosen orders again for packaging processing, the chosen link ends on to the next task operation.

4.2. List Picking Operation Process Design

Chosen according to the picking orders, first optimized according to the time window by the order processing system, namely in the specified time interval processing of a batch of orders, the number of orders for similarity comparison, classification, such as operation, aimed to improving the same goods in the shelf to the chosen order coverage of mouth through the handling of the orders with high similarity of goods or products in the same order to merge into the picking list, in order to reduce the extra storage robot reciprocating motion, improve the picking efficiency.

Under the picking area, operators choose picking mouth for picking orders every time, allocation chosen order based on system optimization according to each single order at the same time, and corresponding chosen box to the number of code labeling. At this point, when choosing picking box, it can order according to the characteristics of the product, such as size, volume, etc., choose the appropriate orders box to save the body material. Then, scans the shelf to system to ensure the labeling has completed, show that nearby warehouse robot which has completed labeling shelves to have taken place according to the requirement.

Put orders input system, workers will picking to optimize the personnel in the chosen terminal in the form of the bottom of the screen to start button calls the picking task, call signal nearby scheduling storage robots according to the system, robot run according to the instruction to picking preparation area in handling with shelf labeling good order to choose appropriate location, shelves, if there has robot waiting in front of the existing in the operation order area, upon the shelves, the robot to replace the original robot position, or according to the in front of the mouth will be chosen each needs first to other need this rack has chosen mouth to operation again returned to the waiting area for choose mouth scheduling to reduce the waiting time for robot.

At the same time, the system dispatch from a recent storage robots according to the chosen order information, robot recognition for shelves according to the instructions and choose the path chosen in the shelf below the lifting shelf according to certain rules, shelves according to the established rules to choose the path to picking, if already ahead in the shelves, the operation is waiting for the waiting in line to leave before a robot to the robot position. When press the start button, the screen display shows above required picking goods information, including name, quantity, location, etc., and the goods display lights up, order shelves response position indicator light is lit up at the same time, the member will be chosen to be the full amount of the goods into the corresponding order after box, press the confirm button beside the crate, the indicator light out, and at the same time the next container and the goods of the light is lit up, corresponding picking rack can be chosen to the next goods until the goods shelves, picking task has been completed, if the other chosen mouth have the shelves of goods demand, the robot run according to the given path to the next selection, waiting for the picking operation, otherwise the robot run to the store according to the established rules, robots waiting for the next scheduling tasks, such as no scheduling tasks automatically back to the empty car parking area, if robots in charging status right now, choosing paths in the charging area waiting in line to the nearest charging. Mouth shelf of all orders will be chosen when the same after completion of all tasks, robots and shelves by gravity and the number of lights on the dual

induction automatic leave, and will be the order shelves to the secondary packaging area split after reduction of customer orders waiting for packaging. List will be chosen according to the operation flow chart shown in figure 3:

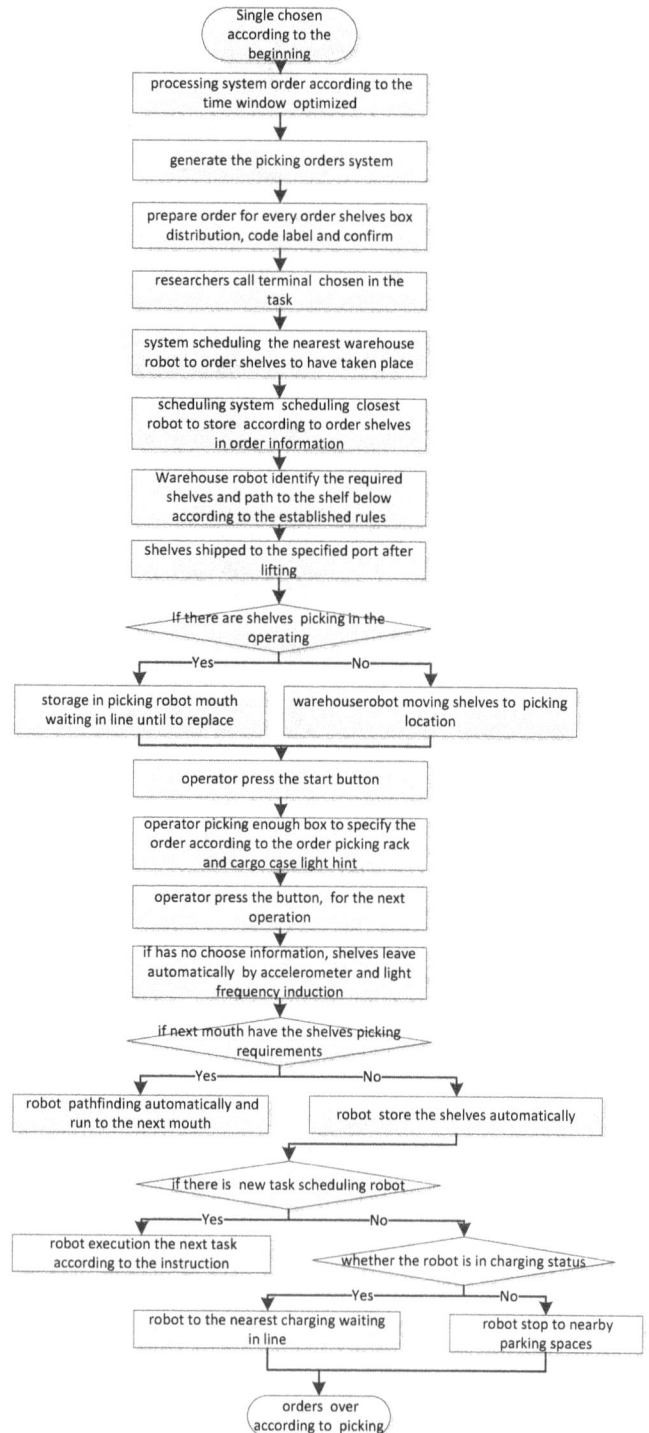

Figure 3. List will be chosen according to the operationflow chart.

5. Conclusion

Picking in e-business distribution center becomes an important factor in the influence of efficient in distribution center. Currently, domestic B2C e-business enterprises have chosen link of low degree of automation, the basic link to manual operation mode. Due to the disadvantages of artificial patterns inherent, makes the lower operational efficiency of the distribution center. The current e-business distribution center selection task for basic patterns is that chosen workers pushing a car and carry them along the storage area to walk S route, at the same time for dozens of order picking, choose car pushed to planting area after walking a circle, then picking orders for two points for a single order.

Picking system based on warehouse robot artificial model has more advantage. Moreover, when the shelves number and order quantity larger makes operation mode in the distribution center area reflect its advantage more, with the enlargement of the scale, the efficiency of the chosen system advantage will also expand. In nowadays e-commerce business development, customer experience becomes the important factor of e-business in an impregnable position[10]. And to improve the customer experience and the efficiency of distribution center operation, especially the efficiency of the low efficiency of the chosen segment is crucial at present.

As a new operation mode, the prophase of fixed investment becomes necessary, the investment will be relatively reduced after a long period of cost-sharing. Currently, labor cost is growing in constant, accumulated over a long period of time, the investment will very much. Traditional model needs to put in storage, storage, picking points to distinguish between equipment, while the picking system based on warehouse robot can do the same area and basic equipment. It will relatively reduce investment of equipment and site input, model for same equipment can improve the utilization rate of equipment. Automated operation pattern has always been diligent direction of the businesses and industries, it not only makes the process simplification, stylized, and the walking reduce operator's mechanized handling action mode makes the process more human. Therefore, e-business based on the storage of robot complied with the trend of the development of the distribution center operation system.

Acknowledgements

This paper is supported by the Funding Project for Technology Key Project of Municipal Education Commission of Beijing (ID:TSJHG201310037036); Funding Project for Beijing key laboratory of intelligent logistics system ;Funding Project of Construction of Innovative Teams and Teacher Career Development for Universities and Colleges Under Beijing Municipality (ID:IDHT20130517), and Beijing Municipal Science and Technology Project (ID:Z131100005413004);Funding Project for Beijing philosophy and social science research base specially commissioned project planning (ID:13JDJGD013).

References

[1] Programme. E-commerce "twelfth five-year" development plan[EB/OL],http://www.miit.gov.cn/n11293472/n11293832/n11293907/n11368223/14527814.html. 2012-03-27.

[2] X.N. Wang. E-commerce logistics management[M].Beijing: Cambridge university press, 2012, 20-21.

[3] ADI company. Kiva Systems will use Blackfin navigation intelligent robots are employed in the warehouse http://www.gongkong.com/webpage/solutions/200910/2009101611212700007.Htm.2009-10-16.

[4] M.Zhang. Parts of new warehouse distribution center layout planning study[D]. Jilin university, 2014.

[5] X.T. Wang. product distribution center planning and research about Qingdao Eurasian jewelry company[D] Ocean university of China, 2013.

[6] PETERSEN, C.G. An evaluation of order picking routing policies[J]. International Journal of Operations &Production Management, 1997, 17(11):1098-1111.

[7] CHOE, K., SHARP, GP., Serfozo, R.S.. Aisle-based order pick systems with batching, zoning and sorting[C]. Progress in Materail Handing Research, 1992, 245-276.

[8] DE KOSTER R., NEUTEBOOM A.J.. The Logistics of Supermarket Chains[M]. Elsevier,2001, Doetinchem.

[9] VAN DEN BERG J.P., GADEMANN A.J.R.N.. Simulation study of an automated storage/retrieval system[J]. International Journal of Production Research, 2000, 38:1339-1356.

[10] Clark, C. 2005. Probabilistic road map sampling strategies for multi-robot motion planning. Journal of Robotics and Autonomous Systems 2005, 53(3-4): 244–264.

10

Research on Remote Monitor and Safety About Substation Based on Wireless Private Network

Wei Lu, Xin Li

Shibei Power Supply Company, Shanghai Municipal Electric Power Company, Shanghai, China

Email address:

13738674523@163.com (Wei Lu)

Abstract: With the technology of security video surveillance applied in unattended electric power substations, security management is the only way to modernize the power grid. The paper proposes a set of integrated solutions combining substation video, security and power supply based on the 4G Wireless Networks. It studies the remote video security system that integrates security and technical precautions with hardware and software combining. The system strengthens the safety and security management of unattended substation relying on the latest 4G private network technology and advanced hardware equipment.

Keywords: 4G Network, Remote Video Surveillance, Security and Technical Precautions, Solar Power Supply

1. Introduction

Nowadays, with the continuous development of China's electric power system ,digital and automation is an inevitable trend of the development of electric power, unmanned substation and its safety management is the only way to modernize the power grid. With the usage of video surveillance system, the staff can inspect the operation [3]. It can be used to obtain the scene of the alarm information and review the accident through video recording.

Recently, the electrical power units in the substations are stolen and damaged throughout the country [1]. In order to ensure the safe operation of power facilities, reduce the loss of property, and ensure the normal operation of urban power network, it is necessary to take decisive measures [4]. High-tech means and effective management are useful methods to continuously reduce the occurrence of such events. At present, a large number of unmanned substations located in the jurisdiction of the Shibei Power Supply Company, Shanghai Municipal Electric Power Company [2]. The paper studies a remote visual security system that integrates security and technical precautions with hardware and software combination.

The paper proposes a set of integrated solutions combining substation video, security and power supply based on the 4G Wireless Networks. It studies the remote video security system that integrates security and technical precautions with hardware and software combining. It concluded that the system can realize the centralized management of substation, make important substation equipment, security monitoring points under real-time monitoring status and reduce the overall cost of operation and maintenance.

2. Current Problems

2.1. Unable to Get the Running Status of the Devices in Time

With a larger number of unmanned substation appears, the daily inspections and switching operations of the substation is still following the tradition of previous work mode because of various objective conditions, which lead to the relevant staff and the management must be present, particularly the substation in the remote areas, and it will cost a lot of manpower and resources on transportation.

2.2. Complicated Construction of Video Surveillance System

It is necessary to lay a great deal of wire for the video surveillance, and it is a difficult task. At the same time to build a monitoring system requires the laying of a large number of video lines, network lines, control lines and power lines, which has brought a huge cost of human and material resources. And ordinary video surveillance can only be viewed in special monitoring center, which cannot be viewed on mobile terminals or other devices, even worse it was subjected to the geographical limitations and networks in emergency situations.

2.3. Image Quality Await to Be Optimized

Ordinary video camera image quality has been unable to meet the requirements of remote monitoring, the image quality will be damaged when the analog camera has been converted into digital, long distance transmission of analog signal, the signal attenuation is serious, the image quality is not good, the image resolution is fixed to PAL or NTSC, which cannot meet the requirements of higher resolution and higher resolution. Video signal dynamic range is small, it is difficult to meet specific application requirements, such as low noise, wide dynamic, high frame rates or other requirements, it also cannot achieve distribution, long distance transmission, sharing or other needs [5].

2.4. Safety Hazards in Scenes

Due to the wide distribution of substation; the environment surrounding is complex and unattended. The security issues of the substation is becoming more and more severer, the issue must be solved including fire prevention[4], security, waterproof and so on, only in this way the safe operation of all the apparatus can be ensured.

3. Research of Anti-theft Technology for Remote Monitoring

3.1. Integration of Remote Monitoring Platform and Security Equipment

SDI HD infrared camera is used in the scene of video surveillance in substation, the shooting effect can up to 720P, and the camera can be remote rotated 360 degree by the monitoring personnel for the camera using cloud control. Site security equipment (such as Infrared burglar, Smoke alarm, Vibration sensor) has been accessed to the host computer. The system uses the remote video terminal client platform to feedback the scene video, security alarm, mobile detection, automatic recording, it also supports remote viewing of PC terminal and mobile terminal equipment [6].

3.2. Data Transmission Through 4G Network

With the maturity of 4G network technology and the coverage of network, the China Telecom's 4G network can provide a more efficient, more stable video transmission network for the system. China Telecom's 4G network is still in trial operation stage, based on the virtual private network services that high-speed packet data telecommunications network had built for mobile customers, by which the corporate customers can get in touch with enterprise intranet Seamless and safely no matter where they are, at the same time the VPND network and the internet network can be isolated. The study based on the integrated of 4G network module and the site monitoring host computer, simultaneously, the 720p HD camera is used as the video capture unit, and transmission of 4G network ensured the high speed, safety, stability and reliability of the network data.

3.3. Solar Backup Power Supply

As the video surveillance and the security system need to maintain uninterrupted operation in 24 hours, in order to ensure that the system is also capable of running in emergency situations, as well as to ensure the storage security of the video hosting computer's data. All apparatus of the system will be used in two ways-powered, one way is mains combines UPS, and another way is solar with the spare battery, while the system utilize the dual power automatic transfer switch to realize the switching between the city electricity and the solar standby power supply, will not cause the malfunction, and meet for the requirement of high reliability. System can be in the field without the cable network and the external power supply [5], the normal operation to ensure that the full 24 hours monitoring.

4. System Structure Introduction

Fig. 1. Remote Monitoring System Configuration.

4.1. Entire Structure Design

Remote anti-theft monitoring system composes of the monitored alarm system, the power supply system and the Server-side system, and the monitored alarm system is contributed by remote monitoring system and site security surveillance system.

Hardware devices of remote monitoring system are composed of high-definition network camera and remote monitor 4G computers.

The devices of site security surveillance system are composed of anti-theft monitoring computer, infrared alarm, smoke alarm, sound and light alarm, vibration alarm, alarm siren, noise reduction devices and so on.

The power supply system composes of UPS host, dual power switch, inverter, solar photovoltaic panels, and battery pack and so on.

The server composes of remote video surveillance server, Video data server, Dedicated 4G VPDN network, VPDN router.

The system transmissions the remote video images and site security surveillance system data to server system through dedicated 4G network by alarm system. Users connect to the server system through a client remote system and can check live video images and alarm information feedback. At the same time, the system supports table PCs and mobile phones monitoring on site.

4.2. System Feature

The system has a reasonable structure design and advanced technology concepts. It's based on live video surveillance and to assist in security anti-theft monitoring. The system using the advanced video compression data, security monitoring and video monitoring combine with 4G network perfectly, meanwhile, the system supports two monitoring modes of C/S and B/S [7].

The design of this system is based on the international protocol standard. All devices can be included in the system as long as they meet the agreement.

Integrated platform has a powerful management function. System combines video surveillance, integrated platform alarm, message alarm in one, achieving an application platform to control all devices in the management system. The system is based on security to assist power production, unified platform for centralized management, real-time monitoring of all front-end equipment and reduce user maintenance costs.

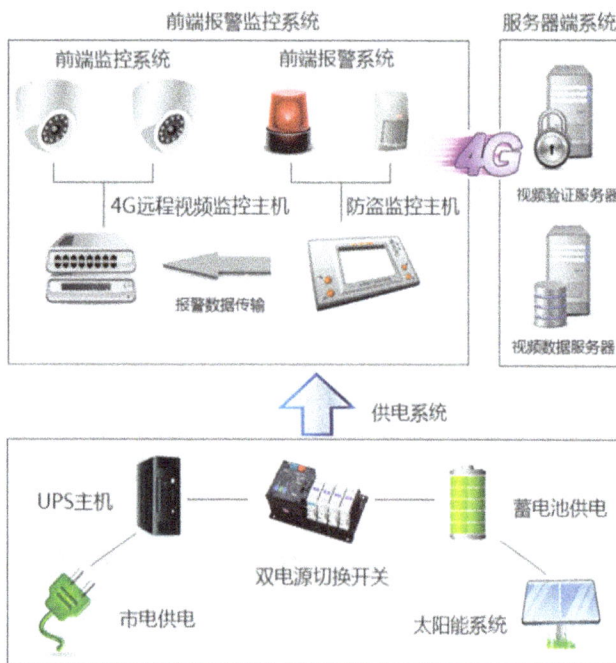

Fig. 2. Configuration of the Safety and Power Supply System.

5. Software Features

5.1. Data Flow Control of the Wireless Network

By using 4G mobile data transmission network, we need to solve the problem of large transport stream, otherwise it will lead to uncontrolled operation and maintenance costs. Research will use the local video store, online video viewing, and emergency data upload program to reduce the transmission of video surveillance traffic. Local video store,

online video viewing:

Traditional video surveillance, the live video monitoring continues to transmit data to a remote server, the server-side data storage, you can connect to the server view back when you need to see the video. So there will be a lot of video data transmission between server and monitoring, resulting the waste of network resource and the data redundant of server [8].

Research will store all monitoring data directly in the local monitoring host disk space, all data is exchanged with other server when monitoring, and server-side unified manages the on-site monitoring equipment. When users need to watch back, using a client platform to connect system server-side. System allocates monitoring disk space through server-side, at the same time data traffic will be produced and the traffic transmission will be reduced.

5.2. Data Save in Incident

If something emergency happened in transformer substation, the system will start the motion detection function. If non-normal mobile phenomenon happened in some specific areas, the system will automatically start recording function. System will select automatically recording function when alarm of system integration arm, it will automatically start recording function when the alarm Detectors (such as infrared alarm) alarm occurs. The recording time can be set in advance; the days of video file can be adjusted.

In this case the video data will be automatically uploaded to a remote server's hard disk through network and notify the administrator unexpected things has happened, so administrator has access to the site video view and it can provide the evidence for events [9].

5.3. Remote Patrol

With the constant improvement of communication network systems and the increasing unmanned substations, although the power sector has real-time monitoring data on the grid, they cannot grasp the site operating conditions and check field devices image. Using remote video inspection technology, we will enable access to the image information within the substation and a variety of alarm information, sector can responses and records the operation of substation equipment running key parts more quickly, intuitively, accurately [10].

Administrators can connect a video camera inside the substation configuration, and understand the station in real time through the client platform.

The resolution of station configuration multiple SDI HD cameras can reach to 720P. That can achieve visualization monitoring and scheduling of substation to ensure unmanned substations' running become more safe and reliable with the help of data transmission of 4G high-speed network.

5.4. Record Video

The system can be set multiple periods of multiple front-end monitoring equipment for recording. Data will be stored in the hard disk of video surveillance system server or front-end devices. Timer mode can be selected as single recording or scheduled recording and each time can be set separate, recording speed can be adjusted.

5.5. Map Showing Substation Location

The system can be positioned by the GPS of front-end monitoring equipment host. It's useful for this system to unified management and visual show of all devices by using GPS technology to combine substation with electronic maps. It's shown in figure.

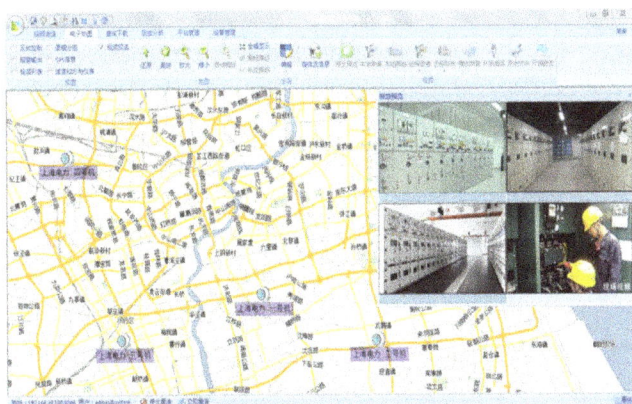

Fig. 3. *Electronic Map of Substation.*

6. Conclusion

Achieving substation automation and modernization of management, continuously improving the safety and reliability of the substation operation and maintenance, improving management efficiency for the planning and construction of video monitoring system and the complete security, anti-technology within the existing substation are the main goals of this article.

Optimizing equipment power supply system by using high-speed secure wireless network technology and integrated substation video surveillance and security equipment. The system can realize the centralized management of substation, make important substation equipment, security monitoring points under real-time monitoring status and reduce the overall cost of operation and maintenance.

References

[1] LIU qingli, LIU jingbo, DENG ying. Video analysis technology in unmanned substations Monitoring. Shanxi Electric Power, 2010, pp.73-76.

[2] LUO lei, ZANG shaohua. Substation remote inspection system wireless communication technology. Power and Technology Science, 2014, pp.92-96.

[3] DING shuwen, WANG cheng. Key Technical Problems Intelligent Substation. For electricity, 2012, pp.9-12+47.

[4] LIU qin, LIU jian. Implementation and Application of unmanned substation Remote Monitoring System Analysis. China Electric Power Education, 2012, pp. 141-142.

[5] LU shen. Remote monitoring system in unmanned substations. Huzhou Teachers College, 2013, pp. 32-35.

[6] SUN jingming. Substation network remote monitoring system application. China Science and Technology Information, 2014, pp.122-124.

[7] PAN yueqing, LUO jingfeng, Monitoring technology in power theft Work. Wireless Internet Technology, 2012, pp.2-7.

[8] MIAO wennan, Car Alarm Monitoring System. 2014, pp.22-24.

[9] ZEN peng, YU haibing, Industrial Wireless Network WIA standard system and technologies. Automation Panorama, 2012, pp.24-27.

[10] XU haiyan, LIU jingwei, The fourth generation of mobile communication technology research. Wireness Internet Technology, 2015, pp.20-21.

11

A Contemporary Overview of the History and Applications of Artificial Life

Jamaluddin Mir[1], Majid Mehmood[2], Malik Touqir Anwar[1], M. Yaqoob Wani[1]

[1]SIT Department, the University of Lahore, Islamabad, Pakistan
[2]Computer Science Department, University of Gujrat, Sialkot, Pakistan

Email address:

mir.jamal@uolisb.edu.pk (J. Mir), majid.mehmood@uogsialkot.edu.pk (M. Mehmood), malik.touqir.anwar910@gmail.com (M. T. Anwar), yaqoobwani@gmail.com (M. Y. Wani)

Abstract: The study of man-made systems which demonstrate certain behaviors that are found to be distinctive characters of natural living systems, found in nature, is called Artificial Life. Artificial life supplements the classic biological science which is focused on the investigation of living organisms by trying to produce life-like characteristics in computer and other such machines. Artificial life is focused on developing an understanding of the fundamental doctrines of life by either creating life-like characteristics in simulations created by computers or by actual physical implementations. Though, the aim of artificial life is concentrated towards both the future and origin of biology, yet the complexity of the subject area requires involvement of other fields of science. The practical as well as the scientific impact of the field of artificial life are equally far reaching.

Keywords: Artificial Life, Artificial Intelligence, Evolution, Cybernetics, Tierra, Self-Replicating Systems

1. Introduction

Since 1980s, the field of artificial life has examined various living systems by an artificial and synthetic approach, with the aim of developing a better understanding of life by means of software. Researchers are of the opinion that the first artificial life model was developed in 1951 by von Neumann [1]. The concept of automata of was described on a cellular level and a self-replicating system was proposed with the intention of developing a universal formal computation system that will be prone to unrestricted evolution [2, 3]. Self-replicating loops were proposed in 1984 [4] which were a lot simpler.

At the same time, the researchers from the field of cybernetics were focusing on the control and communication aspects in these systems [5, 6]. The discipline of cybernetics aims to describe the phenomena in regard to their respective functionalities instead of the substrate. It was proposed by Langton [4] that life demands to be studied as a property of form rather than matter. This is in close relation to the concepts followed by cybernetics. So it can be safely said that ALife is closely related to the field of cybernetics. Furthermore, the fundamental theories of homeostasis [7] and autopoiesis [8] are derived from the field of cybernetics.

Apart from cybernetics, Alife and artificial (AI) are also related where ALife is more focused on systems which can simulate the nature and its laws and AI is aimed at simulating the human intelligence. Another major difference is that of the different modeling strategies used by ALife and AI.

2. ALife and AI

Although ALife and Artificial Intelligence (AI) are quite different from each other, yet both the fields of computer science are somewhat connected by means of ALife's roots in machine learning. The fields of ALife and AI depict a certain amount of similarity as both these fields of computer science rely on similar methodologies, erupting from natural phenomenon. However, there are remarkable differences between AI and ALife. Majority of the classical methods of AI are serial systems that best defined in a top-down manner, comprising of very complex controllers that are centralized and make decisions on the basis of global knowledge of the system. The decisions made by the centralized controller are very much capable of having an impact on any part of the system. Whereas, in contrast to this approach, numerous living systems existing in nature demonstrate highly complex and self-governing behavior which is parallel in nature, depicting the behavior of a distributed network, comprised of low-level agents. The decisions made by each agent are made on the basis of their local environment and local knowledge. As can be seen, this approach is in contrast to the centralized approach of the AI models where on centralized controller makes decisions on the basis of global knowledge. The ALife models are defined in a bottom-up manner which is characteristic of simple agents that are parallel in nature and act in a local manner. These types of models are called agent-based models. The overall behavior of the system emerges as a result of the interactions among the agents.

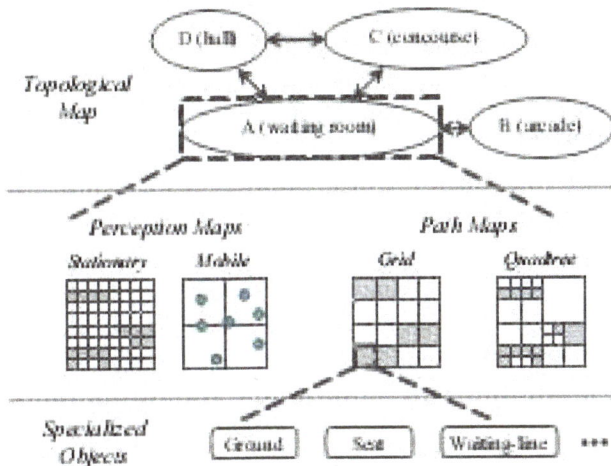

Figure 1. Hierarchical model for ALife.

3. Digital Evolution

Researchers believe that the most efficient way to study the issues about evolving systems is by their implementation in the software. This approach has been central trend in the field of ALife. The earliest significant such implementation in the digital world was Tierra [11]. Simply put, Tierra was a population of computer program that was self-replicating in nature and resided in the computer memory. The genotype comprised of machine code, with each Tierran creature an instance of some genotype. The simulation of the program started with the self-replication of a single ancestor that was placed in the computer memory. The process of self-replication by the ancestor and its subsequent child continues uptil the point when the computer memory is full of such creatures, sharing the same genotype of their ancestor. Among the overall population, older creatures are eliminated so as to create space for new progenies. The population evolves by means of natural selection in case of any errors or mutation as it is said. If a certain error or mutation allows the program to replicate quickly, that particular mutated genotype will tend to spread among the population. As a result, over time, the population becomes extraordinarily diverse. Subsequently, millions of CPU cycles later, the Tierran population now contains several different types of creatures that demonstrate various different relationships in their ecology.

4. Categories of ALife

ALife has been categorized into three extensive but inter-related branches that relate to three different respective methodologies. The majority of the ALife research is in the area of Soft ALife which is aimed at creating simulations depicting life-like behavior. Hard ALife is aimed at developing hardwareimplementations. Wet ALife is aimed at producing living systems from certain bio-chemical materials [9]. In broader sense, therefore, it can be said that ALife is the synthesis and simulation of living systems which is aimed at developing a better understanding of life and life forms by simulations.

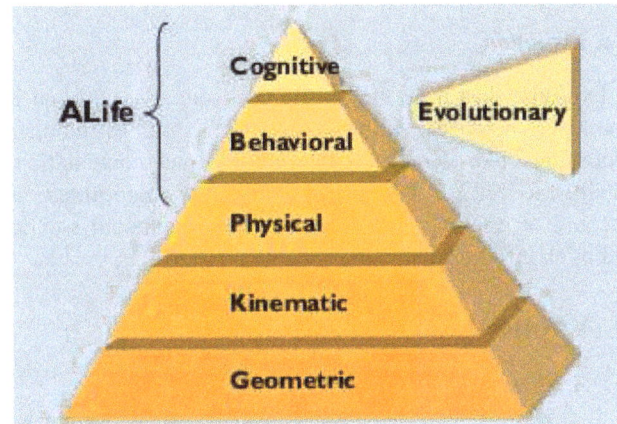

Figure 2. Modeling for Artificial Life.

4.1. Autonomy

Since its inception, ALife has been closely related to the theories of biological autonomy and autopoiesis [10]. Autopoiesis describes the constrained grid of processes which self-maintain the organization in such a manner that it makes them easier to be identified in a chemical substance [11].

4.2. Self-Organization

Self-organization is the ability of systems where local interaction can result in global behaviors for example swarm of bees, flock of birds or traffic pattern [12, 13, 14, 15]. The traditional examples of self-organization in ALife are snowflakes [16] and boids [17] which use pattern formation and cooperative motion [18]. Some researchers argue that the characteristic of self-reflection is a special case of the concept of self-organization because in the process of replication, the object under consideration can conserve as well as duplicate its organization on its own. Self-maintenance is also another exceptional case of self-organization that is more in close relation to homeostasis [19]. Researchers have developed very sophisticated robots which are based on the swarm intelligence. Such feature of self-organization has been a motivation for computational intelligence.

4.3. Adaptation

Any change in an agent or within a system in reaction to the environment which will consequently help that particular system or agent to achieve their goal is called adaptation [15]. Researchers consider adaptation to be the fundamental characteristic of all living systems that is extremely important for autonomy and the very survival of the system or agent. In contrast to AI which is more concerned with predicting and controlling, ALife is more focusing in developing the capability of adaptability in AI [20]. However it must be noted that predictability is as much important as is adaptability, both in natural systems as well as in artificial systems.

4.4. Evolution

Evolution has been one of the key concepts explored by computer sciences with prime focus on evolutionary algorithms like genetic algorithm [21] and computational intelligence [22]. All such evolutionary algorithms are focused on finding the best possible and optimum solution out of an infinitesimally large search space.

5. Applications of ALife

5.1. Artificial Societies

Researchers have described the societies on the basis of relationships between the individuals belonging to the same species. Investigation of all likelihoods of social interactions which used to be very difficult to explore in complex systems has now become possible by the help of computational modeling [23]. The societal modeling concepts have paved the way for developingALife games likeCreatures and TheSims [24].

| Arcade | Main Waiting Room | Concourses and Platforms |

Figure 3. A large scale Artificial Society created by ALife models.

5.2. Computational Biology

Neural networks as well as evolutionary algorithms have found the very basis of modern computations and have proved that concepts derived from biology can be used as the central concepts in other sciences. Neural networks, derived from biology, have been used to develop distributed computation models. Many of these have been used by ALife in different stages and models. The biological immune systems have served as motivation and central logic of numerous new advancements in the field of computer security, as well as in the field of optimization [24].

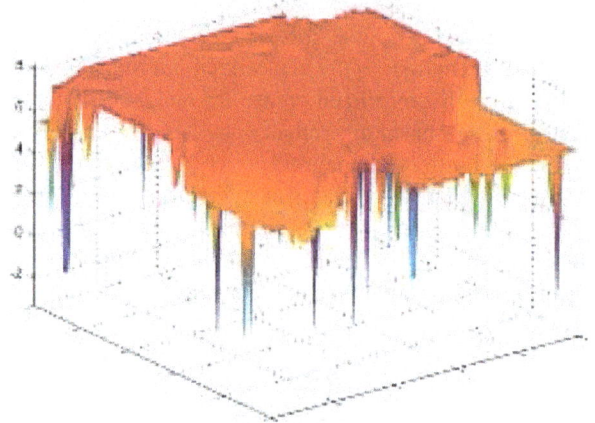

Figure 4. Avida; an ALife software program in which the bits of code depict actual organisms.

6. The Future

Researchers believe that an improved understanding of life will help in enabling us to make intelligent decisions in regard to environmental resources, social life, development of cities, commercial development of biotechnology and many other areas of applied sciences.

The sophisticated nature of the living systems has been the limiting factor when it comes to the scalability of artificial systems that are to be designed. Such concepts can be applied to electric circuits as the relations of the components can be regulated and optimized. The concepts and perceptions of ALife are slowly and gradually infiltrating biological sciences the same was as complex computational modeling is making its way into all fields of biological sciences. If the current trend continues in the field of ALife, soon we would have to revise ourconcepts about real and artificial.

Although artificial life (ALife) is related to artificial intelligence (AI) yet there are very critical differences between the two subject areas, mainly among the modeling strategies used by AI and ALife. Majority of the classic AI models are top-down identified serial systems which use a very sophisticated and centralized controller which is responsible of making decisions on the basis of all the states. The decisions of the controller are capable of affecting any or all thesystem aspects of the whole system. In contrast, as can be observed, majority of the natural systems are parallel or non-linear, having inter-relations based on agents that interact with each other. The decisions of the agents are based on the information available specifically to that particular agent and they are so designed that the decisions of the agents have an impact only on the location of that particular agent. The models followed in ALife are specified by following a bottom-top non-linear systems of agents which interact in their local capacity. As a result of these several local interactions between agents, a global change can be observed.

The synthetic approach of ALife has numerous advantages. The fact that a theory related to life can be expressed by computer code renders it an extremely high level of accuracy and precision. Computer generated models have been observed to simplify the extent of conceptualization that is found to be a prerequisite for constructing outstanding models of natural phenomena.

An important part of the research conducted in the field of Alife is mainly concerned with implementing the research in real life situation, instead of merely carrying out computer simulations. As a result, such products and systems can be engineered which exhibit life-like characteristics. However, a significant part of the research conducted in this field is purely theoretical and aimed to develop a better understanding of life by understanding the very essence of life. The problems that can be solved by ALife can include developing electronic hardware tobio-inspired robotics by making use of evolutionary algorithms which simulate nature inspired processes like the concept of swarm intelligence of bees and flocks of birds by actual swarm of robots.

6.1. Health/Medicine

The growing complexity of the current health-care system has made it ideal for applications from ALife [25]. The ALife applications demand a thorough understanding of very nature of the health-care system which encompasses the understanding of the behavior of the system, responding to the changes in the system, dealing with the monetary issues and, in future, making the whole system adaptable to non-linearity of the real world issues.

6.2. Transport

In most of the metropolitan cities all around the world, the traffic and transportation system poses a large scale complexity and non-linearity. Models and simulations from ALife can be applied to the traffic movement as well as for developing infrastructure network [26]. The number of engineering applications in ALife are only bound to increase in the near future.

6.3. Artificial Societies and Artificial Economies

The ALife models have been vastly applied into what is now described as artificial societies and artificial economies. The agent-based modeling which uses the bottom-up effect among the social systems has been used to generate models which can be simulated in artificial environments. If the bottom-up effect methods of ALife are applied to the markets, the artificial societies evolve into artificial economy. The economic systems are highly complex entities which face several non-linear components and variables. In this regard, the ALife models have a natural ability to deal with all these non-linearities that deal with the non-equilibrium phenomenon.

7. Conclusion

A majority of the real world problems were solved by mathematical programming methods. However, it must be noted that a large portion of the real world problems which were traditionally solved by the optimization methods and mathematical programming techniques were somewhat smaller in scale. Many of the engineering problems in the real world are combinatorial, consequently posing large dimensionality as it is very difficult to modularize all such problems. Since its inception, ALife is slowly integrating into almost all fields of sciences. Specially in all such environments where a human controller is required to make complex decisions based upon several different variables, the ALife based models can simplify the decision making process, thereby making things easier.

References

[1] von Neumann, J. (1951). "The general and logical theory of automata," in Cerebral Mechanisms in Behavior-The Hixon Symposium, 1948 (Pasadena CA: Wiley), 1–41.

[2] von Neumann, J. (1966). The Theory of Self-Reproducing Automata. Champaign, IL: University of Illinois Press.

[3] Mange, D., Stauffer, A., Peparola, L., and Tempesti, G. (2004). "A macroscopic view of self-replication," in Proceedings of the IEEE, Number 12. IEEE. 1929–1945.

[4] Langton, C. G. (1984). Self-reproduction in cellular automata. Physica D 10, 135–144. doi:10.1016/0167-2789(84)90256-2.

[5] Wiener, N. (1948). Cybernetics: Or, Control and Communication in the Animal and the Machine. New York, NY: Wiley and Sons.

[6] Gershenson, C., Csermely, P., Erdi, P., Knyazeva, H., and Laszlo, A. (2014). The past, present and future of cybernetics and systems research. Systema. 1, 4–13.

[7] Ashby, W. R. (1947a). The nervous system as physical machine: with special reference to the origin of adaptive behavior.Mind 56, 44–59. doi:10.1093/mind/LVI.221.44

[8] Varela, F. J., Maturana, H. R., and Uribe, R. (1974). Autopoiesis: the organization of living systems, its characterization and a model. Biosystems 5, 187–196. doi:10.1016/0303-2647(74)90031-8

[9] Rasmussen, S., Chen, L., Nilsson, M., and Abe, S. (2003). Bridging nonliving and living matter. Artif. Life 9, 269–316. doi:10.1162/106454603322392479

[10] Bourgine, P., and Varela, F. J. (1992). "Introduction: towards a practice of autonomous systems," in Toward a Practice of Autonomous Systems: Proceedings of the First European Conference on Artificial Life, eds F. J. Varela and P. Bourgine (Cambridge, MA: MIT Press), xi–xvii.

[11] Maturana, H., and Varela, F. (1980). Autopoiesis and Cognition: The Realization of Living. Dordrecht: Reidel Publishing Company.

[12] Haken, H. (1981). "Synergetics and the problem of selforganization," in Self-Organizing Systems: An Interdisciplinary Approach, eds G. Roth and H. Schwegler (New York, NY: Campus Verlag), 9–13.

[13] Camazine, S., Deneubourg, J.-L., Franks, N. R., Sneyd, J., Theraulaz, G., and Bonabeau, E. (2003). Self-Organization in Biological Systems. Princeton, NJ: Princeton University Press.

[14] Gershenson, C., and Heylighen, F. (2003). "When can we call a system self-organizing?," in Advances in Artificial Life, 7th European Conference, ECAL 2003 LNAI 2801, eds W. Banzhaf, T. Christaller, P. Dittrich, J. T. Kim, and J. Ziegler (Berlin: Springer), 606–614.

[15] Gershenson, C. (2007). Design and Control of Self-Organizing Systems. Mexico City: CopItArxives.

[16] Packard, N. (1986). "Lattice models for solidification and aggregation," in Theory and Application of Cellular Automata, ed. S. Wolfram (Tokyo: World Scientific, Institute for Advanced Study Preprint), 305–310.

[17] Reynolds, C. W. (1987). Flocks, herds, and schools: a distributed behavioral model. Comput. Graph. 21, 25–34. doi:10.1145/37402.37406

[18] Vicsek, T., and Zafeiris, A. (2012). Collective motion. Phys. Rep. 517, 71–140. doi:10.1016/j.physrep.2012.03.004

[19] Williams, H. T. P. (2006). Homeostatic Adaptive Networks. PhD thesis. Leeds, UK: University of Leeds.

[20] Brooks, R. A. (1991). Intelligence without representation. Artif. Intell. 47, 139–160. doi:10.1016/0004-3702(91)90053-M.

[21] Holland, J. H. (1975). Adaptation in Natural and Artificial Systems. Cambridge, MA: The University of Michigan Press.

[22] Gilbert, N., and Conte, R. (eds) (1995). Artificial Societies: The Computer Simulation of Social Life. Bristol, PA: Taylor & Francis, Inc.

[23] Grand, S. (2001). Creation: Life and How to Make it. Cambridge, MA: Phoenix.

[24] Burke, E. K., Kendall, G., Aickelin, U., Dasgupta, D., and Gu, F. (2014). "Artificial immune systems," in Search Methodologies, eds E. K. Burke and G. Kendall (Springer), 187–211.

[25] Plsek, P. E., &Greenhalgh, T. (2001). The challenge of complexity in health care. Bmj, 323(7313), 625-628.

[26] Bauer, D. C., Cannady, J., & Garcia, R. C. (2001). Detecting anomalous behavior: optimization of network traffic parameters via an evolution strategy. In SoutheastCon 2001. Proceedings. IEEE (pp. 34-39). IEEE.

12

Review and Application of Model and Spectral Analysis Based Fault Detection and Isolation Scheme in Actuators and Sensors

H. Bal[2], S. K. Mohanty[1], N. P. Mahalik[2, *], B. B. Biswal[3]

[1]Colllege of Engineering, Biju Patnaik University of Technology, Bhubaneswar, Odisha, India
[2]Department of Industrial Technology, Jordan College of Agricultural Sciences and Technology, California State University, Fresno, California, USA
[3]National Institute of Technology, Rourkela, India

Email address:
skmohanty03@yahoo.com (S. K. Mohanty), nmahalik@gmail.com (N. P. Mahalik), bibhuti.biswal@gmail.com (B. B. Biswal)

Abstract: For condition monitoring of machineries and systems conventional method such as hardware or sensor based error checking scheme were in use. As the automated systems are becoming complex, recently most of the condition-monitoring schemes have been applying sophisticated analytical tools and methods to achieve improved performance. The objective of this paper is to demonstrate model based Fault Detection and Isolation (FDI) schemes for mechatronic systems and devices. First we have reviewed FDI approaches and implementation schemes. Then, we have developed two frameworks: model and spectral signature based for the implementation of FDI schemes. The model based feature estimation and spectral analysis based multiresolution methods are implemented in exemplar devices such as actuators and sensors used in mechatronic systems. Based on the frameworks, the diagnostics and isolation algorithms were developed using MATLAB code. The algorithms are capable of detecting and isolating faults within the systems. The study is comprehensive and the implementation scenarios can be extendible to many types of systems and devices used in the mechatronic domain.

Keywords: FDI, Model-Based, Spectral Analysis, Multiresolution, ANN, FL

1. Introduction

For condition monitoring of machineries and systems, conventional method such as hardware based error checking [1, 15] has been very popular since long. As the automated systems are becoming complex, recently most of the condition-monitoring schemes have been applying sophisticated analytical tools and methods to achieve improved performance [2]. Review on various condition monitoring and FDI approaches can be found from [3]. In this paper we have presented sophisticated analytical methods and approaches in order to develop FDI algorithms which can be applied to a range of systems and devices. A brief review on model-based FDI technique for actuating systems is presented below. Primarily, the model-based approach prepares a model that facilitates generating residuals which are measured and tested statistically in order to compare with the logical normal patterns for decision

making of fault symptoms as illustrated in the Fig. 1 [16-17].

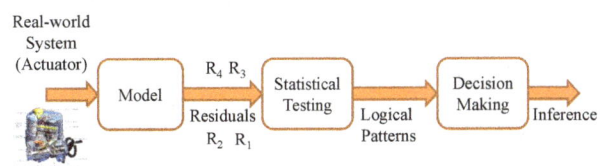

Figure 1. Stages involved in model based FDI approach.

Mechatronic systems are composed of actuators and sensors. We have applied model-based technique to actuators and multi-resolution spectral analysis methods to sensors. The objective is to detect faults in the motor during its operation. First, we reviewed a framework to define appropriate model equations in order to accommodate features capable of finding faults. Fig. 2 shows the schematic of model based FDI scenario. For detail derivation, refer [2-5]. Here, A(s) and B(s) are system matrix, u and y are inputs and outputs respectively. The feature estimation process is based on a task that determines the dependencies between different measurable signals - expressed by mathematical models. Based on measured input/output signals the detection methods generate residuals and then estimate the feature for comparison with the normal features. The changes of features are detected and faults are isolated.

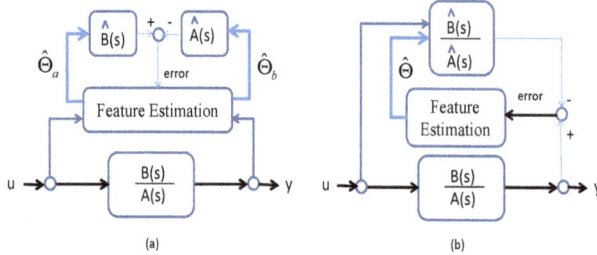

Figure 2. Feature estimation method with minimum errors (a) Equation error, (b) Output error [3, 15, 17].

2. Feature Estimation Based FDI for Actuation Valve Driven by Motor

In this experimental study, we developed a prototype to replicate the motor-actuation based valve control system as shown in Fig. 3. The main valve has two openings: left side opening and right side opening. There are two motors. Each motor drives its corresponding auxiliary valve which is interfaced with the main valve to either open the left side or right side based on the control need. This each motor drives the outflow valve either in positive or negative direction depending upon the amount of opening and closing need. The entire control strategy was implemented based on Distributed Control Scheme (DCS). Under this scenario

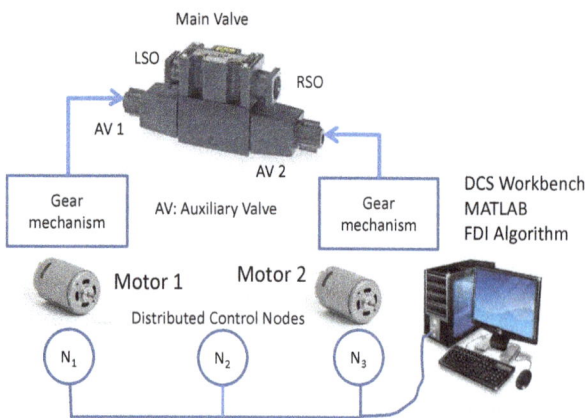

Figure 3. The prototype of motor-actuation based valve control system.

In order to detect the fault it was required to identify the feasible measurands. For this system, the considered measurands are $V(t)$, $I(t)$, $\omega_r(t)$, and $\theta_g(t)$. These measurands can be segmented to auxiliary measurands. For example, considering time variant current, the voltage equation of the electrical subsystem can be written $V(t) - \psi_E$

$\omega_r(t) = RI(t)$; R is the resistance of the coil and ψ_E flux linkage. Thus, flux is derived to be one of the auxiliary measurands featuring faults because the generated rotor torque is proportional to the effective magnetic flux linkage. Note that $\psi_T < \psi_E$ and is given by $\Im_r(t) = \psi_T I(t)$. Similarly, the dynamic equation of the motor can be written as $J_r \omega_r(t) = \psi_T I(t) - \Im_f(t) - \Im_L(t)$, where J is called the moment of inertia. The friction torque becomes $\Im_f(t) = c_f sign\omega_r(t)$ and load torque is $\Im_L(t)$. The expression $\theta_g = \theta_r / v$ is the relationship between the gear ratio v and shaft position (θ_r) and flap position (θ_g). The value of v from the data sheet is 1294. Now the torque of the flap becomes $\Im_L = c_s f(\theta_g)$. Let us formulate a model equation that can reflect the feature estimation scheme. A suggested formulation can be $y(t) = \Delta^T(t)\theta$. The following expressions are written to describe the residuals; each is coupled from at least one measurands [3]. Respective residuals are considered as the image of the respective faults [5]. Several residuals can be combined to represent some form of composite faults. The important variables that are considered for measurement are voltage V; current I, friction c_f , resistance R, and position = ϕ [2-5].

$$F_{R1}(t) = V(t) - RI(t) - \psi_E \omega_r(t)$$
$$F_{R2}(t) = \psi_T I(t) - J_r \omega_r(t) - c_f sign\omega_r(t) - c_s f(\theta_g)$$
$$F_{R3}(t) = V(t) - R / \psi_T (J_r \omega_r(t) + c_s f(\theta_g)$$
$$+ c_f sign\omega_r(t) + \psi_E \omega_r(t))$$
$$F_{R4}(t) = \theta_g(t) - \theta_r(t) / v$$

2.1. Results

Previously we had implemented the FDI algorithm in DSP (Digital Signal Processor) platform, a centralized system, and presented the results in [4]. This work is a step forward to earlier work in that the control architecture and strategy have been implemented based on DCS (Distributed Control Systems) scheme rather than centralized. That is the FDI algorithm was embedded within the DCS architecture using fieldbus type technology. Once the model is developed, the feature estimation algorithm is written to identify abnormality in the features or parameters. In order to achieve this, two sub-processes are needed: classification and decision. ANN (Artificial Neural Network) based classification is very popular for nonlinear systems. The decision algorithm was implemented using FL (Fuzzy Logic) methods. Fig. 4 illustrates the FDI processes that include both learning and run-time phase.

Fig.5. shows the signals measured at the outputs using sensors. It also shows the feature estimations and the derived residuals. The residuals are estimated based on the equations presented above. Five different faults corresponding to resistance, friction, voltage, position, and current were deliberately incorporated into the system and subsequently some of the features were observed via estimation using the model equations. Finally, the fault signatures such as residuals were generated to isolate and localize the exact fault. Some of the experimental results are shown in Table 1. We also conducted the success rate of identifying the faults. The success rate in this laboratory based demonstration system was 100%.

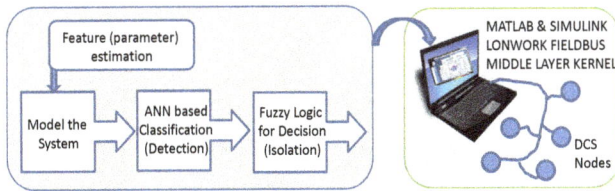

Figure 4. *Experimental set up and algorithm development phases.*

Table 1. *Feature deviation due to faults.*

Faults introduced	Estimation of some features					Residuals generated in percentage (%)			
	$\hat{R}$	$\hat{c}_f$	$\hat{V}$	$\hat{\phi}_g$	$\hat{I}$	F_{r1}	F_{r2}	F_{r3}	F_{r4}
R	+	+	+	+	+	94	77	86	78
c_f	+	+	+	+	+	79	81	85	87
V	+	+	+	+	+	98	98	90	84
ϕ	+	+	+	+	+	88	92	90	99
I	+	+	+	+	+	96	90	93	93

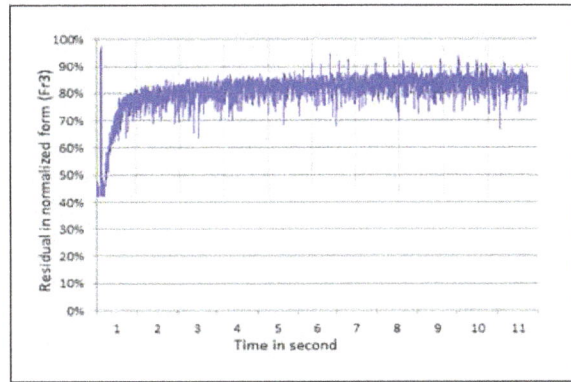

(a)

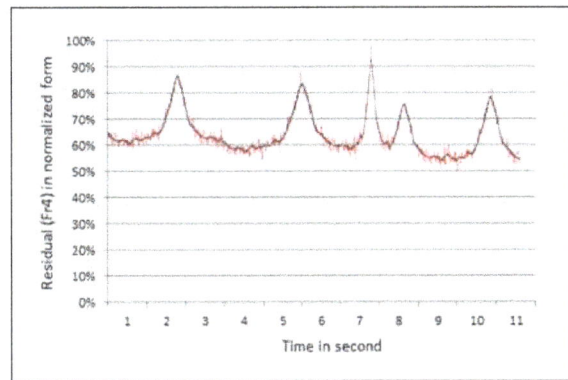

(b)

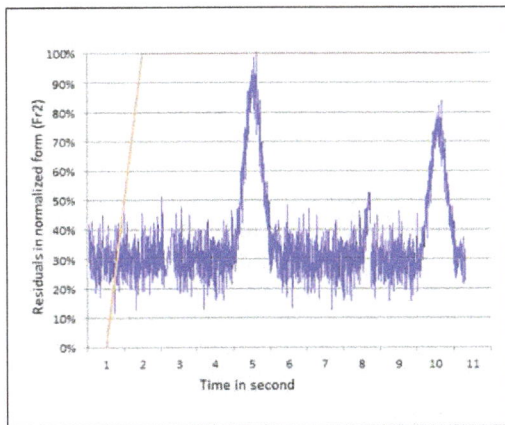

(c)

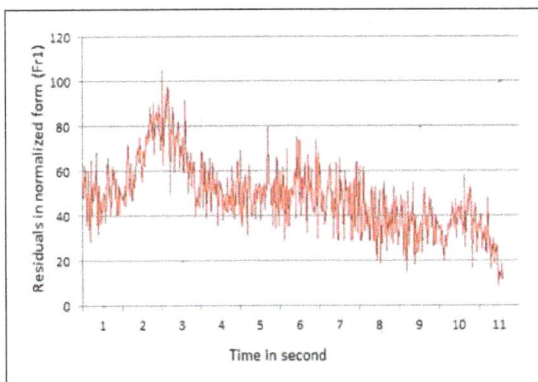

(d)

Figure 5. *Feature estimation based FDI results (Residuals when the valve leaks): Residuals at 15-50% during normal operation. (a) leakage fault, (d) sensor offset positive, (e) sensor offset negative; (B) Residuals for positive offset on supply air.*

The estimation in all experiment showed positive (+). Using the normalization method, we have set the threshold value (minimum) of the residual as 75%. The faults were introduced manually. The employed control scheme is closed loop based. Both ANN and FL algorithms) were used to classify [6] and isolate [5] the faults, respectively.

2.2. Classification and Decision Using ANN and FL

The FDI phases are described in the Fig. 3. It can be stated that ANN forms the basis of classification, [4] while FL serves as the decision tool. [2]. Both classification and decision together are treated as residual evaluation phases of the FDI [6-7]. Residual evaluation is thus a part of fault isolation [8]. The goal of residual evaluation for FDI is to device whether a fault in a system under consideration has occurred avoiding wrong decision that cause false alarm. In order to reduce the false alarm rate, classification and decision by combination of ANN and FL, respectively are suggested. As a validation of the residual evaluation we conducted several experiments with the above actuator based valve control system. As another example, the Fig.5 shows some other faults corresponding to appropriate residuals. In particular, it is necessary to understand the leakage fault of the valve when it is completely closed. The figure is self-explanatory and the results presented in Fig.4 and Fig.5 are based on the FDI algorithm represented in Fig.3. The classification and decision aspect of the algorithm is realized using ANN and FL toolset of the MATLAB software [9]. The mapping for a given input to an output using FL can be formulated by using a process called fuzzy inference. Fuzzy inference system is having wide range of applications in the fields such as automatic control, data classification, decision analysis, expert system and computer vision [10]. MATLAB has been designed with three functions namely plotfis, plotmf and gensurf which provides a high level view of the fuzzy inference system. All the fuzzy inference system information is contained in the MATLAB object known as FIS (Fuzzy inference system) structure. This structure is stored inside each GUI tool. This structure is examined by access functions such as getfis and setfis.

3. Second Experiment with Wheelchair

In considering advancements in computing platform and digital technology, this work studies FDI in actuating systems [11]. Actuators are a class of technological systems that always entail greater reliability. This work presents a study on two types of FDI approaches. They are model-based and spectral analysis.

3.1. Model-Based FDI

In the second experiment for the evaluation of the feature estimation based fault detection is a wheel chair shown in the Fig.6. Fig.7. shows schematic of the steering system.

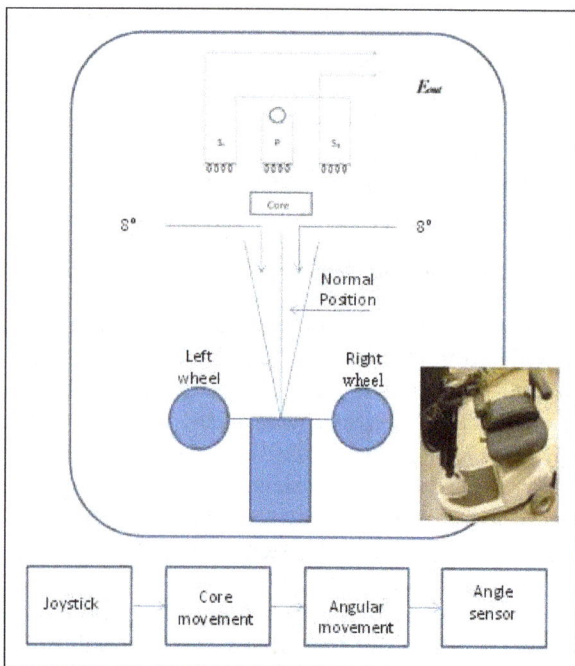

Figure 6. A 3-wheel based rehabilitation chair [18].

The objective is to observe the driving activities while the

wheel chair changes the lane. This behavior of the lane change is extremely important to a handicapped person sitting the chair. The faults in the steering actuator must be detected and isolated. If the behavior of the lane change is abnormal then the steering operation must be switched to the standby actuating system.

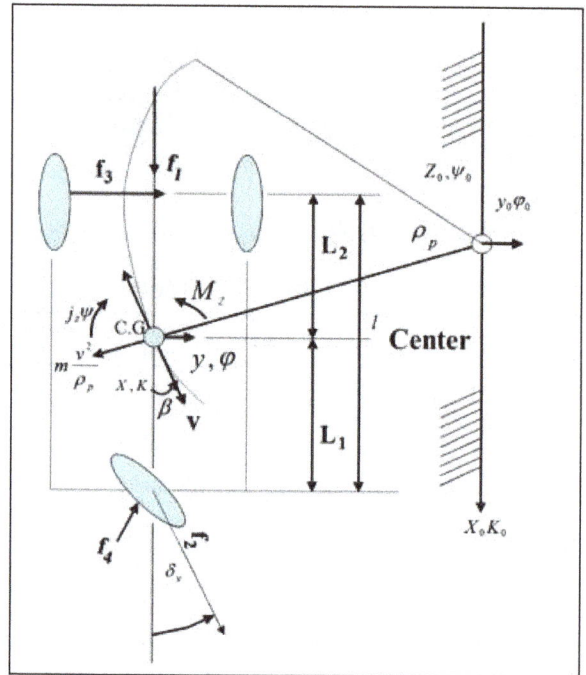

Figure 7. Geometrical drawings of the steering behavior [3].

Based on theoretical modeling it is perceived that the characteristic velocity will be the preferred choice as regards to a feature of this system [3-5]. Here the lane-change behavior is reflected through three behaviors such as normal-steering, under-steering and over-steering. The wheel chair has 3 wheels as shown in the Fig.7. By referring [3], the dynamic model equations of this wheel chair were presented in detail in [5] and [4]. The important model equations used in this paper can be written from [3].

Since this is a lane change scenario, the dynamics can be modeled considering lateral movement. The following terms are defined.

a: position error b: lateral position error
c: vertical position error ψ : yaw angle

Δ : steering angle Δa : wheel angle
F: slip angle m: mass
I_a: moment of inertia V: driving velocity
S_1: front wheel stiffens S_2: rear wheel stiffness
L_1: front axle length to C.G L_2: rear axle length to C.G
L: front and rear axle length i_{st}: gear ratio
ρ : radius F_1: force acting on rear tire

First, consider the characteristic equation as of a vehicle as follows.

$$S^2 + C_1 S + C_0 = 0$$

Where,

$$C_1 = \frac{1}{I_a mv}[(I_a + ml^2_{\,1})S_1 + (I_a + ml^2_{\,2})S_2]$$

$$C_0 = \frac{1}{I_a mv}[S_1 S_2 (L_1 + L_2)^2 + mv^2 (S_2 L_2 - S_1 L_1)]$$

The characteristic velocity [3] can be written as follows.

$$V^2_{\,ch}(t) = \frac{S_1(t)S_2(t)^2}{m(S_2(t)L_2 - S_1(t)L_1)}$$

The above characteristic expression that is used for stability is as follows.

$$\frac{V^2}{V_{ch}^{\;2}} < 1$$

The input and output model equations [3] are as follows.

$$\begin{bmatrix} \ddot{b} \\ \ddot{\psi} \end{bmatrix} = \begin{bmatrix} \dfrac{s_1'}{mi_{st}} \\ \dfrac{s_1'L_1}{I_a i_{st}} \end{bmatrix} \Delta_a = \begin{bmatrix} \dfrac{mv + s_2 + s_1'}{mv} & \dfrac{s_1 L_1}{mv} \dfrac{s_1 L_1}{mv} \\ \dfrac{s_2 L_2 s_1' L_1}{I_a v} & \dfrac{s_1' L^2_{\,1}}{I_a v} \dfrac{s_2 L^2_{\,2}}{I_a v} \end{bmatrix} \begin{bmatrix} \dot{b} \\ \dot{\psi} \end{bmatrix}$$

$$\begin{bmatrix} \dot{b}(t) \\ \dot{\psi}(t) \end{bmatrix} = \begin{bmatrix} 1 & 0 \\ 0 & 1 \end{bmatrix} \begin{bmatrix} \dot{b} \\ \dot{\psi} \end{bmatrix}$$

The steering wheel angle during natural steering will be

$$\frac{\dot{\psi}(t)}{\Delta a(t)} = \frac{1}{i_{st}L} \frac{v(t)}{1 + \left(\dfrac{v(t)}{v_{ch}(t)}\right)^2}$$

With the measured steering wheel angle at the input, the velocity and the yaw rate at the output the characteristics velocity can be written as follows.

$$[V_{ch}^{\;2}(t)]\left[1 - \frac{\Delta a(t)V(t)}{\dot{\psi}(t)i_{st}L}\right] = -V^2(t)$$

3.2. Results

The overall objective of this research was to develop advanced diagnostic and prognostic method by making the systems more predictable and available through developing model-based dynamic health management algorithms. The developed algorithm utilizing model equations to estimate the features [12] (both variables and parameters) by using sensor signals and refining it to a data infrastructure for FDI usage based in DCS implementation strategy. Thus the research not only demonstrates FDI scheme, but its implementation methods using advanced fieldbus type technology systems and architecture (e.g., DCS). The architecture can serve as basis for developing a framework for identifying fault symptoms before it occurs.

The work involves techniques for managing uncertainty and predictive modeling of future behavior. The steering behavior depends on the velocity. We set the nominal velocity as reference. 50% of nominal velocity either side was set as understeering and oversteering. Several sensors were calibrated and compensated beforehand in order to measure the measurands such as angle, acceleration, yaw rate and normal velocity. As mentioned in the equation, the effect of instability scenario was also introduced in the modeling. The wheelchair's lane change behavior was studied multiple times. Similar to actuation valve, the residuals were generated and estimation was produced. Fig.8 shows the results corresponding to lane change behavior and its predictive model.

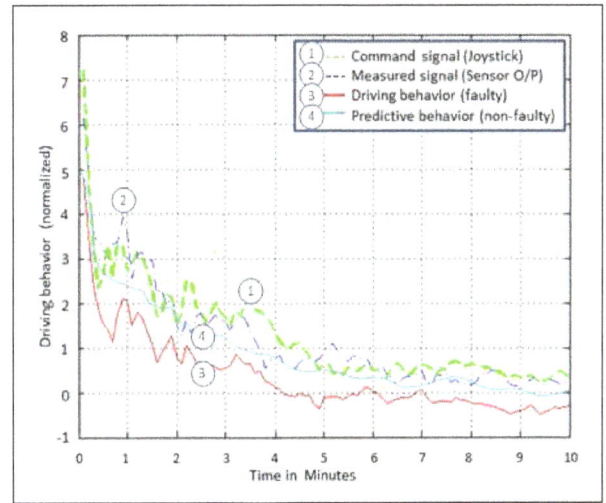

Figure 8. Results showing command signal (Joystick, measured signal (Sensor O/P), faulty driving behavior, and predictive driving behavior (non-faulty based on model based FDI).

3.3. Spectral Analysis based FDI

The wheelchair has many components responsible for its operation. The bottom part of the Fig. 6 depicts the movement of the joystick and the resulting output. One such essential component is the LVDT sensor which plays significant role in lane change behavior. It is an electromechanical transducer which is coupled to the wheelchair and it converts the joystick movement into the core movement. LVDT is encountered with different kind of fault, out of which the most common are overvoltage faults, core not moving fault and inductance or coil fault. Over- or under-voltage occurs when the applied voltage is in excess that cannot be handled by the primary coil of the LVDT.

Non-movement of core is associated with flux equivalency and corresponding differential voltages at both the coils. In this study, we have modeled the system where all these faults can have one index or signature that what we call inefficiency of the coil. The considered fault occurrence in an LVDT results from the inefficiency of the coils. It is assumed that the joystick makes slight deviations of about 8° with respect to the expected line while moving along a straight path. Due to this core movement there is an angle involved which is detected with the help of an angle sensor. The output of an angle sensor is a random signal as shown in Fig.9(a).

Scalogram of the non-faulty random signal we get is shown in Fig.9b.

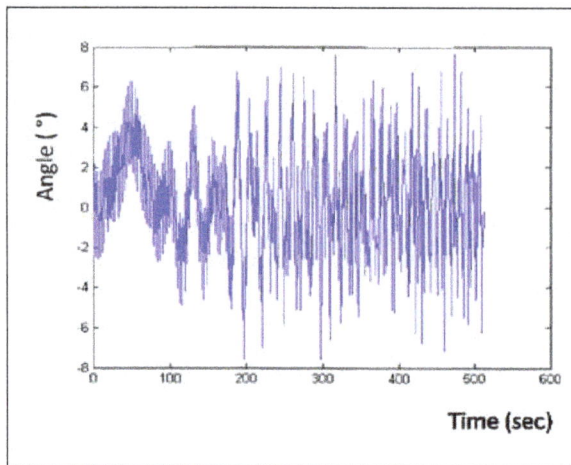

(a)

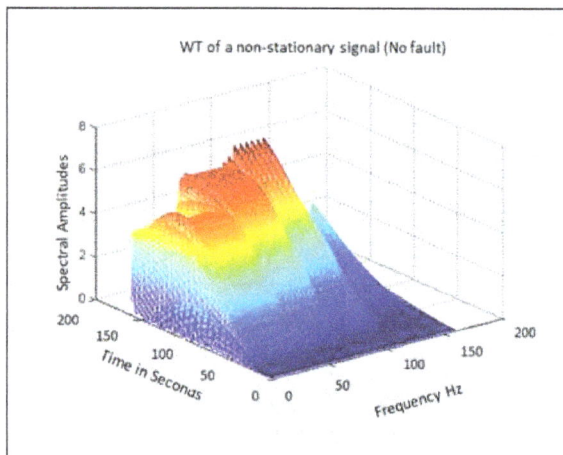

(b)

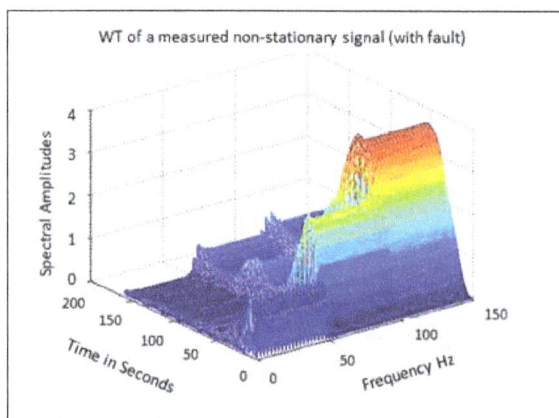

(c)

Figure 9. *Results showing spectral analysis based FDI [(a) The random signal expected at the output of an angle sensor, (b) Scalogram of the LVDT signal showing all the frequency components, (c) Scalogram of the random signal obtained from the angle sensor output superimposed with the LVDT fault].*

The fault in the core takes place when the command is given by the joystick but the relative movement of the core stops. In our simulation study the random signal from an angle sensor and the dither signal are superimposed and its WT is computed using the MATLAB. The Fig.9(c) shows the superimposition of the random signal obtained from an angle sensor output and the LVDT fault (core not moving). It is clear from the figure that the WT is able to record the change in the frequency components, as well as their time localization.

4. Conclusions

This paper provides validation of different FDI schemes based on the results obtained through several experimental studies [13]. The studies were carried out for model based feature estimation and spectral analysis based multiresolution signal decomposition methods, respectively. First the feature estimation based FDI for actuating valve driven by two motors was illustrated. In this case the experimental study was carried out for an electrical motor. Several experiments were conducted with actuator based valve control system for the validation of residual evaluation. As a second example, in this paper a rehabilitation wheel chair was considered for feature estimation based fault detection evaluation. The objective was to detect and isolate the faults in steering actuators. The classification and confirmation of the faults are implemented with the help of Artificial Neural Network (ANN) and Fuzzy Logic (FL) [14]. Finally, the validation of multiresolution technique of a sensor was demonstrated. The failure modes were considered. The scalograms of the output signals with the faults signatures were examined and detected.

Acknowledgements

The authors acknowledge Prof. R. Isernann for developing the model based approaches for several dynamical systems. The model-based work is based on much of Prof. R. Isernann's earlier work.

References

[1]   Juan Dai, Chen, C.L.P., Xiao-Yan Xu, Peng Hu, Condition monitoring on complex machinery for predictive maintenance and process control, IEEE International Conference on Systems, Man and Cybernetics, 2008. SMC 2008., vol., no., pp.3595,3600, 12-15 Oct. 2008

[2]   Dogan Gökhan and Murat Başaran, Condition monitoring of speed controlled induction motors using wavelet packets and discriminant analysis, Journal of Expert Systems with Applications, Volume 38 Issue 7, pp 8079-8086, 2011, Pergamon Press, Inc. Tarrytown, NY, USA

[3]   Rolf Isermann, Model-based fault-detection and diagnosis – status and applications, Annual Review in Control, 29 (2005) 71–85

[4] Mohanty, S.K.; Mahalik, N., "Some Studies on FDI in Actuating Systems," Process Automation, Control and Computing (PACC), 2011 International Conference on , vol., no., pp.1,6, 20-22 July 2011, doi: 10.1109/PACC.2011.5979009

[5] Mohanty, S, "Fault Detection in Mechatronic Systems", PhD Thesis, 2006, Utkal University, India

[6] Nadia Ben Amor, Multifont Arabic Characters Recognition Using Hough Transform and HMM/ANN Classification, Journal of multimedia, vol. 1, no. 2, May 2006

[7] Magdi. A. Koutb, M. Nabila, El-Rabaie, Hamdi. A. Awad, Ibrahim. A. Abd El-Hamid, Neural Fuzzy Fault Detection And Isolation In Greenhouses, IEEE Control System Magazine, pp. 28-47,

[8] Emmanuel Mazars, Imad M. Jaimoukha, and Zhenhai Li, Computation of a Reference Model for Robust Fault Detection and Isolation Residual Generation, Journal of Control Science and Engineering, 2008

[9] http://serdis.dis.ulpgc.es/~ii-rf/Manuales/Matlab/Matlab%205%20-%20Reference%20Manual.PDF

[10] H. T. Mok, C. W. Chan, Online fault detection and isolation of nonlinear systems based on neuro-fuzzy networks, Engineering Applications of Artificial Intelligence, Vol. 21, Iss. 2, pp 171-181, 2008

[11] Inseok Hwang, Sungwan Kim, Youdan Kim and Chze Eng Seah, A Survey of Fault Detection, Isolation, and Reconfiguration Methods, IEEE Transaction on Control Systems Technology, Vol 18, No. 3, 2010.

[12] A. Ashokan and D. Sivakumar, Intergration of Fault Detection and Isolation Control In A Multi-Input Multi-Output System, Journal Automation & System Engineering,

[13] D. Füssel and R. Isermann, Hierarchical motor diagnosis utilizing structural knowledge and a self-learning neuro-fuzzy-scheme. IEEE Trans. on Ind. Electronics, Vol. 74, No. 5, pp. 1070-1077, 2000

[14] S.M. El-Shal, A.S. Morris, A fuzzy expert system for fault detection in statistical process control of industrial processes, IEEE Transactions on Applications and Reviews, Systems, Man, and Cybernetics, Part C, Vol. 30, Iss. 2, pp. 281 – 289, 2000

[15] Balle, P.; Isermann, Rolf, "Fault detection and isolation for nonlinear processes based on local linear fuzzy models and parameter estimation," American Control Conference, 1998. Proceedings of the 1998, vol.3, no., pp.1605,1609 vol.3, 21-26 Jun 1998, doi: 10.1109/ACC.1998.707277

[16] Jie Chen and R.J. Patton, "Robust Model-Based Fault Diagnosis for Dynamic Systems", Book, Springer Science Business Media, New York., 1999

[17] B. Freyermuth, R. Isermann (Ed.), "Fault Detection, Supervision and Safety for Technical Processes" IFAC Symposia Series, 1991.

[18] Harbilas Bal and N. P. Mahalik, "Fault detection (FDI) in actuators, 32nd Annual Central California Research Symposium, California State University, April 6, Fresno, California.

13

Robust Nonlinear Adaptive Controller Design for Horizontal Position Control of a Rotary Wing Autonomous Vehicle Using Backstepping Method

Tushar Kanti Roy

Department of Electronics & Telecommunication Engineering, Rajshahi University of Engineering & Technology, Rajshahi, Bangladesh

Email address:
roy.kanti03@gmail.com

Abstract: In this paper, a new approach of designing robust adaptive backstepping controller for horizontal position control of a rotary wing autonomous unmanned vehicle (RAUV) with consideration of parametric uncertainties and external disturbances is proposed. Based on this new approach, the proposed RAUV controller is adaptive to the parametric uncertainties and robust to the external disturbances. To prove the convergence of different tracking error to zero, a control Lyapunov function (CLF) is formulated in every step of the design process of controller and which is guaranted through the negative definiteness of the derivative of CLF. At last, a numerical evaluation is performed on a highly fedility nonlinear simulation model to justify the usefulness of the proposed controller. The performance of the designed controller is also compared with a classical PID controller. Simulation results demonstrate that the proposed controller provides an improved performance for the closed-loop system in the presence of parametric and external uncertainties within the UAH model over the existing controller.

Keywords: Adaptive Robust Backstepping Controller, Control Lyapunov Function, External Disturbance, Rotary Wing Autonomous Vehicle, Parametric Uncertainty

1. Introduction

There are variety of rotary wing unmanned autonomous vehicles (RUAV), but among these the unmanned autonomous helicopters (UAHs) constitute one of the most versatile and agile platforms for the development of autonomous flight systems. A small-scale unmanned helicopter can operate in different flight modes, such as vertically take-off/landing, hovering, longitudinal or lateral flight, and bank to turn which gives them the advantage of effective observation from various positions. Hovering and vertically take-off are necessarily needed. Nowadays, there are new trends of UAH controller design due to their high level of agility, maneuverability and capability of operating in adverse weather conditions. To achieve these performances of an UAH, the trajectory tracking and disturbance rejection capability need to be significantly improved. But the problem is that it is a naturally unstable system with nonlinear dynamics. The main difficulties of an UAH are higher nonlinearities which arise from the cross-couplings due to the tail rotor, main rotor, engine and dynamic uncertainties [1]. Besides, it is an underactuated mechanical system with six degrees of freedom (6-DOF) because it has only four control inputs. To cope such problems as mentioned above, different approaches have been proposed by the researchers in literature [2]-[4]. Thus, in order to improve the tracking performance of an UAH in the presence of parametric and external uncertainties within the system a robust adaptive backstepping controller is proposed in this paper.

Different conventional controllers are available to stabilize to flight of an UAH which are designed on the linear approximation around an operating point [5], [6]. But these controller are not suitable when operating point is changed due to any external or inter uncertainties. Thus, recently, various advance nonlinear control techniques have been applied to the control the UAH for different operating points under large disturbance [7]-[11]. A robust H_∞ control method of the longitudinal and lateral dynamics of the BELL 205 helicopters in the presence of model uncertainties is presented in [12]. A robust feedback method is propoed in [13] to reject the external wind gust, where the external wind gust is assumed to be the sum of a fixed number of sinusoids with unknown amplitudes, frequencies and phases. To control of an autonomous scale helicopter under the consideration of parameter uncertainties and uniform time varying three-dimensional wind gusts a backstepping method is proposed in [14]. A nonlinear H$_\infty$ horizontal position controller for hovering and landing flight of a RUAV in the presence of horizontal wind gusts is proposed in [15]. Similar control approach is used in [16] to control the altitude and attitude for a hovering flight of an UAH in the presence of vertical wind gusts. A nonlinear robust small-gain control method is proposed in [17] to control the vertical motion of an autopilot helicopter during landing flight condition in the presence of parametric uncertainties on the plant and actuator model. The hover flight control of a helicopter by considering the 1-DOF system using the neural adaptive technique was proposed in [18]. Though, these controller are designed by considering the uncertainties with the system of an UAH but how to control the longitudinal and lateral dynamics of an UAH are not clear in this papers. Moreover, the longitudinal and lateral dynamics control to improve the trajectory tracking performance of an UAH under consideration of parametric uncertainties along with external disturbances are still uncovered.

The main aim of this paper is to design a robust adaptive bacstepping controller to control the horizontal position of an UAH under the consideration of parametric uncertainties and external disturbances. Note that backstepping is a recursive nonlinear control design method, which provides an alternative to the feedback linearization. The advantage of this technique over feedback linearization technique is that it can gain from the stabilizing nonlinear terms rather than eliminating them. For the controlling purposes, the forward and sideward motions of an UAH, together with roll and pitch motions are controlled by lateral and longitudinal cyclic stick inputs via the flapping motion of the main rotor. Finally, a nonlinear simulation model is used to evaluate the effectiveness of the designed controller for hovering flight control and compared to that of an existing PID controller.

The rest part of this paper is organized as follows. Section 2 briefly introduces the mathematical model of UAH to design the controller. The control problem is formulated in Section 3. Section 4 presents the design procedure of the proposed controller. Section 5 discusses the simulation results. Finally, the conclusion of the work is presented in Section 6.

2. Dynamical Model of UAH

The UAH has the specific characteristic as compared to fixed wing aircraft such as, it can move vertically, float in the air, turn in place, move forward and laterally and can perform these movements in combinations. Due to these characteristics, the dynamic modeling of an UAH is a very

complex problem. The motion states and control inputs of an UAH in the 6-DOF form can be represented as follows:

$$x = \{u, w, q, \theta, v, p, r, \phi, \Psi\}$$
$$u_c = \{\delta_{col}, \delta_{lat}, \delta_{lat}, \delta_{ped}\}$$

where u, v and w represent linear velocity in body frame; p, q and r denote roll, pitch and yaw rates; and ϕ, θ and Ψ represent the roll, pitch and yaw angle, respectively of an UAH. A single main rotor helicopter has four independent control inputs such as δ_{lat}, δ_{lon}, δ_{col} and δ_{ped} which denote the deflection of the lateral cyclic, longitudinal cyclic, main rotor collective pitch and tail rotor collective pitch, respectively. The collective commands control the magnitude of the main rotor and tail rotor thrust and other two control commands control the inclination of the tip-path-plane (TPP) on the longitudinal and lateral direction. The equations which describing the net forces of the UAH can be expressed as,

$$m(\dot{u} - vr + wq) = X - mg \sin \theta \qquad (1)$$

$$m(\dot{v} - wp + ur) = Y + mg \sin \phi \cos \theta \qquad (2)$$

$$m(\dot{w} + uq - vp) = Z + mg \cos \phi \cos \theta \qquad (3)$$

The equations which describe the moments of the UAH can be expressed as,

$$L = I_{xx}\dot{p} - I_{xz}\dot{r} + qr(I_{zz} - I_{yy}) - I_{xz}pq \qquad (4)$$

$$M = I_{yy}\dot{q} + pr(I_{xx} - I_{zz}) + I_{xz}(p^2 - r^2) \qquad (5)$$

$$N = -I_{xz}\dot{p} + I_{zz}\dot{r} + pq(I_{yy} - I_{xx}) + I_{xz}qr \qquad (6)$$

In order to complete the system modeling, the following three equations are essential which relate the Euler angle rates to the angular velocity [19].

$$\dot{\phi} = p + q \sin \phi \tan \theta + r \cos \phi \tan \theta \qquad (7)$$

$$\dot{\theta} = q \cos \phi - r \sin \phi \qquad (8)$$

$$\dot{\Psi} = \frac{q \sin \phi + r \cos \phi}{\cos \theta} \qquad (9)$$

The longitudinal and lateral cyclic tilt of the main rotor disk is controllable through the cyclic pitch. Therefore, the longitudinal and lateral flapping dynamics can be represented by the following first-order equations [20].

$$\dot{a}_1 = -q - \frac{a_1}{\tau} + \frac{1}{\tau}(\frac{\partial a}{\partial u}u + A_{lon}\delta_{lon}) \qquad (10)$$

$$\dot{b}_1 = -p - \frac{b_1}{\tau} + \frac{1}{\tau}(\frac{\partial b}{\partial u}v + A_{lat}\delta_{lat}) \qquad (11)$$

where δ_{lat} and δ_{lon} are the lateral and longitudinal cyclic control inputs, a1 and b1 are the lateral and longitudinal flapping angles and Alon and Blat are the effective steady-state longitudinal and lateral gains from the cyclic inputs to the main rotor flapping angles. The terms $A_u = \dfrac{\partial a_1}{\partial u}$ and $B_v = \dfrac{\partial b_1}{\partial v}$ are constants and represent the longitudinal and lateral Dihedral effect. The Dihedral effect is the change of tip-path-plane (TPP) tilts due to the longitudinal and lateral velocities [21]. The Dihedral effect is modeled by the following equation.

$$\frac{\partial a_1}{\partial u} = -\frac{\partial b_1}{\partial v} = \frac{2}{\Omega R_b}\left(\frac{8C_T}{a\sigma} + \sqrt{\frac{C_T}{2}}\right) \quad (12)$$

where R_b is the main rotor radius, σ solidity ratio, 'a' lift curve slope and C_T thrust coefficient. Since the rotor is symmetric, so the consideration is $A_u = -B_v$. In order to design the controller, the linearization is essential to derive a simplified working model, due to the inherent instability under hover and slow flight conditions. So, after linearizing the equations (1)-(8) the following parameterized model of decoupled longitudinal and lateral dynamics can be obtained,

$$\begin{bmatrix} \dot{u} \\ \dot{q} \\ \dot{\theta} \\ \dot{a}_1 \end{bmatrix} = \begin{bmatrix} X_u & X_q & -g & X_a \\ M_u & M_q & 0 & M_a \\ 0 & 1 & 0 & 0 \\ \dfrac{A_u}{\tau} & -1 & 0 & -\dfrac{1}{\tau} \end{bmatrix} \begin{bmatrix} u \\ q \\ \theta \\ a_1 \end{bmatrix} + \begin{bmatrix} 0 \\ 0 \\ 0 \\ \dfrac{A_{lon}}{\tau} \end{bmatrix}\delta_{lon} \quad (13)$$

$$\begin{bmatrix} \dot{v} \\ \dot{p} \\ \dot{\phi} \\ \dot{b}_1 \end{bmatrix} = \begin{bmatrix} Y_v & Y_p & g & Y_b \\ L_v & L_p & 0 & L_b \\ 0 & 1 & 0 & 0 \\ \dfrac{B_v}{\tau} & -1 & 0 & -\dfrac{1}{\tau} \end{bmatrix} \begin{bmatrix} v \\ p \\ \phi \\ b_1 \end{bmatrix} + \begin{bmatrix} 0 \\ 0 \\ 0 \\ \dfrac{A_{lat}}{\tau} \end{bmatrix}\delta_{lat} \quad (14)$$

where $X_u = \dfrac{1}{m}\dfrac{\partial X}{\partial u}$, $M_u = \dfrac{1}{I_{yy}}\dfrac{\partial M}{\partial u}$ are the force and moment derivatives normalized by the mass of the helicopter or respective moment of inertia. The pitching flap-stiffness constant is represented by M_a that can be computed as follows $M_a = \dfrac{mM_z}{I_{yy}} + \dfrac{K_\beta}{I_{yy}}$, where M_z is the height of the rotor hub above the fuselage center of gravity, Iyy is the pitching moment of inertia and K_β is the rotor blade spring stiffness. Similarly the lateral flap-stiffness constant L_b can be computed as follows $L_b = \dfrac{mM_z}{I_{xx}} + \dfrac{K_\beta}{I_{xx}}$. The proposed linear model as described by equations (13)-(14), has been successfully espoused for control applications in a large number of small-scale unmanned helicopters [22]-[27]. The nonlinear robust adaptive nonlinear controller will be designed using backstepping technique based on equations (13)-(14). However, before design the controller the control problem formulation is discussed in the following section.

3. Control Problem Formulation

From the longitudinal dynamics model as described by equation (13), it can be seen that the longitudianl flapping a_1 is a function of u and q. Similarly, from the lateral dynamics model, it can be seen that the lateral flapping b_1 is a function of v and p. Under this condition, to continue the design procedure of the proposed controller is not possible. From [28], it is clear that the effect of lateral and longitudinal forces produced by the flapping angles can be neglected as they have a minimal effect on the translational dynamics as compared to the propulsion forces produced by the stability derivatives X_θ and X_φ. Moreover, according to the control purposes, the dynamics of an UAH should be separated into two interconnected subsystems. The first subsystem accounts for the longitudinal dynamics and second subsystem for lateral dynamics. Now, after neglecting the effect of parameters X_a, Y_b, X_q and Y_p the longitudinal-lateral dynamics will have a strict feedback form which is suitable for the proposed controller. Again, as the mass of an UAH is continuously varying so the parameters of an UAH is not fixed. Moreover, the dynamics of the UAH will be affected by the external wind gusts. Thus, under the above assumptions, the simplified model equations of the longitudinal dynamics can be written as follows:

$$\dot{x} = u$$
$$\dot{u} = X_u\,u - g\,\theta \quad (15)$$
$$\dot{\theta} = q$$
$$\dot{q} = \delta\,M_u\,u + M_q\,q + M_a\,a_1 + d_1$$

Similarly, the simplified model equations of the lateral dynamics can be written as follows:

$$\dot{y} = v$$
$$\dot{v} = Y_v\,v + g\,\phi \quad (16)$$
$$\dot{\phi} = p$$
$$\dot{p} = \eta L_v\,v + L_p\,p + L_b\,b_1 + d_2$$

where d_1 and d_2 are the external disturbances. Based on equations (15)-(16), the proposed robust adaptive backstepping controller design procedure is shown in the following section.

4. Controller Design

In this section, the design procedure of the proposed controller for the longitudinal and lateral dynamics of an UAH is presented based on the appropriate decoupled model as described by equations (15)-(16). The objective of this control is to regulate the several physical quantities (e.g. position, attitude etc.) for improving the flight condition of an UAH.

4.1. Longitudinal Dynamics

In this subsection, the design procedure of the adaptive robust backstepping controller is shown based on the Lyapunov function for the longitudinal dynamics as described by equation (15) under consideration of internal and external uncertainties. The design procedure is divided into four steps which is evaborately discussed as follows.

Design step 1: First, for the longitudinal position tracking objective, let define the longitudinal position tracking error as

$$z_1 = x - x_d, \quad \dot{z}_1 = \dot{x} - \dot{x}_d, \quad \dot{z}_1 = u \qquad (17)$$

Here u is assumed as a virtual control variable and its desired value u_d is a stabilizing function for equation (17). Let z_2 be an another error variable representing the difference between the actual u and its desired value u_d, i.e.,

$$z_2 = u - u_d, \quad u = z_2 + u_d$$

Therefore, interms of z2 the equation (17) can be written as

$$\dot{z}_1 = z_2 + u_d \qquad (18)$$

At this stage a virtual control law should be designed for ud in such a way that which would make $z_1 \to 0$ as $t \to \infty$.

Now, consider the first CLF as follows:

$$W_1 = \frac{1}{2} z_1^2$$

whose time derivative after substituting the value of $\dot{z}_1$ is

$$\dot{W}_1 = z_1(z_2 + u_d) \qquad (19)$$

Now an appropriate virtual control law ud need to be selected in such a way that which would make $\dot{W}_1 \le 0$. Under this condition, the stabilizing function is chosen as

$$u_d = -k_1 z_1 \text{ with k1>0} \qquad (20)$$

where k_1 is a scalar parameter which can be used to tune the output response. Then equation (19) can be written as

$$\dot{W}_1 = -k_1 z_1^2 + z_1 z_2 \qquad (21)$$

From equation (21), it can be seen that if z2=0 then

$$\dot{W}_1 = -k_1 z_1^2 \le 0.$$

The second coupling term of equation (21) will be cancelled in the next step. Now the time derivative of u_d which is essential in the next step can be written as

$$\dot{u}_d = -k_1 u \qquad (22)$$

As $\dot{z}_1 = u$.

Design step 2: In this step, the error dynamics for $z_2 = u - u_d$ is derived whose time derivative can be written as follows:

$$\dot{z}_2 = x_u u - g\theta + k_1 u \qquad (23)$$

In which θ is viewed as an another vitual control variable. Now define a virtual control law θ_d and let z_3 be an another error variable which is representing the difference between actual control and virtual control i.e., $z_3 = \theta - \theta_d$ and after taking time derivative it can be written as

$$\dot{z}_2 = (x_u + k_1)u - g(z_3 + \theta_d) \qquad (24)$$

Now choose a second CLF as follows:

$$W_2 = W_1 + \frac{1}{2} z_2^2 \qquad (25)$$

whose time derivative by inseting equations (21) and (24) can be written as

$$\dot{W}_2 = -k_1 z_1^2 + z_2\{z_1 + (x_u + k_1)u - g\theta_d\} - gz_2 z_3 \qquad (26)$$

Now an appropriate stabilizing function θ_d can be selected in such a way to cancel out the terms related to z_1, z_2 and u, while the term involving z_3 cannot be removed and this is

$$\theta_d = g^{-1}\{z_1 + (x_u + k_1)u + k_2 z_2\} \qquad (27)$$

Then equation (26) can be written as

$$\dot{W}_2 = -k_1 z_1^2 - k_2 z_2^2 - gz_2 z_3 \qquad (28)$$

From equation (28), it is clear that if $z_3 = 0$ then $\dot{W}_2 = -k_1 z_1^2 - k_2 z_2^2 \le 0$ which is negative definite. In order to complete the next step, the time derivative of θ_d is essential which can be written as

$$\dot{\theta}_d = g^{-1}\{u + (x_u + k_1)(x_u u - g\theta) + k_2 \dot{z}_2\} \qquad (29)$$

Design step 3: Here the error dynamics for $z_3 = \theta - \theta_d$ is derived whose time derivative is

$$\dot{z}_3 = \dot{\theta} - \dot{\theta}_d \qquad (30)$$

By substituting the values of $\dot{\theta}$ and $\dot{\theta}_d$ into equation (30), it can be written as

$$\dot{z}_3 = q - g^{-1}\{u + (x_u + k_1)(x_u u - g\theta)\} \\ - g^{-1}k_2\{(x_u + k_1)u - g(z_3 + \theta_d)\} \tag{31}$$

In which q is viewed as the virtual control input. Now define a stabilizing control law qd and let z4 be an error variable representing the difference between actual and virtual control input, i.e., $z_4 = q - q_d$ and interm of this error variable the equation (31) can be written as

$$\dot{z}_3 = z_4 + q_d - g^{-1}\{u + (x_u + k_1)(x_u u - g\theta)\} \\ - g^{-1}k_2\{(x_u + k_1)u - g(z_3 + \theta_d)\} \tag{32}$$

Now choose another CLF as

$$W_3 = W_2 + \frac{1}{2}z_3^2 \tag{33}$$

whose time derivative is

$$\dot{W}_3 = -k_1 z_1^2 - k_2 z_2^2 - g z_3 z_2 + z_3[z_4 + q_d \\ - g^{-1}\{u + (x_u + k_1)(x_u u - g\theta)\} - g^{-1}k_2\{(x_u + k_1)u \\ - g(z_3 + \theta_d)\}]$$

At this stage, the stabilizing q_d need to be selected in such a way that cancel out the terms related to z_1, z_2, z_3 and u, while the term involving z_4 cannot be removed. Thus, the stabilizing function q_d is

$$q_d = g z_2 + g^{-1}\{u + (x_u + k_1)(x_u u - g\theta)\} \\ + g^{-1}k_2\{(x_u + k_1)u - g(z_3 + \theta_d)\}] - k_3 z_3 \tag{34}$$

After that selection,

$$\dot{W}_3 = -k_1 z_1^2 - k_2 z_2^2 - k_3 z_3^2 + z_3 z_4 \tag{35}$$

From equation (35), it is clear that if $z_4 = 0$ then equation (35) simplified to

$$\dot{W}_3 = -k_1 z_1^2 - k_2 z_2^2 - k_3 z_3^2 \le 0 \tag{36}$$

In order to complete the final step, the time derivative of q_d can be written as

$$\dot{q}_d = f \tag{37}$$

where

$$f = g\dot{z}_2 + g^{-1}\{\dot{u} + (x_u + k_1)(x_u \dot{u} - g\dot{\theta})\} \\ + g^{-1}k_2\{(x_u + k_1)\dot{u} - g(\dot{z}_3 + \dot{\theta}_d)\}] - k_3 \dot{z}_3$$

The derivation of longitudinal dynamics control law along with the stability and robustness analysis of the system is shown in the following step.

Design step 4: The dynamics of final error can be obtained as following by taking time derivative of z4

$$\dot{z}_4 = \delta M_u u + M_q q + d_1 + M_a a_1 - f \tag{38}$$

In equation (38), the actual control input appears. By incorporating the estimation error of unknown parameter δ, equation (38) can be rewritten as

$$\dot{z}_4 = (\tilde{\delta} + \hat{\delta})M_u u + M_q q + d_1 + M_a a_1 - f \tag{39}$$

where $\tilde{\delta} = \delta - \hat{\delta}$ is the estimation error of unknown parameter δ. The objective is to design the actual control input a1 such that z_1, z_2, z_3, and z_4 converge to zero as t→∞. The presence of the parameter estimation error suggests the following form of the CLF

$$W_4 = W_3 + \frac{1}{2}z_4^2 + \frac{1}{2\gamma_1}\tilde{\delta}^2 \tag{40}$$

The time derivative of W_4 becomes

$$\dot{W}_4 = \dot{W}_3 + z_4 \dot{z}_4 - \frac{1}{\gamma_1}\tilde{\delta}\dot{\hat{\delta}} \tag{41}$$

where γ_1 is a positive design constant which determines the convergence speed of the estimation. By substituting the values of $\dot{W}_3$ and $\dot{z}_4$ into equation (41), it can be written as

$$\dot{W}_4 = -k_1 z_1^2 - k_2 z_2^2 - k_3 z_3^2 + z_4\{z_3 + \hat{\delta}M_u u + M_q q \\ + M_a a_1 - f + d_1\} - \gamma_1^{-1}\tilde{\delta}(\dot{\hat{\delta}} - \gamma_1 M_u u)$$

Now the final control law and adaptation law are chosen in such a way that which would make $\dot{W}_4 \le 0$. Thus, the final control law and adaptation law are chosen as follows:

$$a_1 = -\frac{\{z_3 + \hat{\delta}M_u u + M_q q - f + k_4 z_4 + \Gamma \operatorname{sgn}(z_4) + \hat{d}_1\}}{M_a}$$

$$\dot{\hat{\delta}} = \gamma_1 z_4 M_u u \tag{42}$$

where $\hat{d}_1$ is an estimate parameter which represents a best guess for the unknown external disturbance d_1 and sgn is the signum function which can be written as

$$\operatorname{sgn}(z_4) = \begin{cases} +1 & if \quad z_4 > 0 \\ 0 & if \quad z_4 = 0 \\ -1 & if \quad z_4 < 0 \end{cases} \tag{43}$$

The estimation error on d_1 is assumed to be bounded by knowing the constant, Γ, i.e., $\| d_1 - \hat{d}_1 \| \le \Gamma$. Then by using the Schwartz inequality the derivative of W4 becomes

$$\dot{W}_4 \le -k_1 z_1^2 - k_2 z_2^2 - k_3 z_3^2 - k_4 z_4^2 - \| z_4 \| \left[\Gamma - \| d_1 - \hat{d}_1 \| \right]$$

Since $\| d_1 - \hat{d}_1 \| \le \Gamma$, so $\dot{W}_4 \le 0$.

4.2. Lateral Dynamics

In this subsection, the robust adaptive backstepping controller is designed based on the Lyapunov function for the lateral dynamics in the presence of parametric and external uncertainties. Again, consider a CLF which is used to augment the estimated parameter error,

$$W_{10} = W_7 + \frac{1}{2}z_8^2 + \frac{1}{2\gamma_l}\tilde{\eta}^2 \qquad (44)$$

Using the similar procedure as mentioned in the previous subsection, the following control input can be obtained for the lateral dynamics

$$b_1 = -L^{-1}{}_b\{z_7 + \hat{\eta} L_v v + L_p p - f(z_6, z_7, v, \phi, \phi_d) + $$

$$k_8 z_8 + \Gamma_1 \operatorname{sgn}(z_8) + \hat{d}_2\} \qquad (45)$$

$$\dot{\hat{\eta}} = \gamma_2 L_v v z_8 \qquad (46)$$

Finally, we get the following equation.

$$\dot{W}_{10} = -k_5 z_5^2 - k_6 z_6^2 - k_7 z_7^2 - k_8 z_8^2 \leq 0 \qquad (47)$$

Thus, it can be proved that the system is Lyapunov stable and the errors are asymptotically converging to an arbitrarily small neighborhood of zero. Note that the detailed design procedure of the proposed controller for lateral dynamics is not illustrated in this paper. Simulation studies are conducted in the following section to show the effectiveness of this proposed controller.

Remark 1: Control inputs in the controller design process are set to be longitudinal and lateral flapping angles. They will be converted later into longitudinal cyclic and lateral cyclic for implementation.

5. Simulation Results

The performance of the designed controller has been conducted on a nonlinear simulation model using the MATLAB simulink. To show the superiority of the proposed robust adaptive backstepping controller over an existing controller, the performance is also compared with a classical PID controller. The simulation is conducted in the case where the desired positions are set to x = 0 m and y = 0 m. The roll angle trim φref is initialized at 4.50 to compensate the effect of tail rotor thrust.

The corresponding system responses with both controllers are shown in Fig. 1 to Fig. 3. The horizontal position responses of an UAH with both controllers is shown in Fig. 1. It is clear that the position tracking error is almost zero with the proposed controller (solid blue line) but it is relatively large with the PID controller (solid green line) and it is oscillating.

From the simulation result, it is clear that the proposed controller is more robust than the PID control under the condition of parametric and external disturbances within the

system of an UAH. The corresponding velocity response of an UAH is shown in Fig. 2, from where it is clear that horizontal velocities settle to approximately 0 m/s at about 3 s after the start of the simulation in both y direction and x direction from the beginning of the simulation. But for the PID controller, it can be seen that they are oscillating and are not completely damped.

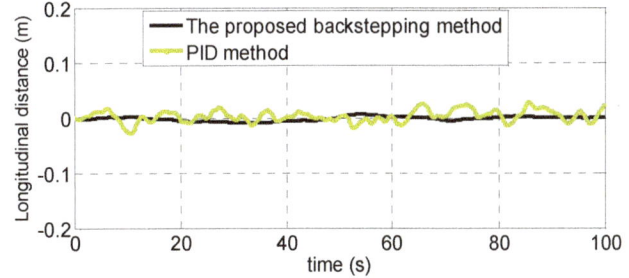

(a). Longitudinal distance of UAH.

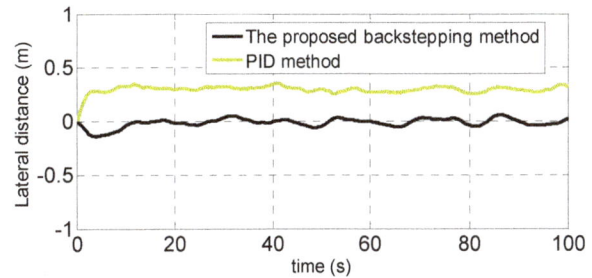

(b). Lateral distance of UAH.

Fig. 1. *Horizontal positions response using robust adaptive backstepping and PID controllers.*

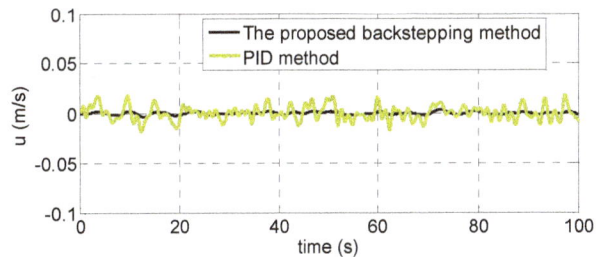

(a). Longitudinal velocity of UAH.

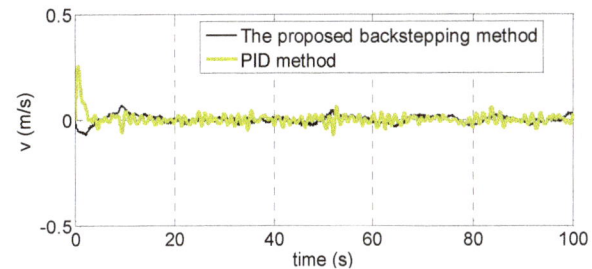

(b). Lateral velocity of UAH.

Fig. 2. *Horizontal velocities response using robust adaptive backstepping and PID controllers.*

Thus, from the simulation results, it is obvious that the proposed controller is able to achieve the desired horizontal positions in the presence of parametric and external disturbances by providing more stabilize hovering flight.

The corresponding roll and pitch angle responses of an UAH during hover flight are shown in Fig. 3. From where, it can be seen that due to the proposed controller, the roll angle settles to the desired value at 4.5° wihtin few second, but for the PID controller, it settles in between about 2.5° to 3.90°. It can be seen that the angle responses are more stable with the proposed controller than the existing controller in terms of settling time and oscillations.

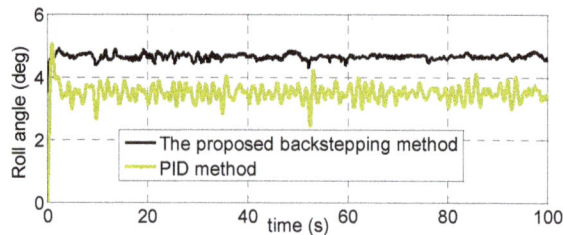

(a). Roll angle of UAH.

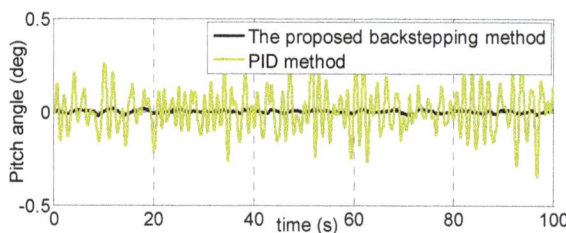

(b). Pitch angle of UAH.

Fig. 3. *Roll and Pitch angles response using the adaptive robust backstepping and PID controllers.*

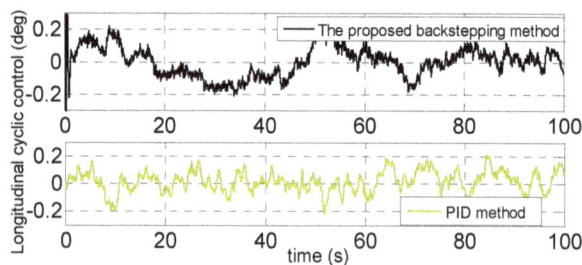

(a). Control signal of longitudinal dynamics.

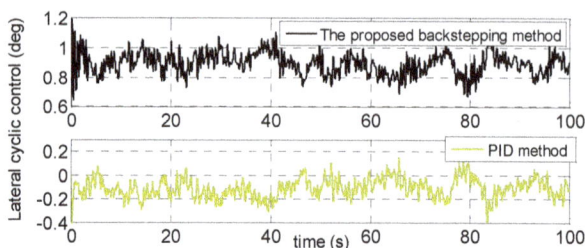

(b). Control signal of lateral dynamics.

Fig. 4. *Control signals using the robust adaptive backstepping and PID controllers.*

The control signals for the both controllers is shown in Fig. 4, from where it can be seen that control signals do not

exceed the physical constraints of the helicopter. From the simulation results, it is clear that the designed controller is more effective than an existing controller in terms of settling time and damping of oscillation.

6. Conclusion

In this paper, a backstepping method to design a robust adaptive controller for an UAH is proposed to enhance the stabilization of horizontal position control of a hovering flight. Based on the proposed formulation, the designed controller is adaptive to the unknown parameters and robust to the external disturbances. From the theoretical and numerical evaluations, it can be concluded that the proposed controller has the capability to stabilize the longitudinal and lateral dynamics as it can track the pre-defined reference trajectory despite the presence of parametric and external uncertainties within the UAH system model. Future work will be devoted on the implementation of the controller to a real system, e.g., flight test under the consideration of parametric and external uncertainties to prove the feasibility in the real life.

References

[1] T. K. Roy, M. Garrat, H. R. Pota and H. Teimoori, "Hover flight control of a small helicopter using robust backstepping and PID," in Proceedings of the 10th World Congress on Intelligent Control and Automation, 6-8 July 2012.

[2] L. Guo, C. Melhuish, and Q. Zhu, "Towards neural adaptive hovering control of helicopters," in Proc. IEEE Int. Conf. Control Applications, Glasgow, U.K., Sep. 2002, pp. 54-58.

[3] T. K. Roy, "Position control of a small helicopter using robust backstepping," in Proceedings of the IEEE 7th International Conference on Electrical and Computer Engineering, pp. 787-790, 20-22 December 2012.

[4] T. K. Roy and A. A. Suman, "Adaptive backstepping controller for altitude control of a small-scale helicopter by considering the ground effect compensation," in Proc. of the IEEE2nd International Conference on Informatics, Electronics and Vision (ICIEV), pp. 1-5, 2013, Bangladesh.

[5] H. Shim, T. J. Koo, and F. Hoffmann, "A comprehensive study of control design for an autonomous helicopter," in Proceedings of the 37th International Conf. on Decision and Control, pp. 3653-3658, 1998.

[6] X. H. Xia, and Y. J. Ge, "Finite-horizon optimal linear control for autonomous soft landing of small-scale helicopter," In Proceedings of the International Conf. on Information and Automation, pp. 1160-1164, 2010.

[7] T. K. Roy, and M. Garrat, "Altitude control of an unmanned autonomous helicopter via robust backstepping controller under horizontal wind gusts," in Proceedings of the IEEE 7th International Conf. on Electrical and Computer Engineering, pp. 771-774, 20-22 December 2012.

[8] T. K. Roy, M. Garrat, H. R. Pota, and H. Teimoori, "Robust altitude control of an unmanned autonomous helicopter using backstepping," in Proceedings of the 10th WCICA, pp. 1650-1654, 6-8 July 2012.

[9] T. K. Roy and H. M. Rasel, "Altitude control of an unmanned autonomous helicopter via adaptive backstepping controller under horizontal wind gusts," IJCEE, vol. 5, no. 1, pp. 1-7, 2013.

[10] T. K. Roy and A. A. Suman, "Adaptive backstepping controller for altitude control of a small-scale helicopter by considering the ground effect compensation," in Proc. of the IEEE2nd International Conference on Informatics, Electronics and Vision (ICIEV), May 2013, pp. 1-5, Bangladesh.

[11] T. K. Roy, M. A. Mahmud, M. Z. I. Sarkar, and M. F. Pervej, "Robust backstepping controller for yaw dynamics control of an unmanned autonomous helicopter," in Proc. of the International Forum on Strategic Technology (IFOST), 2014.

[12] D. J. Wallcer, M.C. Turner, AJ. Smerlas, ME. Strange, and A.W. Gubbels, "Robust control of the longitudinal and lateral dynamics of the BELL 205 helicopter," Proceedings of the American Control Conference, June 1999.

[13] K. A. Danapalasingam, J.–J. Leth, A. la C.-Harbo and M. Bisgaard "Robust helicopter stabilization in the face of wind disturbance" 49th IEEE Conference on Decision and Control, December 15-17, 2010.

[14] T. Cheviron, F. Plestan, and A. Chriette, "A robust guidance and control scheme of an autonomous scale helicopter in presence of wind gusts," International Journal of Control, Vol. 82, No. 12, December 2009.

[15] T. K. Roy, M. Garrat, H. R. Pota and H. Teimoori, "Robust control for longitudinal and lateral dynamics of small scale helicopter, in Proceedings of the 31st CCC, 25-27 July, 2012.

[16] S. Suresh, P. Kashyab and M. Nabi, "Automatic take-off control system for helicopter an H∞ apporach," 11th Conf. Control, Automation, Robotics and Vision, 7-10 December, 2010.

[17] A. Isidori, L. Marconi and A. Serrani, "Robust nonlinear motion control of a helicopter," IEEE Transactions on Automatic Control, Vol. 48, No. 3 March 2003.

[18] L. Guo, C. Melhuish, and Q. Zhu, "Towards neural adaptive hovering control of helicopters," in Proc. IEEE Int. Conf. Control Applications, Glasgow, U.K., Sep. 2002, pp. 54-58.

[19] D. McLean. Automatic Flight Control Systems. Prentice Hall, 1990.

[20] T. K. Roy, "Robust adaptive control for longitudinal and lateral dynamics of a small scale helicopter," IJCCNet, vol. 1, no. 3, pp. 22 – 34, 2012.

[21] T. K. Roy, "Robust backstepping control for small helicopter" Master thesis, November 2012, The University of New South Wales, Australia.

[22] M. Tischler and R. Remple, "Aircraft and rotorcraft system identification: engineering methods with flight-test examples," in AIAA, 2006.

[23] T. K. Roy, "Horizontal position control of a small scale helicopter using adaptive backstepping controller," IJCCNet, vol. 2, no. 2, pp. 1 – 14, May 2013.

[24] T. K. Roy, M. Garrat, H. R. Potat and H. Teimoori, "Robust Backstepping Control for Longitudinal and Lateral Dynamics of Small Scale Helicopter," Journal of University of Science and Technology of China, Vol. 42, No. 7, Jul. 2012.

[25] T. K. Roy, "Longitudinal and lateral dynamics control of a small scale helicopter using adaptive backstepping controller under horizontal wind gusts," Asian Transaction Engineering, vol. 3, no. 1, March 2013.

[26] G. Cai, B. Chen, K. Peng, M. Dong, and T. Lee, "Modeling and control system design for a UAV helicopter," in 14th Mediterranean Conference on Control and Automation, June 2006, pp. 1–6.

[27] T. K. Roy, "Longitudinal and lateral dynamics control of a small scale helicopter using adaptive backstepping controller under horizontal wind gusts," *Asian Transaction on Engineering*, vol. 3, no. 1., 2013.

[28] I. A. Raptis and K. P. Valavanis, "Velocity and heading tracking control for small-scale unmanned helicopters," in Proceedings of American Control Conference, 2011.

14

Constructions of Implications Satisfying the Order Property on a Complete Lattice

Yuan Wang[1, *], Keming Tang[1], Zhudeng Wang[2]

[1]College of Information Science and Technology, Yancheng Teachers University, Yancheng, People's Republic of China
[2]School of Mathematics and Statistics, Yancheng Teachers University, Yancheng, People's Republic of China

Email address:

yctuwangyuan@163.com (Yuan Wang), tkmchina@126.com (Keming Tang), zhudengwang2004@163.com (Zhudeng Wang)
*Corresponding author

Abstract: In this paper, we further investigate the constructions of fuzzy connectives on a complete lattice. We firstly illustrate the concepts of left (right) semi-uninorms and implications satisfying the order property by means of some examples. Then we give out the formulas for calculating the upper and lower approximation implications, which satisfy the order property, of a binary operation.

Keywords: Fuzzy Logic, Fuzzy Connective, Implication, Order Property

1. Introduction

In fuzzy logic systems (see [1-2]), connectives "and", "or" and "not" are usually modeled by t-norms, t-conorms, and strong negations on $[0,1]$ (see [3]), respectively. Based on these logical operators on $[0,1]$, the three fundamental classes of fuzzy implications on $[0,1]$, i.e., R-, S-, and QL-implications on $[0,1]$, were defined and extensively studied (see [4-8]). But, as was pointed out by Fodor and Keresztfalvi [9], sometimes there is no need of the commutativity or associativity for the connectives "and" and "or". Thus, many authors investigated implications based on some other operators like weak t-norms [10], pseudo t-norms [11], pseudo-uninorms [12], left and right uninorms [13], semi-uninorms [14], aggregation operators [15] and so on.

Uninorms, introduced by Yager and Rybalov [16] and studied by Fodor et al. [17], are special aggregation operators that have proven useful in many fields like fuzzy logic, expert systems, neural networks, aggregation, and fuzzy system modeling. This kind of operation is an important generalization of both t-norms and t-conorms and a special combination of t-norms and t-conorms [17]. However, there are real-life situations when truth functions cannot be associative or commutative. By throwing away the commutativity from the axioms of uninorms, Mas et al. introduced the concepts of left and right uninorms on $[0, 1]$ in [18] and later in a finite chain in [19], Wang and Fang [13, 20] studied the left and right uninorms on a complete lattice. By removing the associativity and commutativity from the axioms of uninorms, Liu [14] introduced the concept of semi-uninorms and Su et al. [21] discussed the notion of left and right semi-uninorms on a complete lattice. On the other hand, it is well known that a uninorm (semi-uninorm, left and right uninorms) U can be conjunctive or disjunctive whenever $U(0, 1) = 0$ or 1, respectively. This fact allows to use uninorms in defining fuzzy implications [13-14, 22-23].

Constructing fuzzy connectives is an interesting topic. Recently, Wang [24] laid bare the formulas for calculating the smallest pseudo-t-norm that is stronger than a binary operation and the largest implication that is weaker than a binary operation, Su and Wang [25] investigated the constructions of implications and coimplications on a complete lattice and Wang et al. [26-28] studied the relations among implications, coimplications and left (right) semi-uninorms on a complete lattice. Moreover, Wang et al. [27, 29-30] investigated the constructions of implications and coimplications satisfying the neutrality principle.

In this paper, based on [24-30], we study the constructions

of implications satisfying the order property on a complete lattice. After recalling some necessary definitions and examples about the left (right) semi-uninorms and implications on a complete lattice in Section 2, we give out the formulas for calculating the upper and lower approximation implications, which satisfy the order property, of a binary operation in Section 3.

The knowledge about lattices required in this paper can be found in [31].

Throughout this paper, unless otherwise stated, L always represents any given complete lattice with maximal element 1 and minimal element 0; J stands for any index set.

2. Left (Right) Semi-Uninorms and Implications

In this section, we recall some necessary definitions and examples about the left (right) semi-uninorms and implications on a complete lattice.

Definition 2.1 (Su et al. [21]). A binary operation U on L is called a left (right) semi-uninorm if it satisfies the following two conditions:

(U1) *there exists a left (right) neutral element, i.e., an element $e_L \in L$ ($e_R \in L$) satisfying $U(e_L, x) = x$ ($U(x, e_R) = x$ for all $x \in L$,*

(U2) *U is non-decreasing in each variable.*

If a left (right) semi-uninorm U is associative, then U is the left (right) uninorm [13] on L.

If a left (right) semi-uninorm U with the left (right) neutral element $e_L \in L$ ($e_R \in L$) has a right (left) neutral element $e_R \in L$ ($e_L \in L$), then $e_L = U(e_L, e_R) = e_R$. Let $e = e_L = e_R$. Here, U is the semi-uninorm [14].

For any left (right) semi-uninorm U on L, U is said to be left-conjunctive and right-conjunctive if $U(0, 1) = 0$ and $U(1, 0) = 0$, respectively. U is said to be conjunctive if both $U(1, 0) = 0$ and $U(1, 0) = 0$ since it satisfies the classical boundary conditions of AND.

U is said to be strict left-conjunctive and strict right-conjunctive if U is conjunctive and for any $x \in L, U(x, 1) = 0 \Leftrightarrow x = 0$ and $U(1, x) = 0 \Leftrightarrow x = 0$, respectively.

Definition 2.2 (Wang and Fang [13]). A binary operation U on L is called left (right) arbitrary $\vee$-distributive if

$$U(\vee_{j \in J} x_j, y) = \vee_{j \in J} U(x_j, y) \quad \forall x_j, y \in L$$

$$\left(U(x, \vee_{j \in J} y_j) = \vee_{j \in J} U(x, y_j) \quad \forall x, y_j \in L \right); \quad (1)$$

left (right) arbitrary $\wedge$-distributive if

$$U(\wedge_{j \in J} x_j, y) = \wedge_{j \in J} U(x_j, y) \quad \forall x_j, y \in L$$

$$\left(U(x, \wedge_{j \in J} y_j) = \wedge_{j \in J} U(x, y_j) \quad \forall x, y_j \in L \right). \quad (2)$$

If a binary operation U is left arbitrary $\vee$-distributive ($\wedge$-distributive) and also right arbitrary $\vee$-distributive ($\wedge$-distributive), then U is said to be arbitrary $\vee$-distributive ($\wedge$-distributive).

Noting that the least upper bound of the empty set is 0 and the greatest lower bound of the empty set is 1, we have

$$U(0, y) = U(\vee_{j \in \Phi} x_j, y) = \vee_{j \in \Phi} U(x_j, y) = 0$$

$$\left(U(x, 0) = U(x, \vee_{j \in \Phi} y_j) = \vee_{j \in \Phi} U(x, y_j) = 0 \right) \quad (3)$$

for any $x, y \in L$ when U is left (right) arbitrary $\vee$-distributive,

$$U(1, y) = U(\wedge_{j \in \Phi} x_j, y) = \wedge_{j \in \Phi} U(x_j, y) = 1$$

$$\left(U(x, 1) = U(x, \wedge_{j \in \Phi} y_j) = \wedge_{j \in \Phi} U(x, y_j) = 1 \right) \quad (4)$$

for any $x, y \in L$ when U is left (right) arbitrary $\wedge$-distributive.

For the sake of convenience, we introduce the following symbols:

$U_s^{e_L}(L)$: the set of all left semi-uninorms with the left neutral element e_L on L;

$U_s^{e_R}(L)$: the set of all right semi-uninorms with the right neutral element e_R on L;

$U_{cs}^{se_L}(L)$: the set of all strict left-conjunctive left semi-uninorms with the left neutral element e_L on L;

$U_{cs}^{e_R s}(L)$: the set of all strict right-conjunctive right semi-uninorms with the right neutral element e_R on L;

$U_{\vee cs}^{se_L}(L)$: the set of all strict left-conjunctive left arbitrary $\vee$-distributive left semi-uninorms with the left neutral element e_L on L;

$U_{cs\vee}^{e_R s}(L)$: the set of all strict right-conjunctive right arbitrary $\vee$-distributive right semi-uninorms with the right neutral element e_R on L.

Example 2.1 (Su et al. [21]). Let $e_L \in L$,

$$U_{sW}^{e_L}(x, y) = \begin{cases} y \text{ if } x \geq e_L, \\ 0 \text{ otherwise,} \end{cases} \quad U_{sM}^{e_L}(x, y) = \begin{cases} y \text{ if } x \geq e_L, \\ 1 \text{ otherwise,} \end{cases}$$

$$U_{csM}^{e_L}(x, y) = \begin{cases} 0 & \text{if } x = 0 \text{ or } y = 0, \\ y & \text{if } 0 < x \leq e_L, y \neq 0, \\ 1 & \text{otherwise,} \end{cases}$$

where x and y are elements of L. Then $U_{sW}^{e_L}$ and $U_{sM}^{e_L}$ are, respectively, the smallest and greatest elements of

$U_s^{e_L}(L)$. By Example 2 and Theorem 8 in [26], we see that $U_{cs}^{se_L}(L)$ and $U_{\vee cs}^{se_L}(L)$ are two join-semilattices with the greatest element $U_{csM}^{e_L}$.

Example 2.2. Let $e_L \in L$,

$$U_{csW}^{se_L}(x,y) = \begin{cases} y & \text{if } x \geq e_L, \\ \wedge\{a \in L \mid a \neq 0\} & \text{if } 0 < x \text{ not} \geq e_L, y = 1, \\ 0 & \text{otherwise.} \end{cases}$$

When $e_L \neq 0$ and $\wedge\{a \in L \mid a \neq 0\} \neq 0$, it is straightforward to verify that $U_{csW}^{se_L}$ is a strict left-conjunctive left semi-uninorm with the left neutral element e_L. If $U \in U_{cs}^{se_L}(L)$, then

$$U(x,y) \geq \begin{cases} U(e_L,y) = y & \text{when } x \geq e_L, \\ \wedge\{a \in L \mid a \neq 0\} & \text{when } 0 < x \text{ not} \geq e_L, y = 1, \\ 0 & \text{otherwise,} \end{cases}$$

i.e., $U \geq U_{csW}^{se_L}$. Thus, $U_{csW}^{se_L}$ is the smallest element of $U_{cs}^{se_L}(L)$.

Moreover, assume that $\vee\{a \in L \mid a \text{ not} \geq e_L\} \text{ not} \geq e_L$. For any $x_j \in L \, (j \in J)$, if $\vee_{j \in J} x_j \geq e_L$, then there exists $j_0 \in J$ such that $x_{j_0} \geq e_L$,

$$\begin{aligned} U_{csW}^{se_L}(\vee_{j \in J} x_j, y) &= y = U_{csW}^{se_L}(x_{j_0}, y) \\ &= \vee_{j \in J} U_{csW}^{se_L}(x_j, y) \quad \forall y \in L; \end{aligned} \tag{5}$$

if $0 < \vee_{j \in J} x_j \text{ not} \geq e_L$, then $x_j \text{ not} \geq e_L$ for any $j \in J$ and there exists $j_0 \in J$ such that $0 < x_{j_0} \text{ not} \geq e_L$,

$$\begin{aligned} U_{csW}^{se_L}(\vee_{j \in J} x_j, 1) &= \wedge\{a \mid a \neq 0\} = U_{csW}^{se_L}(x_{j_0}, 1) \\ &= \vee_{j \in J} U_{csW}^{se_L}(x_j, 1); \end{aligned} \tag{6}$$

$$\begin{aligned} U_{csW}^{se_L}(\vee_{j \in J} x_j, y) &= 0 = U_{csW}^{se_L}(x_{j_0}, y) \\ &= \vee_{j \in J} U_{csW}^{se_L}(x_j, y) \quad y \neq 1; \end{aligned} \tag{7}$$

if $\vee_{j \in J} x_j = 0$, then $x_j = 0$ for any $j \in J$,

$$U_{csW}^{se_L}(\vee_{j \in J} x_j, y) = 0 = \vee_{j \in J} U_{csW}^{se_L}(x_j, y) \quad \forall y \in L. \tag{8}$$

Therefore, $U_{csW}^{se_L}$ is left arbitrary $\vee$-distributive and the smallest element of $U_{\vee cs}^{se_L}(L)$.

Example 2.3. Let $e_R \in L$,

$$U_{sW}^{e_R}(x,y) = \begin{cases} x & \text{if } y \geq e_R, \\ 0 & \text{otherwise,} \end{cases} \qquad U_{sM}^{e_R}(x,y) = \begin{cases} x & \text{if } y \geq e_R, \\ 1 & \text{otherwise,} \end{cases}$$

$$U_{csM}^{e_R}(x,y) = \begin{cases} 0 & \text{if } x = 0 \text{ or } y = 0, \\ x & \text{if } 0 < y \leq e_R, x \neq 0, \\ 1 & \text{otherwise,} \end{cases}$$

$$U_{csW}^{e_R s}(x,y) = \begin{cases} x & \text{if } y \geq e_R, \\ \wedge\{a \in L \mid a \neq 0\} & \text{if } 0 < y \text{ not} \geq e_R, x = 1, \\ 0 & \text{otherwise.} \end{cases}$$

where x and y are elements of L. By Example 2.6 in [26], we know that $U_{sW}^{e_R}$ and $U_{sM}^{e_R}$ are, respectively, the smallest and greatest elements of $U_s^{e_R}(L)$. By Example 3 and Theorem 8 in [26], we see that $U_{cs}^{e_R s}(L)$ and $U_{cs\vee}^{e_R s}(L)$ are two join-semilattices with the greatest element $U_{csM}^{e_R}$.

Similarly, When $e_R \neq 0$ and $\wedge\{a \in L \mid a \neq 0\} \neq 0$, $U_{csW}^{e_R s}$ is the smallest element of $U_{cs}^{e_R s}(L)$. Moreover, if $\vee\{a \in L \mid a \text{ not} \geq e_R\} \text{ not} \geq e_R$, then $U_{csW}^{e_R s}$ is the smallest element of $U_{cs\vee}^{e_R s}(L)$.

Definition 2.3 (Fodor and Roubens [1], Baczynski and Jayaram [4], Bustince et al. [6], De Baets and Fodor [22]). *An implication* I *on* L *is a hybrid monotonous (with decreasing first and increasing second partial mappings) binary operation that satisfies the corner conditions* $I(0,0) = I(1,1) = 1$ *and* $I(1,0) = 0$.

An implication I *is said to satisfy the order property with respect to* e *(w. r. t. e, for short) when* $x \leq y$ *if and only if* $I(x,y) \geq e$ *for any* $x, y \in L$.

Implications are extensions of the Boolean implication $\rightarrow$ ($P \rightarrow Q$ meaning that P is sufficient for Q).

Note that for any implication I on L, due to the monotonicity, the absorption principle holds, i.e., $I(0,x) = I(x,1) = 1$ for any $x \in L$.

For the sake of convenience, we introduce the following symbols:

$I(L)$: the set of all implications on L;

$I_\wedge(L)$: the set of all right arbitrary $\wedge$-distributive implications on L;

$I^{ope}(L)$: the set of all implications which satisfy the order property w. r. t. e on L;

$I_\wedge^{ope}(L)$: the set of all right arbitrary $\wedge$-distributive implications which satisfy the order property w.r.t. e on L.

Clearly, $I(L)$, $I_\wedge(L)$, $I^{ope}(L)$ and $I_\wedge^{ope}(L)$ are all meet-semilattices. By Example 2.4 in [25], we know that $I_\wedge(L)$ is not a join-semilattice.

Definition 2.3. Let U be a binary operation on L. Define $I_U^L, I_U^R \in L^{L \times L}$ as follows:

$$I_U^L(x,y) = \vee\{z \in L \mid U(z,x) \leq y\} \quad \forall x, y \in L, \tag{9}$$

$$I_U^R(x,y) = \vee\{z \in L \mid U(x,z) \leq y\} \quad \forall x, y \in L. \tag{10}$$

Here, I_U^L *and* I_U^R *are, respectively, called the left and right residuum of the binary operation* U .

By virtue of Theorems 4.4 and 4.5 in [13], we know that U and I_U^R satisfy the following right residual principle:

$$U(x, z) \le y \Leftrightarrow z \le I_U^R(x, y) \quad \forall x, y, z \in L \qquad (11)$$

when a binary operation U is right arbitrary $\vee$ -distributive; U and I_U^L satisfy the following left residual principle:

$$U(z, x) \le y \Leftrightarrow z \le I_U^L(x, y) \quad \forall x, y, z \in L \qquad (12)$$

when U is left arbitrary $\vee$ -distributive.

When U is non-decreasing in each variable, it is easy to see that I_U^L and I_U^R are all decreasing in the first variable and increasing in the second one by Definition 2.3.

Example 2.4. For some left and right semi-uninorms in Examples 2.1-2.3, a simple computation shows that

$$I_{U_{csW}^{se_L}}^L(x, y) = \begin{cases} 0 & \text{if } x = 1 \text{ and } y = 0, \\ 1 & \text{if } x \le y, \\ \vee\{a \in L \mid a \text{ not} \ge e_L\} & \text{otherwise,} \end{cases}$$

$$I_{U_{csM}^{e_L}}^L(x, y) = \begin{cases} 1 & \text{if } x = 0 \text{ or } y = 1, \\ e_L & \text{if } 0 < x \le y < 1, \\ 0 & \text{otherwise,} \end{cases}$$

$$I_{U_{csW}^{e_{Rs}}}^R(x, y) = \begin{cases} 0 & \text{if } x = 1 \text{ and } y = 0, \\ 1 & \text{if } x \le y, \\ \vee\{a \in L \mid a \text{ not} \ge e_R\} & \text{otherwise,} \end{cases}$$

$$I_{U_{csM}^{e_R}}^R(x, y) = \begin{cases} 1 & \text{if } x = 0 \text{ or } y = 1, \\ e_R & \text{if } 0 < x \le y < 1, \\ 0 & \text{otherwise,} \end{cases}$$

where x and y are elements of L . By the virtue of Theorem 8 in [26], we see that $I_{U_{csM}^{e_L}}^L$ is the smallest element of both $I^{ope_L}(L)$ and $I_\wedge^{ope_L}(L)$.

When $e_L \ne 0$ and $\vee\{a \in L \mid a \text{ not} \ge e_L\} \text{ not} \ge e_L$, it is easy to see that $I_{U_{csW}^{se_L}}^L$ is the greatest element of $I^{ope_L}(L)$.

Moreover, assume that $\wedge\{a \in L \mid a \ne 0\} \ne 0$. For any $y_j \in L\, (j \in J)$, if $\wedge_{j \in J} y_j = 0$, then there exists $j_0 \in J$ such that $y_{j_0} = 0$,

$$\begin{aligned} I_{U_{csW}^{se_L}}^L(x, \wedge_{j \in J} y_j) &= I_{U_{csW}^{se_L}}^L(x, 0) = I_{U_{csW}^{se_L}}^L(x, y_{j_0}) \\ &= \wedge_{j \in J} I_{U_{csW}^{se_L}}^L(x, y_j) \quad \forall x \in L; \end{aligned} \qquad (13)$$

if $x \le \wedge_{j \in J} y_j$, then $x \le y_j$ for any $j \in J$,

$$I_{U_{csW}^{se_L}}^L(x, \wedge_{j \in J} y_j) = 1 = \wedge_{j \in J} I_{U_{csW}^{se_L}}^L(x, y_j) \quad \forall x \in L; \qquad (14)$$

if $0 < \wedge_{j \in J} y_j \text{ not} \ge x$, then $0 < y_j$ for any $j \in J$ and there exists $j_0 \in J$ such that $0 < y_{j_0} \text{ not} \ge x$,

$$I_{U_{csW}^{se_L}}^L(x, y_j) \ge \vee\{a \in L \mid a \text{ not} \ge e_L\}; \qquad (15)$$

$$\begin{aligned} I_{U_{csW}^{se_L}}^L(x, \wedge_{j \in J} y_j) &= \vee\{a \in L \mid a \text{ not} \ge e_L\} \\ &= I_{U_{csW}^{se_L}}^L(x, y_{j_0}) = \wedge_{j \in J} I_{U_{csW}^{se_L}}^L(x, y_j). \end{aligned} \qquad (16)$$

Therefore, $I_{U_{csW}^{se_L}}^L$ is the greatest element of $I_\wedge^{ope_L}(L)$.

$I_{U_{csM}^{e_L}}^L$ Similar conclusions hold for $I^{ope_R}(L)$ and $I_\wedge^{ope_R}(L)$.

3. Constructing the Implications Satisfying the Order Property

Recently, Su and Wang [25] have studied the constructions of implications and coimplications and Wang et al. [27, 29-30] further investigated the constructions of implications and coimplications satisfying the neutrality principle on a complete lattice.

This section is a continuation of [25, 27, 29-30]. We will study the constructions of the upper and lower approximation implications which satisfy the order property.

It is easy to verify that if $J \ne \Phi$, then

$$I_j \in I^{ope_L}(L) \, \forall j \in J \implies \wedge_{j \in J} I_j \in I^{ope_L}(L). \qquad (17)$$

When $e_L \ne 0$ and $\vee\{a \in L \mid a \text{ not} \ge e_L\} \text{ not} \ge e_L$, we see that $I^{ope_L}(L)$ is also a complete lattice with the smallest element and greatest element $I_{U_{csW}^{se_L}}^L$ by Example 2.4. Thus, for a binary operation A on L , if there exists $I \in I^{ope_L}(L)$ such that $A \le I$, then

$$\wedge\{I \mid A \le I, I \in I^{ope_L}(L)\} \qquad (18)$$

is the smallest implication that is stronger than A and satisfies the order property w. r. t. e_L on L . Here, we call it the upper approximation implication, which satisfies the order property w. r. t. e_L , of A and write as $[A]_I^{ope_L}$. Similarly, if there exists $I \in I^{ope_L}(L)$ such that $I \le A$, then

$$\vee\{I \mid I \le A, I \in I^{ope_L}(L)\} \qquad (19)$$

is the largest implication that is weaker than A and satisfies the order property w. r. t. e_L on L . Here, we call it the lower approximation implication, which satisfies the order property w. r. t. e_L , of A and write as $(A)_I^{ope_L}$.

Likewise, for a binary operation A on L , we may introduce the following symbols:

$[A]_I^{ope_R}$: the upper approximation implication, which satisfies the order property w. r. t. e_R, of A ;

$(A]_I^{ope_R}$: the lower approximation implication, which satisfies the order property w. r. t. e_R, of A ;

$[A]_I^{ope_L \wedge}$ ($[A]_I^{ope_R \wedge}$): the upper approximation right arbitrary $\wedge$ -distributive implication, which satisfies the order property w. r. t. e_L (e_R), of A ;

$(A]_I^{ope_L \wedge}$ ($(A]_I^{ope_R \wedge}$): the lower approximation right arbitrary $\wedge$ -distributive implication, which satisfies the order property w. r. t. e_L (e_R), of A .

Definition 3.1 (see Su and Wang [25]). Let A be a binary operation on L. Define the upper approximation implicator A_{ui} and the lower approximation implicator A_{li} of A as follows:

$$A_{ui}(x, y) = \vee\{A(u, v) \mid u \geq x, v \leq y\} \quad \forall x, y \in L, \qquad (20)$$

$$A_{li}(x, y) = \wedge\{A(u, v) \mid u \leq x, v \geq y\} \quad \forall x, y \in L. \qquad (21)$$

Theorem 3.1 (see Su and Wang [25]). Let $A, B \in L^{L \times L}$. Then the following statements hold:

$$A_{li} \leq A \leq A_{ui}. \qquad (22)$$

$(A \vee B)_{ui} = A_{ui} \vee B_{ui}$ and

$$(A \wedge B)_{li} = A_{li} \wedge B_{li}. \qquad (23)$$

A_{ui} and A_{li} are hybrid monotonous.

If A is are hybrid monotonous, then $A_{ui} = A_{li} = A$.

Theorem 3.2. Let $A \in L^{L \times L}$.

(1) If A is right arbitrary $\vee$ -distributive, then A_{ui} is also right arbitrary $\vee$ -distributive,

$$(I_A^R)_{li} = I_{A_{ua}}^R, (I_A^R)_{ui} \leq I_{A_{la}}^R, \qquad (24)$$

$$A_{ua}(x, (I_A^R)_{li}(x, y)) \leq y \quad \forall x, y \in L. \qquad (25)$$

(2) If A is right arbitrary $\wedge$ -distributive, then A_{li}

is also right arbitrary $\wedge$ -distributive.

(3) If A is left arbitrary $\vee$ -distributive, then,

$$(I_A^L)_{li} = I_{A_{ua}}^L, (I_A^L)_{ui} \leq I_{A_{la}}^L, \qquad (26)$$

$$A_{ua}((I_A^L)_{li}(x, y), x) \leq y \quad \forall x, y \in L. \qquad (27)$$

Proof. We only prove that statement (1) holds.

Assume that A is a right arbitrary $\vee$ -distributive binary operation on L. Clearly, A_{ua} is also right arbitrary $\vee$ -distributive. By Definition 3.1, the monotonicity of A and

I_A^R, and the right residual principle, we have that

$$
\begin{aligned}
I_{A_{ua}}^R(x, y) &= \vee\{z \in L \mid A_{ua}(x, z) \leq y\} \\
&= \vee\{z \in L \mid \vee\{A(u, v) \mid u \leq x, v \leq z\} \leq y\} \\
&= \vee\{z \in L \mid \vee\{A(u, z) \mid u \leq x\} \leq y\} \\
&= \vee\{z \in L \mid A(u, z) \leq y \; \forall u \leq x\} \\
&= \vee\{z \in L \mid z \leq I_A^R(u, y) \; \forall u \leq x\} \\
&= \vee\{z \in L \mid z \leq \wedge_{u \leq x} I_A^R(u, y)\} \qquad (28) \\
&= \wedge_{u \leq x} I_A^R(u, y) \; \forall x, y \in L,
\end{aligned}
$$

$$
\begin{aligned}
(I_A^R)_{li}(x, y) &= \wedge\{I_A^R(u, v) \mid u \leq x, v \geq y\} \\
&= \wedge\{I_A^R(u, y) \mid u \leq x\} = \wedge_{u \leq x} I_A^R(u, y) \; \forall x, y \in L.
\end{aligned} \qquad (29)
$$

Thus, $(I_A^R)_{li} = I_{A_{ua}}^R$. Similarly, we have that

$$(I_A^R)_{ui}(x, y) = \vee\{I_A^R(u, y) \mid u \geq x\} \; \forall x, y \in L, \qquad (30)$$

$$
\begin{aligned}
A_{la}(x, z) &= \wedge\{A(u_1, v) \mid u_1 \geq x, v \geq z\} \\
&= \wedge\{A(u_1, v) \mid u_1 \geq x\} \; \forall x, z \in L,
\end{aligned} \qquad (31)
$$

$$
\begin{aligned}
&(I_{A_{la}}^R)(x, y) \\
&= \vee\{z \in L \mid \wedge\{A(u_1, z) \mid u_1 \geq x\} \leq y\} \; \forall x, y \in L.
\end{aligned} \qquad (32)
$$

If $u \geq x$, let $z = I_A^R(u, y)$, then

$$
\begin{aligned}
A(u, z) &= A(u, \vee\{c \in L \mid A(u, c) \leq y\}) \\
&= \vee\{A(u, c) \mid A(u, c) \leq y\} \leq y, \qquad (33) \\
&\wedge\{A(u_1, z) \mid u_1 \geq x\} \leq A(u, z) \leq y.
\end{aligned}
$$

So, $(I_A^R)_{ui}(x, y) \leq (I_{A_{la}}^R)(x, y)$ for any $x, y \in L$, i.e., $(I_A^R)_{ui} \leq I_{A_{la}}^R$.

Moreover, we know that A_{ua} is right arbitrary $\vee$ -distributive and hence

$$
\begin{aligned}
A_{ua}(x, (I_A^R)_{li}(x, y)) &= A_{ua}(x, I_{A_{ua}}^R(x, y)) \\
&= A_{ua}(x, \vee\{z \in L \mid A_{ua}(x, z) \leq y\}) \qquad (34) \\
&= \vee\{A_{ua}(x, z) \mid A_{ua}(x, z) \leq y\} \leq y \; \forall x, y \in L.
\end{aligned}
$$

The theorem is proved.

Below, we give out the formulas for calculating the upper and lower approximation implications which satisfy the order property.

Theorem 3.3. Suppose that $A \in L^{L \times L}$, $e_L \neq 0$ and $\vee\{a \in L \mid a \text{ not} \geq e_L\} \text{not} \geq e_L$.

(1) If $A \leq I_{U_{csW}^{se_L}}^L$, then $[A]_I^{ope_L} = I_{U_{csM}^{e_L}}^L \vee A_{ui}$;

if $A \geq I_{U_{csM}^{e_L}}^L$, then $(A]_I^{ope_L} = I_{U_{csW}^{se_L}}^L \wedge A_{li}$.

(2) If $\wedge\{a \in L \mid a \neq 0\} \neq 0$, $A \geq I_{U_{csM}^{e_L}}^L$ and A is right

arbitrary $\wedge$-*distributive, then*

$$[A]_I^{ope_L \wedge} = I_{U_{csW}^{se_L}}^L \wedge A_{li}. \qquad (35)$$

Moreover, if A is non-decreasing in its first variable, then $(A]_I^{ope_L \wedge} = I_{U_{csW}^{se_L}}^L \wedge A$.

Proof. Assume that $\vee\{a \in L \mid a \text{ not} \geq e_L\} \text{not} \geq e_L$ and $e_L \neq 0$. Then $I_{U_{csM}^{e_L}}^L$ and $I_{U_{csW}^{se_L}}^L$ are, respectively, the smallest and greatest elements of $I^{ope_L}(L)$ by Example 2.4.

(1) If $A \leq I_{U_{csW}^{se_L}}^L$, let $I_1 = I_{U_{csM}^{e_L}}^L \vee A_{ui}$, then $A \leq I_1$ and

$$I_{U_{csM}^{e_L}}^L \leq I_1 \leq I_{U_{csW}^{se_L}}^L \qquad (36)$$

Thus, $I_1(0,0) = I_1(1,1) = 1$ and $I_1(1,0) = 0$. If $x \leq y$, then $I_1(x,y) \geq I_{U_{csM}^{e_L}}^L(x,y) \geq e_L$; if $I_1(x,y) \geq e_L$, then $I_{U_{csW}^{se_L}}^L(x,y) \geq I_1(x,y) \geq e_L$ and so $x \leq y$, i.e., I_1 satisfies the order property w. r. t. e_L. By Theorem 3.1 (3) and the hybrid monotonicity of $I_{U_{csM}^{e_L}}^L$, we know that I_1 is hybrid monotonous. So, $I_1 \in I^{ope_L}(L)$. If $A \leq I$ and $I \in I^{ope_L}(L)$, then $A_{ui} \leq I_{ui} = I$ and $I_1 = I_{U_{csM}^{e_L}}^L \vee A_{ui} \leq I$. Therefore,

$$[A)_I^{ope_L} = I_{U_{csM}^{e_L}}^L \vee A_{ui}. \qquad (37)$$

If $A \geq I_{U_{csM}^{e_L}}^L$, let $I_2 = I_{U_{csW}^{se_L}}^L \wedge A_{li}$, then $I_2 \leq A$,

$$A_{li} \geq (I_{U_{csM}^{e_L}}^L)_{li} = I_{U_{csM}^{e_L}}^L, I_{U_{csM}^{e_L}}^L \leq I_2 \leq I_{U_{csW}^{se_L}}^L. \qquad (38)$$

Thus, we can prove in an analogous way that $I_2 \in I^{ope_L}(L)$ and $(A]_I^{ope_L} = I_{U_{csW}^{se_L}}^L \wedge A_{li}$.

(2) When $\wedge\{a \in L \mid a \neq 0\} \neq 0$, $I_{U_{csM}^{e_L}}^L$ and $I_{U_{csW}^{se_L}}^L$ are, respectively, the smallest and greatest elements of $I_\wedge^{ope_L}(L)$ by Example 2.4. Let $I_3 = I_{U_{csW}^{se_L}}^L \wedge A_{li}$. If $A \geq I_{U_{csM}^{e_L}}^L$, then $I_3 \in I^{ope_L}(L)$ by statement (1). Noting that A is right arbitrary $\wedge$-distributive, we can see that A_{li} is also right arbitrary $\wedge$-distributive by Theorem 3.2 (1). So, I_3 is right arbitrary $\wedge$-distributive, i.e., $I_3 \in I_\wedge^{ope_L}(L)$. By the proof of statement (1), we know that $(A]_I^{ope_L \wedge} = I_{U_{csW}^{se_L}}^L \wedge A$.

Moreover, if A is non-decreasing in its first variable, then $A_{li} = A$ by Theorem 3.1 (4) and so

$$(A]_I^{ope_L \wedge} = I_{U_{csW}^{se_L}}^L \wedge A. \qquad (39)$$

The theorem is proved.
Analogous to Theorem 3.3, we have the following theorem.

Theorem 3.4. Suppose that $A \in L^{L \times L}$, $e_R \neq 0$ and $\vee\{a \in L \mid a \text{ not} \geq e_R\} \text{not} \geq e_R$.

(1) If $A \leq I_{U_{csW}^{e_{R^s}}}^R$, then $[A)_I^{ope_R} = I_{U_{csM}^{e_R}}^R \vee A_{ui}$;

if $A \geq I_{U_{csM}^{e_R}}^R$, then $(A]_I^{ope_R} = I_{U_{csW}^{e_{R^s}}}^R \wedge A_{li}$.

(2) If $\wedge\{a \in L \mid a \neq 0\} \neq 0$, $A \geq I_{U_{csM}^{e_R}}^R$ and A is right arbitrary $\wedge$-distributive, then

$$(A]_I^{ope_R \wedge} = I_{U_{csW}^{e_{R^s}}}^R \wedge A_{li}. \qquad (40)$$

Moreover, if A is non-decreasing in its first variable, then $(A]_I^{ope_R \wedge} = I_{U_{csW}^{e_{R^s}}}^R \wedge A$.

4. Conclusions and Future Works

Constructing fuzzy connectives is an interesting topic. Recently, Wang et al. [24-25, 27, 29-30] investigated the constructions of implications and coimplications on a complete lattice. In this paper, motivated by these works, we give out the formulas for calculating the upper and lower approximation implications, which satisfy the order property, of a binary operation.

In a forthcoming paper, we will investigate the relationships between left (right) semi-uninorms and implications on a complete lattice.

Acknowledgements

This work is supported by Science Foundation of Yancheng Teachers University (16YCKLQ006), the National Natural Science Foundation of China (61379064) and Jiangsu Provincial Natural Science Foundation of China (BK20161313).

References

[1] J. Fodor and M. Roubens, "Fuzzy Preference Modelling and Multicriteria Decision Support", Theory and Decision Library, Series D: System Theory, Knowledge Engineering and Problem Solving, Kluwer Academic Publishers, Dordrecht, 1994.

[2] G. J. Klir and B. Yuan, "Fuzzy Sets and Fuzzy Logic, Theory and Applications", Prentice Hall, New Jersey, 1995.

[3] E. P. Klement, R. Mesiar and E. Pap, "Triangular Norms", Trends in Logic-Studia Logica Library, Vol. 8, Kluwer Academic Publishers, Dordrecht, 2000.

[4] M. Baczynski and B. Jayaram, "Fuzzy Implication", Studies in Fuzziness and Soft Computing, Vol. 231, Springer, Berlin, 2008.

[5] M. Baczynski and B. Jayaram, "QL-implications: some properties and intersections", Fuzzy Sets and Systems, 161, 158-188, 2010.

[6]   H. Bustince, P. Burillo and F. Soria, "Automorphisms, negations and implication operators", Fuzzy Sets and Systems, 134, 209-229, 2003.

[7]   F. Durante, E. P. Klement, R. Mesiar and C. Sempi, "Conjunctors and their residual implicators: characterizations and construction methods", Mediterranean Journal of Mathematics, 4, 343-356, 2007.

[8]   Y. Shi, B. Van Gasse, D. Ruan and E. E. Kerre, "On dependencies and independencies of fuzzy implication axioms", Fuzzy Sets and Systems, 161, 1388-1405, 2010.

[9]   J. Fodor and T. Keresztfalvi, "Nonstandard conjunctions and implications in fuzzy logic", International Journal of Approximate Reasoning, 12, 69-84, 1995.

[10] J. Fodor, "Srict preference relations based on weak t-norms", Fuzzy Sets and Systems, 43, 327-336, 1991.

[11] Z. D. Wang and Y. D. Yu, "Pseudo-t-norms and implication operators on a complete Brouwerian lattice", Fuzzy Sets and Systems, 132, 113-124, 2002.

[12] Y. Su and Z. D. Wang, "Pseudo-uninorms and coimplications on a complete lattice", Fuzzy Sets and Systems, 224, 53-62, 2013.

[13] Z. D. Wang and J. X. Fang, "Residual operators of left and right uninorms on a complete lattice", Fuzzy Sets and Systems, 160, 22-31, 2009.

[14] H. W. Liu, "Semi-uninorm and implications on a complete lattice", Fuzzy Sets and Systems, 191, 72-82, 2012.

[15] Y. Ouyang, "On fuzzy implications determined by aggregation operators", Information Sciences, 193, 153-162, 2012.

[16] R. R. Yager and A. Rybalov, "Uninorm aggregation operators", Fuzzy Sets and Systems, 80, 111-120, 1996.

[17] J. Fodor, R. R. Yager and A. Rybalov, "Structure of uninorms", Internat. J. Uncertainly, Fuzziness and Knowledge-Based Systems, 5, 411-427, 1997.

[18] M. Mas, M. Monserrat and J. Torrens, "On left and right uninorms", Internat. J. Uncertainly, Fuzziness and Knowledge-Based Systems, 9, 491-507, 2001.

[19] M. Mas, M. Monserrat and J. Torrens, "On left and right uninorms on a finite chain", Fuzzy Sets and Systems, 146, 3-17, 2004.

[20] Z. D. Wang and J. X. Fang, "Residual coimplicators of left and right uninorms on a complete lattice", Fuzzy Sets and Systems, 160, 2086-2096, 2009.

[21] Y. Su, Z. D. Wang and K. M. Tang, "Left and right semi-uninorms on a complete lattice", Kybernetika, 49, 948-961, 2013.

[22] B. De Baets and J. Fodor, "Residual operators of uninorms", Soft Computing, 3, 89-100, 1999.

[23] M. Mas, M. Monserrat and J. Torrens, "Two types of implications derived from uninorms", Fuzzy Sets and Systems, 158, 2612-2626, 2007.

[24] Z. D. Wang, "Generating pseudo-t-norms and implication operators", Fuzzy Sets and Systems, 157, 398-410, 2006.

[25] Y. Su and Z. D. Wang, "Constructing implications and coimplications on a complete lattice", Fuzzy Sets and Systems, 247, 68-80, 2014.

[26] X. Y. Hao, M. X. Niu and Z. D. Wang, "The relations between implications and left (right) semi-uninorms on a complete lattice", Internat. J. Uncertainly, Fuzziness and Knowledge-Based Systems, 23, 245-261, 2015.

[27] X. Y. Hao, M. X. Niu, Y. Wang and Z. D. Wang, "Constructing conjunctive left (right) semi-uninorms and implications satisfying the neutrality principle", Journal of Intelligent and Fuzzy Systems, 31, 1819-1829, 2016.

[28] M. X. Niu, X. Y. Hao and Z. D. Wang, "Relations among implications, coimplications and left (right) semi-uninorms", Journal of Intelligent and Fuzzy Systems, 29, 927-938, 2015.

[29] Z. D. Wang, M. X. Niu and X. Y. Hao, "Constructions of coimplications and left (right) semi-uninorms on a complete lattice", Information Sciences, 317, 181-195, 2015.

[30] Z. D. Wang, "Left (right) semi-uninorms and coimplications on a complete lattice", Fuzzy Sets and Systems, 287, 227-239, 2016.

[31] G. Birkhoff, "Lattice Theory", American Mathematical Society Colloquium Publishers, Providence, 1967.

15

Fuzzy Fine Tuning Therapies for Intelligence of High Speed Electronic Packaging Equipment

He Yunbo[*]**, Hu Yongshan, Chen Xin, Gao Jian, Yang Zhijun, Chen Yun, Tang Hui, Ao Yinhui, Zhang Yu**

Guangdong Provincial Key Lab. of Computer Integrated Manufacturing Systems, Key Laboratory of Mechanical Equipment Manufacturing and Control Technology of Ministry of Education, School of Electromechanical Engineering, Guangdong University of Technology, Guangzhou, P.R. China

Email address:

heyunbo@gdut.edu.cn (He Yunbo)

[*]Corresponding author

Abstract: It's always difficult to satisfy both strict requirements of dynamic performance and settling performance of high-precision and high-acceleration point-to-point motions on electronic packaging equipment, i.e., Die bonders, Wire bonders, Flip-chip bonders, Wafer bumping machines, and so on. Focusing on this difficulty, Fuzzy logic is promoted in this paper to fine tune feed forward coefficient parameters of motion systems on packaging equipment to satisfy multiple targets of dynamic and settling performance requirements. This intelligent fine tuning approach, with achievable multiple motion targets, provides a therapy to automate and optimize motion system performance on packaging equipment, help on development and mass production of semiconductor packaging equipment, and improve machine intelligence.

Keywords: Dynamic Performance, Settling Performance, Electronic Packaging Equipment, Fuzzy Logic, Fine Tuning, Machine Intelligence

1. Introduction

Electronic packaging machines (i.e., Die bonders, Wire bonders, Flip-chip bonders, Wafer bumping machines, and so on) require their moving axes to move with high speed and high precision during mass production of electronic and semiconductor manufacturing. Due to the large production volume, similar movements are repeated by thousands of hundreds of times 24 hours a day. Even a small improvement in the motion control performance brings faster and finer therapies in manufacturing, which result in significant economic gains. To achieve optimized dynamic tracking and settling motion control performance, maximum feedback system bandwidth and feedforward control (FFC) are quoted. FFC is important to provide large control output in dynamic tracking, so that motors and actuators can be driven according to fast motion profile to achieve and maintain high acceleration. With FFC in dynamic stages, settling performance can be affected and shaped accordingly. To achieve optimum dynamic tracking and settling performance, performance fine tuning is important and frequently addressed [1-8].

People, especially for those control engineers and scientists who are familiar with control systems and experienced in fine tuning, prefer manual fine tuning of motion performance. Yet, electronic packaging machinesare frequently used by machine operators, technicians and maintenance engineers in production. Above users normally have no pre knowledge of motion systems and fine tuning. Auto tuning or self tuning is a must for them. Currently in semiconductor equipment industry, speed command and acceleration command of a motion profile are selected as FFC resources (or inputs) to simplify implementation of FFC modeling for convenience. And, two proportional coefficients (proportional coefficient of velocity command kvel and that of acceleration command kacc) are utilized to shape dynamic tracking performance by searching or fine tuing.

Searching mechanism can be described as below: The default FFC coefficients Kvel and Kacc, and their up/low searching ranges are initially set, as well as a combined performance target, which includes dynamic and static performance indicators and their individual weightings. Then, a best Kacc is found by fixing a default Kvel and searching for a minimum combined performance target with captured motion performances through machine running round by round with incremental Kacc between its low and up ranges; then fix the best Kacc and repeat above machine running and performance searching process to search for a best Kvel. Searching mechanism is far worse than tuning. Drawbacks of searching mechanism are described as follows:

1. Much time consumption and small range of Up/Low limits of FFC coefficients Kvel/Kacc. Obviously, much time is needed when searching step by step between low and up ranges of Kvel/kacc. Low/up ranges cannot be wide, otherwise much time should be consumed.

2. Strict default values of Kvel/Kacc. Before searching starts, an experienced engineer or expert is needed to set the default Kvel/Kacc values to around the best. Tuning result may not be good if default Kvel/Kacc values are not good enough.

3. Uncontrollable searching result. There are too many objectives and weightings to build up the final integrated single performance target. After searching, we just know that the final Kvel/Kacck may be good as the final integrated performance target is the best. But for every individual performance objective, such as DPE (max Dynamic Position Error) and PUS/POS (static Position UnderShoot/OverShoot), and DAC in settling, we don't know whether they are in specs when the searching result, the final set of Kvel/Kacc is applied. In other words, we cannot separately control these multiple objectives. We have to integrate all these objectives to one single performance target with different weightings for every performance index. Different Kvel/Kacc and different individual performances may lead to the same integrated performance target. This searching method may lead to uncontrollable Kvel/Kacc, as well as non-repetitive performance.

4. Fixed tuning resolution/step. The searching method cannot identify whether the default value is near or far from the best value during searching. It cannot accelerate or decelerate the incremental steps in searching process. For example, given a default value for Kacc 270 and suppose the best value is 272, by searching method, the program must search from the low limit (i.e. 0) to up limit (i.e. 400) step by step with a fixed incremental step, say 3. After over 100 times of machine running, performance capturing, calculations and comprisions, with much time consumed, it can just find a value (either 270 or 273) as the best searching value. It's still a little different from 272, which is actually the best.

5. Little flexibility to apply to new products/users/purposes. The up/low limits, default values, objectives, and weightings must be carefully set by an experienced engineer before the searching mechanism is submitted for use in mass production. This method is not convenient to apply to new products and users directly. And also, it's hard to expand this method for other purposes such as to find the best DPE, Amax (maximum Acceleration) or DAC settling range in a motion.

An intelligent tuning method is needed to actually tune the FFC coefficients toward controllable motion performance, and to save tuning time. Stearns H. (Dept. of Mech. Eng., Univ. of California, Berkeley, Berkeley, CA) published a paper on feedforward controller tuning. A fixed-structure feedforward controller of a wafer stage of a photolithography machine is tuned. The feedforward controller is tuned with the objective of minimizing a cost function that is a quadratic function of tracking error. Simulation and experimental results are presented which show that the iterative tuning method effectively reduces tracking error. [4] An automation technique is presented by Timothy P. K. (Drexel University), which yields high-performance, low-cost optoelectronic alignment and packaging through the use of intelligent control theory and system-level modeling. The approach is based on model-based control, to build an a priori knowledge model, specific to the assembled package's optical power propagation characteristics, and use this to set the initial "feed-forward" conditions of the automation system. The research was preliminary, algorithms were only verified through simulation and the author was starting hardware implementation and testing, with donated equipment from Kulicke & Soffa, Willow Grove, PA. [9] Some publications involved in Fine tuning. [1-2, 4-8, 10-11] Fuzzy logic feedforward fine tuning was roughly introduced by the authors. [12]

Before fuzzy logic is introduced, we must find the rules how FFC parameters influence motion performance while motion profile (i.e., motion distance, motion maximum velocity, and motion maximum acceleration) and other motion conditions are pre-determined. It's difficult to theoretically map FFC parameters to the dynamic motion tracking system. Fortunately we can find the trend that FFC parameters affect actual motion after careful study on quite a lot of motions with different FFC parameters and different profile movement conditions. Fuzzy logic should tune FFC parameters along with the correct direction towards expected motion tracking performance. It's also important to appropriately adjust the tuning incremental steps under different circumstances. That is, according to tracking performance, if the present FFC value is far from the best value (too big or too small), the tuning program shall automatically accelerate tuning to approach the best value quickly; and if the present value is around the best one, the tuning program shall decrease steps to catch the nearby best value accurately. In a word, an "intelligent" tuning method is needed to keep the tuning process fast and accurate, besides, tuning shall be smooth and robust enough. Fuzzy logic suits this tuning process a lot because it is intelligent enough to deal well with complexity of this dynamic tuning process without modeling.

2. Fuzzy Logic Fine Tuning

2.1. How Is Fuzzy Logic Introduced to FFC Tuning

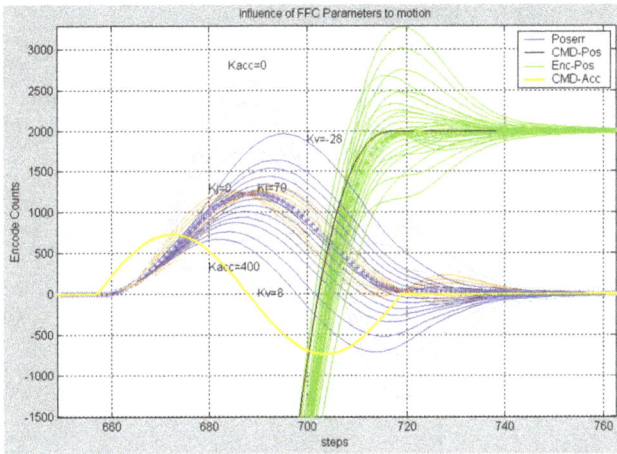

Figure 1. Influence of FFC Parameters to Motion Performance.

Figure 1 is the simulation result of FFC coefficient influence to a typical motion control system on electronic packaging equipment. Simulation illustrates the typical influence of FFC coefficient parameters to motion performance. The dynamic segment of the Position Error curve shifts upper or lower if Kvel is gradually increased or decreased, shown as the blue curves in Fig. 2. Kacc (shown as gray curves in Fig. 2) is like a sine wave to make both the curve's maximum and minimum values larger or smaller, while performances at the beginning and end of the dynamic stage of motion keep no change.

The tuning idea is to simultaneously tune Kvel and Kacc by fuzzy logic according to DPE (Dynamic tracking Position Error) and PUS (Position UnderShoot).

Fuzzy rules of tuning Kvel and Kacc are introduced in Table 1 from watching DPE (Dynamic Position Error) and PUS (Position UnderShoot). It must be pointed out that here PUS is a little bit different from usual PUS. PUS may also refer to negative POS if it's mainly overshoot at the end of the motion. And here DPE means the difference of the actual maximum dynamic position error to expected DPE.

This is a typical double input (fuzzy variables DPE and PUS) double output (fuzzy variables Kvel and Kacc) fuzzy inference problem. Fuzzy subsets of DPE, PUS, Kvel and Kacc quote {NB NM NS ZO PS PM PB}. Where, N means Negative; P means Positive; B refers to Big; M refers to Medium; S refers to Small. The 49 fuzzy inference rules are expressed in the following and Table 1.

If DPE is NB and PUS is NB, then Kvel is PB and Kacc is ZO;

If DPE is NB and PUS is NM, then Kvel is PB and Kacc is ZO;

…

If DPE is PB and PUS is PB, then Kvel is NB and Kacc is ZO.

Table 1. Fuzzy rules of Kvel and Kacc tuning.

Kvel/Kacc		PUS						
		NB	NM	NS	ZO	PS	PM	PB
DPE	NB	PB/ZO	PB/ZO	PB/ZO	PB/PS	ZO/PM	ZO/PB	ZO/PB
	NM	PM/NS	PM/ZO	PM/ZO	PS/PS	ZO/PS	ZO/PM	ZO/PB
	NS	PS/NB	PS/NS	PS/ZO	PS/PS	ZO/PS	ZO/PM	ZO/PB
	ZO	ZO/NB	ZO/NS	ZO/NS	ZO/ZO	ZO/PS	ZO/PS	NS/PB
	PS	ZO/NB	ZO/NS	ZO/NS	NS/ZO	NS/ZO	NS/ZO	NM/PM
	PM	ZO/NB	ZO/NM	ZO/NS	NS/NS	NM/ZO	NM/ZO	NM/PM
	PB	ZO/NB	ZO/NB	NM/NM	NM/NM	NM/NS	NB/ZO	NB/ZO

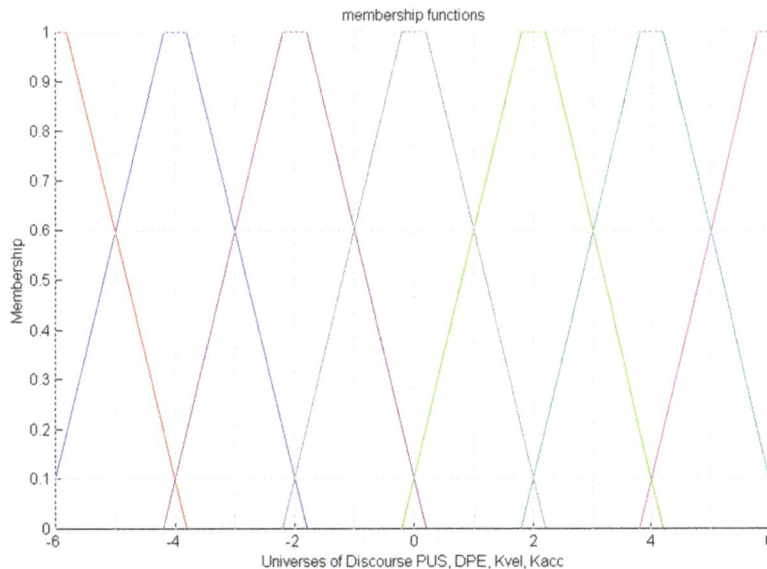

Figure 2. Trapezoid Memberships.

To simplify fuzzy inference process, the universes of discourses PUS, DPE, Kvel and Kacc quote {-6, -5, -4, -3, -2, -1, 0, 1, 2, 3, 4, 5 ,6}. Differences of these four discourses are arranged in quantification factors for PUS and DPE and scaling factors for Kvel and Kacc. Trapezoid membership functions for NB, NM, NS, ZO, PS, PM and PB are quoted and shown in Figure 2.

2.2. Fuzzy Inference

49 Fuzzy relations are as follows:

$$R_{Kvel1} = (NB_{DPE} \times PB_{Kvel}) \circ (NB_{PUS} \times PB_{Kvel}) = R_{DPE2Kvel1} \cap R_{PUS2Kvel1} \tag{1}$$

$$R_{Kacc1} = (NB_{DPE} \times ZO_{Kacc}) \circ (NB_{PUS} \times ZO_{Kacc}) = R_{DPE2Kacc1} \cap R_{PUS2Kacc1} \tag{2}$$

...

$$R_{Kvel49} = (PB_{DPE} \times NB_{Kvel}) \circ (PB_{PUS} \times NB_{Kvel}) = R_{DPE2Kvel49} \cap R_{PUS2Kvel49} \tag{3}$$

$$R_{Kacc49} = (PB_{DPE} \times ZO_{Kacc}) \circ (PB_{PUS} \times ZO_{Kacc}) = R_{DPE2Kacc49} \cap R_{PUS2Kacc49} \tag{4}$$

The memberships of two fuzzy inputs $\overrightarrow{DPE_i}$ and $\overrightarrow{PUS_j}$ quote $[0, ..., 0, 1(i), 0, ..., 0]$ and $[0, ..., 0, 1(j), 0, ..., 0]$, respectively. Then after fuzzification (or fuzzy quantification),

$$\vec{K}vel_1 = (\overrightarrow{DPE_i} \circ R_{DPE2Kvel1}) \cap (\overrightarrow{PUS_j} \circ R_{PUS2Kvel1}) \tag{5}$$

$$\vec{K}acc_1 = (\overrightarrow{DPE_i} \circ R_{DPE2Kacc1}) \cap (\overrightarrow{PUS_j} \circ R_{PUS2Kacc1}) \tag{6}$$

...

$$\vec{K}vel_{49} = (\overrightarrow{DPE_i} \circ R_{DPE2Kvel49}) \cap (\overrightarrow{PUS_j} \circ R_{PUS2Kvel49}) \tag{7}$$

$$\vec{K}acc_{49} = (\overrightarrow{DPE_i} \circ R_{DPE2Kacc49}) \cap (\overrightarrow{PUS_j} \circ R_{PUS2Kacc49}) \tag{8}$$

Then the 2 fuzzy output variables, $\vec{K}vel$ and $\vec{K}acc$, are deduced in the following:

$$\vec{K}vel\Big|_{\substack{\overrightarrow{DPE_i} \\ \overrightarrow{PUS_j}}} = \bigcup_{k=1}^{49} \vec{K}vel_k \tag{9}$$

$$\vec{K}acc\Big|_{\substack{\overrightarrow{DPE_i} \\ \overrightarrow{PUS_j}}} = \bigcup_{k=1}^{49} \vec{K}acc_k \tag{10}$$

Centroid Defuzzification Technique (the center of gravity procedure) is quoted for computing the crisp value of Kvel and Kacc,

$$Kvel\Big|_{\substack{\overrightarrow{DPE_i} \\ \overrightarrow{PUS_j}}} = \sum_{k=1}^{length(\vec{K}vel)} \frac{\mu(k)\vec{K}vel(k)}{\mu(k)} \tag{11}$$

$$Kacc\Big|_{\substack{\overrightarrow{DPE_i} \\ \overrightarrow{PUS_j}}} = \sum_{k=1}^{length(\vec{K}acc)} \frac{\mu(k)\vec{K}acc(k)}{\mu(k)} \tag{12}$$

From above inference, fuzzy tables of Kvel and Kacc are given as Table 2 and Table 3.

Table 2. Fuzzy table of Kvel.

Kvel		PUS
		{-6, -5, -4, -3, -2, -1, 0, 1, 2, 3, 4, 5, 6}
DPE	-6	{5.2, 5.0, 5.2, 5.0, 4.8, 4.5, 4.2, 2.1, 0.6, 0.0, 0.0, 0.0, 0.0}
	-5	{4.4, 4.4, 4.4, 4.4, 4.1, 3.4, 3.0, 2.2, 0.8, 0.0, 0.0, 0.0, 0.0}
	-4	{3.7, 3.7, 3.7, 3.7, 3.7, 3.0, 2.0, 1.3, 0.6, 0.0, 0.0, 0.0, 0.0}
	-3	{3.0, 3.0, 3.0, 3.0, 3.0, 3.0, 2.0, 1.0, 0.3, 0.0, 0.0, 0.0, 0.0}
	-2	(2.0, 2.0, 2.0, 2.0, 2.0, 2.0, 2.0, 1.0, 0.3, 0.0, -0.6, -0.7, -0.6}
	-1	{1.0, 1.0, 1.0, 1.0, 1.0, 1.0, 1.0, 1.0, 0.3, 0.0, -0.7, -2.1, -2.1}
	0	{0.3, 0.3, 0.3, 0.3, 0.0, 0.0, 0.0, 0.0, 0.0, -0.3, -0.6, -2.1, -4.2}
	1	{0.0, 0.0, 0.0, 0.0, -0.3, -1.0, -1.0, -1.0, -1.0, -1.0, -1.3, -2.4, -3.7}
	2	{0.0, 0.0, 0.0, 0.0, -0.3, -1.0, -2.0, -2.0, -2.0, -2.0, -2.0, -2.7, -3.4}
	3	{0.0, 0.0, 0.0, 0.0, -0.3, -1.0, -2.0, -3.0, -3.0, -3.0, -3.0, -3.0, -3.7}
	4	{0.0, 0.0, -0.6, -0.8, -0.6, -1.3, -2.0, -3.0, -3.7, -3.7, -3.7, -3.7, -3.7}
	5	{0.0, 0.0, -0.8, -2.0, -2.0, -2.0, -2.7, -3.0, -3.7, -4.4, -4.4, -4.4, -4.4}
	6	{0.0, 0.0, -0.6, -2.0, -3.4, -3.3, -3.4, -3.7, -3.7, -4.4, -5.2, -5.0, -5.2}

Table 3. Fuzzy table of Kacc.

Kacc		PUS
		{-6, -5, -4, -3, -2, -1, 0, 1, 2, 3, 4, 5, 6}
DPE	-6	{-0.3, -0.3, -0.3, 0.0, 0.3, 1.0, 2.0, 3.0, 3.7, 4.1, 4.8, 5.0, 5.2}
	-5	{-1.0, -1.0, -0.3, 0.0, 0.3, 1.0, 2.0, 3.0, 3.0, 3.4, 4.1, 4.4, 5.0}
	-4	{-2.0, -1.3, -0.6, -0.3, 0.0, 1.0, 2.0, 2.3, 2.3, 3.0, 3.7, 4.4, 5.2}
	-3	{-3.0, -2.2, -1.3, -1.0, 0.0, 1.0, 1.7, 2.0, 2.3, 3.0, 3.7, 4.4, 5.0}
	-2	(-4.2, -3.0, -2.0, -1.0, 0.0, 0.7, 1.4, 1.7, 2.0, 3.0, 3.7, 4.1, 4.8}
	-1	{-4.5, -3.3, -2.0, -1.0, -0.7, 0.0, 0.7, 1.0, 2.0, 3.0, 3.0, 3.4, 4.5}
	0	{-4.8, -3.3, -2.0, -1.7, -1.4, -0.7, 0.0, 1.0, 2.0, 2.0, 2.0, 3.0, 4.2}
	1	{-4.5, -3.3, -2.3, -2.0, -1.7, -1.0, 0.0, 1.0, 1.0, 1.0, 1.3, 2.4, 3.7}
	2	{-4.8, -3.3, -2.3, -2.3, -2.0, -1.0, 0.0, 0.0, 0.0, 0.3, 0.6, 2.0, 3.4}
	3	{-4.5, -3.4, -3.0, -3.0, -2.0, -1.0, -1.0, -1.0, -0.3, 0.0, 0.8, 2.0, 3.3}
	4	{-4.8, -4.1, -3.7, -3.0, -2.0, -2.0, -2.0, -1.3, -0.6, -0.3, 0.4, 2.0, 3.4}
	5	{-5.0, -4.4, -4.1, -3.4, -3.0, -3.0, -2.7, -2.0, -1.3, -1.0, 0.4, 2.0, 2.0}
	6	{-5.2, -5.0, -4.8, -4.1, -3.7, -3.7, -3.4, -2.7, -2.0, -1.0, 0.4, 0.8, 0.6}

Quantification factors for PUS and DPE and scaling factors for Kvel and Kacc are important and shall be carefully selected. In FFC tuning, Variable Step Size Method is quoted in order to acquire a fast and accurate tuning result.

2.3. Experimental Results and Analysis

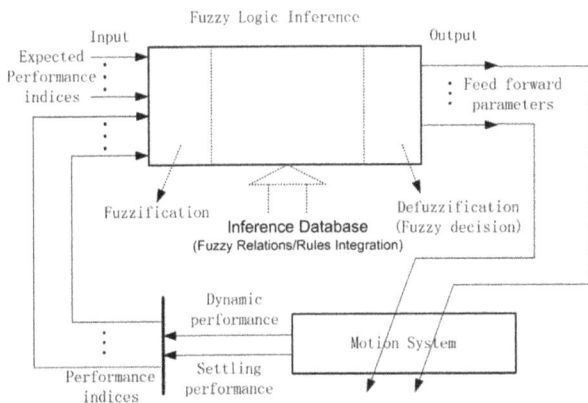

Figure 3. Illustration of fuzzy logic feed forward tuning process.

The Fuzzy FFC Tuning process is illustrated in Figure 3. And the tuning algorithm is programmed and tested on wire bonding machines.

2.3.1. Time Consumption Test

Table 4. Time consumption comparison of two FFC tuning methods.

	MC1	MC2	MC3	MC4
Fuzzy FFC Tuning	1 min	1 min 14 sec	1 min 5 sec	1 min 50 sec
Traditional Auto Tuning	10-12 minutes			

As shown in Table 4, Fuzzy FFC tuning saves 80% tuning time.

2.3.2. Repeatability Test

Repeatability tests were conducted on MC1 (3 times) and MC2 (5 times), as shown in Table 5 and Table 6. Standard Deviations are within 10% of mean values. The results show good repeatability of fuzzy FFC tuning.

Table 5. Repeatability Test of Fuzzy FFC Tuning on MC1.

	Motion 1		Motion 2		Motion 3		Motion 4		Motion 5		Motion 6	
	Kvel	Kacc	Kvel	Kacc	Kvel	Kacc	Kvel	Kacc	Kvel	Kacc	Kvel	Kacc
Test 1	19	173	17	145	15	183	14	230	15	152	18	171
Test 2	19	174	19	154	15	183	14	233	15	150	18	175
Test 3	19	175	17	145	15	183	14	230	15	149	18	174
Mean	19	174	18	148	15	183	14	231	15	150	18	173
STDev	0	1	1.2	5.2	0	0	0	1.7	0	1.5	0	2.1
STD%	0	0.6	6.5	3.5	0	0	0	0.7	0	1	0	1.2

Table 6. Repeatability Test of Fuzzy FFC Tuning on MC2.

	Motion 1		Motion 2		Motion 3		Motion 4		Motion 5		Motion 6	
	Kvel	Kacc	Kvel	Kacc	Kvel	Kacc	Kvel	Kacc	Kvel	Kacc	Kvel	Kacc
Test 1	18	254	18	226	14	261	14	358	17	234	17	260
Test 2	18	256	18	222	13	266	14	360	16	228	18	259
Test 3	18	252	18	222	14	261	14	356	17	228	18	256
Test 4	18	246	18	223	14	260	14	356	17	234	18	250
Test 5	19	249	17	212	14	263	14	355	17	230	19	251
Mean	18	251	18	221	14	262	14	357	17	231	18	255
STDev	0.4	4	0.4	5.3	0.4	2.4	0	2	0.4	3	0.7	4.5
STD%	2.5	1.6	2.5	2.4	3.2	0.9	0	0.6	2.7	1.3	3.9	1.8

2.3.3. *Performance Test*

Performances are captured on above 4 machines. As an example, Table 7 shows the performance indices tuned on MC1. Extracted from both dynamic and static performances, motion performance indices (DPE: Maximum Dynamic Position Error; Sserr: Static settling position error; Ts:

Settling time; Pos: Position overshoot; Pus: Position undershoot; Dve: Maximum Dynamic velocity error; Vo: Velocity overshoot; Vu: Velocity undershoot.) meet with packaging process requirements for every specific motion of six typical motions.

Table 7. *Performance indices of Fuzzy Tuning on MC1.*

	Kvel/Kacc	DPE	Sserr	Ts	Pos	Pus	Dve	Vo	Vu
Motion 1	19/173	1542	-29.2	0.13	20.4	2.1	77	6.7	3.2
Motion 2	17/145	225.5	-23	0.13	9.5	6.5	32.6	8.5	2.2
Motion 3	15/183	806	-20.6	0.13	5.1	23.7	70.4	6.7	1.6
Motion 4	14/226	251.5	-26.4	0.13	13.8	5.8	15.4	3.4	2.2
Motion 5	15/151	185.8	-23.7	0.13	7.3	6.8	30.2	7.1	2.1
Motion 6	15/178	1528.9	-9.3	0.13	11.9	21.8	87.6	7.3	1.6

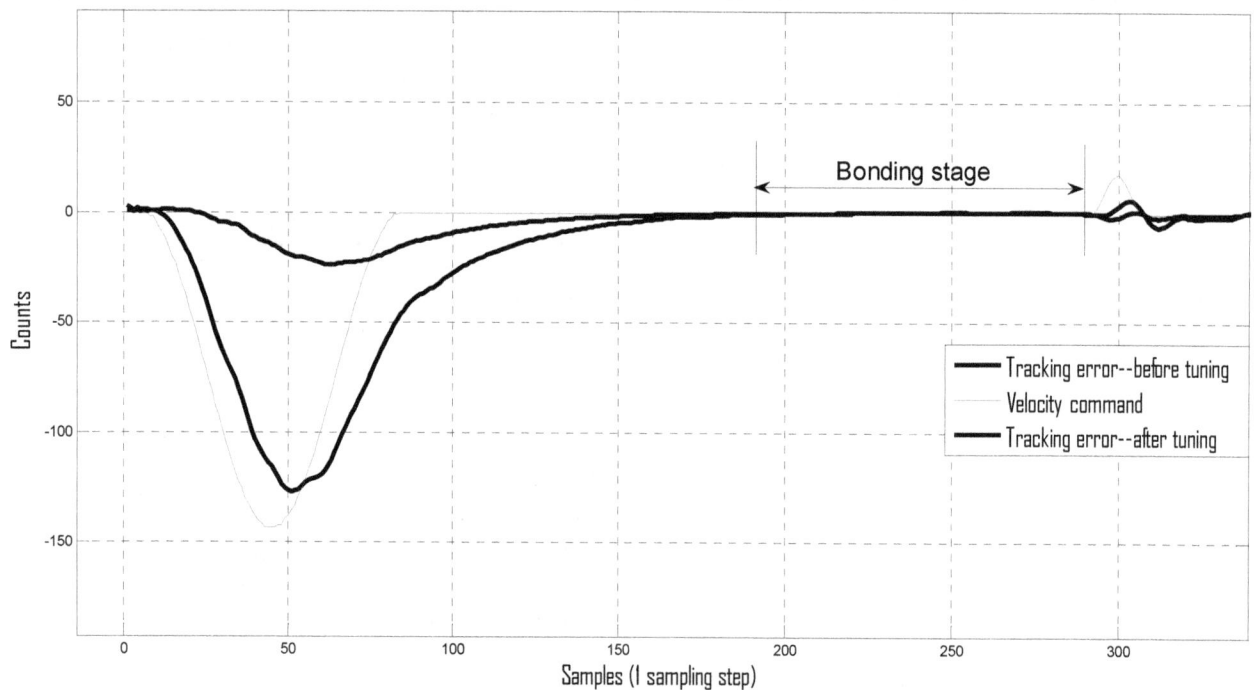

Figure 4. *Tracking errors before and after tuning.*

Figure 4 shows tuning results of tracking position errors. Compared to position tracking errors before tuning, they are improved (diminished) a lot at both dynamic tracking and settling stages. Tracking performance tuned helps to reduce offsets along XY direction during bonding, and save waiting time for settling before bonding.

Experiments show that Fuzzy logic approach is a reliable and time-saving FFC tuning method. And, it's robust enough.

3. Conclusion

Feed forward coefficient parameters of motion systems on packaging equipment are tuned by fuzzy logic to satisfy multiple targets of dynamic and settling performance requirements. The major contributions of Fuzzy FFC Tuning are to save tuning time and get controllable dynamicmotion

performance. This intelligent tuning therapy is used to shape dynamic tracking, so as to help on performance in the settling stage. Experimental results demonstrate that fuzzy logic is a feasible and effective solution to motion performance fine tuning on electronic packaging equipment.

Acknowledgements

Authors show their acknowledgments to supporters of this paper: the National Basic Research Program of China (973 Program No.2011CB013104), the Applied Science and Technology Research and Development Project of Guangdong Province (No. 2015B010133005), and the Scientific Innovation Key Project from Guangdong Educational Department (No. 2012CXZD0020, No. 2012A080303004).

References

[1] Iuliana R., Maarten S., Rogier E., "Adaptive Iterative Learning Control for High Precision Motion Systems," IEEE TRANSACTIONS ON CONTROL SYSTEMS TECHNOLOGY, VOL. 16, NO. 5, SEPTEMBER 2008, pp. 1075-1082.

[2] Aurelio P., Antonio V., "An Iterative Approach for Noncausal Feedforward Tuning", Proceedings of the 2007 American Control Conference, Marriott Marquis Hotel at Times Square, New York City, USA, July 11-13, 2007, pp. 1251-1256.

[3] Marcel H., Daan H., Maarten S., "MIMO feed-forward designin wafer scanners using a gradient approximation-based algorithm" Control EngineeringPractice18(2010), pp. 495–506.

[4] Stearns H., Mishra S., Tomizuka M., "Iterative Tuning of Feedforward Controller with Force Ripple Compensation for Wafer Stage," 10th IEEE International Workshop, AMC '08, Trento, Italy, 2008, pp. 234-239.

[5] Morteza M., "Self-tuning PID controller to three-axis stabilization of a satellite with unknown parameters", International Journal of Non-Linear Mechanics 49 (2013), pp. 50–56.

[6] Meric C., SerdarI., "A novel auto-tuning PID control mechanism for nonlinear systems", ISA Transactions 58 (2015), pp. 292–308.

[7] Onur K., MujdeG., IbrahimE., EnginY., TufanK., "Online tuning of fuzzy PID controllers via rule weighing based on normalized acceleration", Engineering Applications of Artificial Intelligence 26 (2013) , pp. 184–197.

[8] Osama E., Mohammad E., Nabila M., "Development of Self -Tuning Fuzzy Iterative Learning Control for Controlling a Mechatronic System," International Journal of Information and Electronics Engineering, Vol. 2, No. 4(2012), pp. 565-569.

[9] Timothy P. K., Allon G., Shubham K. B., "Model-based Optoelectronic Packaging Automation," IEEE Journal of Selected Topics in Quantum Electronics, Vol. 10, No. 3 (2004), pp. 445-454.

[10] Frank B., Tom O., Maarten S., "Accuracy Aspects in Motion Feedforward Tuning," *2014 American Control Conference (ACC)*, Portland, Oregon, USA, 2014, pp. 2178-2183.

[11] Dennis B., Niels V. D., "Combined Input Shaping and Feedforward Control for Flexible Motion Systems," 2012 American Control Conference, Fairmont Queen Elizabeth, Montréal, Canada, 2012, pp. 2473-2478.

[12] Yunbo H., Xin C., etc., Invited paper, 17th Electronics Packaging Technology Conference, EPTC 2015 Singapore, "Development of High Performance Bonding Machines with Improved Motion Control and Intelligent Fine Tuning Algorithm," 2-4 December 2015.

16

MAC Protocols Design for Smart Metering Network

Yue Yang[1], Yanling Yin[2], Zixia Hu[1]

[1]Electrical Engineering, University of Washington, Seattle, Washington, USA
[2]Electrical Engineering, Harbin Engineering University, Harbin, USA

Email address:
yueyang@uw.edu (Yue Yang), yinyanling@hrbeu.edu.cn (Yanling Yin), huzixia1984@gmail.com (Zixia Hu)

Abstract: The new generation of power metering system - i.e. Advanced Metering Infrastructure (AMI) - is expected to enable remote reading, control, demand response and other advanced functions, based on the integration of a new two-way communication network, which will be referred as Smart Metering Network (SMN). In this paper, we focus on the design principles of multiple access control (MAC) protocols for SMN. First, we list several AMI applications and its benefits to the current power grid and user experience. Next, we introduces several features of SMN relevant to the design choice of the MAC protocols, including the SMN architecture and candidate communication technologies. After that, we propose some performance evaluation metrics, such as scalability issue, traffic types, delay and etc, and give a survey of the associated research issues for the SMN MAC protocols design. In addition, we also note progress within the new IEEE standardization task group (IEEE 802.11ah TG) currently working to create SMN standards, especially in the MAC protocols aspect.

Keywords: MAC Protocols, Smart Metering Network, Advanced Metering Infrastructure, WiFi, 802.11ah, Smart Meters

1. Introduction

Traditionally, the collection of power data from end-customer premises has been accomplished by using conventional power meters. Even if such meters could be remotely read as in Automated Meter Reading systems (AMR), such capabilities were limited to one-way, infrequent data upload. The next generation – Advanced Metering Infrastructure (AMI) - is based on a two-way communication network, Smart Metering Network (SMN), which is comprised with the communication transceivers integrated into smart meters. Therefore, in addition to the remote reading function, AMI may also support remote control, load management and demand response by sending down-link control information from the utility. We provide a high-level comparison of these three generations of power metering systems in the Fig.1.

Generation	System	Comm. Network	Terminal Device	Function
Last	Traditional Power Metering System	No Comm. Network	Conventional Meters	Manual Reading
Current	Automatic Meter Reading (AMR)	One-way Comm. Network	Conventional Meters Coupled with One-way Comm. Infrastructure	Remote Reading
Current/ Next	Advanced Metering Infrastructure (AMI)	Two-way Comm. Network (SMN)	Smart Meters	Remote Reading and Control, Load Management, Demand Response, Fault Detection

Figure 1. Comparison among three Generations of Power Metering Systems.

Since the AMI crucially depends on the two-way networks, the communication aspects of SMN design have begun to draw attentions [5], [23], [24], [7], [17]. Some of those discuss the choice of communication architectures [6] that are appropriate for the various AMI applications, so as to achieve the traditional goals of communication reliability, efficiency and security. On the other hand, some papers talk about the pros and cons of multiple different communication technologies, such as power line communications [25], Zigbee [10] and WiFi [26], and compare their suitability to the AMI applications. Additionally, another key point, which largely determines the efficiency and performance of data communications in SMN is the Multiple Access Control (MAC) protocols, which is exactly the focus of this paper. First, in Section III, we discuss some SMN unique features that will significantly impact the choice of MAC protocols. After that, the MAC performance metrics and their associated research issues are presented in detail in Section IV. The entire paper is concluded in Section V.

2. AMI Applications and Benefits

We briefly summarize several representative applications of AMI and their resulting benefits.

2.1. Outage Detection and Management

Traditionally, outage may only be detected by a report from customers and other monitoring at the control center. With the aid of SMN, the utility can devise a more efficient way to detect the outage and improve response time. For example, once the voltage drops below a threshold for a duration, the Smart Meter may report a warning message and enable the control center to detect and locate the source of an impending outage event.

2.2. Demand Response

With SMN, the customers will be able to acquire their own real-time energy usage data and use dynamic pricing information sent by the utility, to make better decisions or manage automatically on use of the electric appliances within the premises, so as to conserve energy and cost.

2.3. Power Quality Monitoring

The status of monitoring of the distribution segment (from the distribution substation to customers) is currently very limited. AMI enables collection of various types of real-time data at many more points - such as feeder voltage and current, power quality within a customer premise etc. This will lead to more efficient and accurate power quality monitoring.

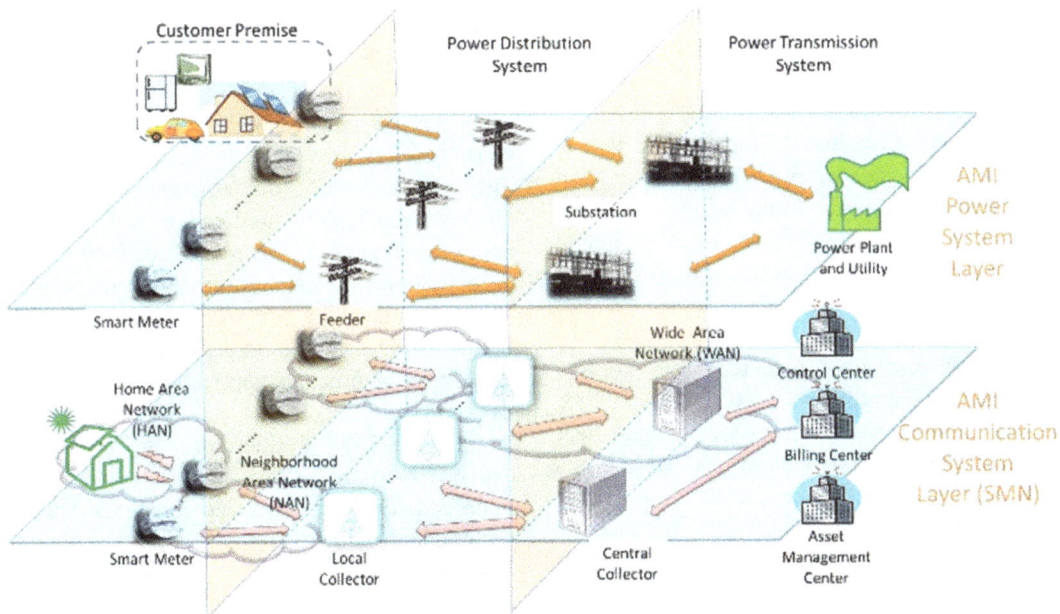

Figure 2. AMI Power and Communication System Architectures.

2.4. Remote Connect and Disconnect

With the aid of SMN, a utility may remotely enable or disable the energy delivery to certain customers, so as to smooth consumption peaks or automatically react to some emergency events.

The above is only a partial list of applications and benefits from AMI. In summary, with the help of the bi-directional communication network, AMIs can improve operational efficiency, reliability and security.

3. Smart Metering Network Features

3.1. Smart Metering Network System Architecture

Fig.2 presents an SMN system based on the Smart Grid Architecture given in [1]. A short overview of its main components follows:

Smart Meter (SM): This device has three different roles. First, the Smart Meter is a multi-utility instrument measuring electric power consumption (and possibly in future, gas, water and heat). It can thus act as an energy control center, i.e. as a point of aggregation for usage information collected by using a Home Area Network (HAN) that connects home appliances. Finally, the Smart Meter also serves as the gateway between HAN and external network; it reports on energy consumption, sends out urgent data, receives remote commands from the utility and is responsible for security of the above transactions. It is noted that SMs nodes are fixed as they are deployed in customer premises. They are usually powered from the main supply and hence power-saving issue is not as important as in traditional battery-powered Wireless Sensor Network nodes.

Home Area Network (HAN): This is composed of multiple inter-connected electric appliances, such as air conditioner, dish washer, plug-in hybrid electric vehicles (PHEV), etc. and the Smart Meter. All the components inside HAN share information or deliver control commands to each other. For example, the dish washer may send a signal to Smart Meter, requesting it to send a `postpone' command to the charging PHEV, so that it may operate without incurring excessive energy cost at that time.

Local Collector: Between the Smart Meter and Utility Center, there could be multiple layers of intelligent electronic devices (IEDs) that acts as data concentrators. For example, a data collection node closer to customer premises - named as Local Collector - collects SM data from multiple premises and relays it to the Central Collector. Additional functions at the Local Collector may include simple data processing and distributed decision and intelligence using its own data. The network segment between Smart Meter and Local Collector is called Neighborhood Area Network (NAN), while that above Local Collector belongs to Wide Area Network (WAN).

Central Collector: A centralized data repository for the entire region operated by the utility that acts as the interface with Control Center, Billing Center and Asset Management Center. These centers may use this data to conduct analysis and evaluate system status, and make decisions or deliver control commands to other components.

The mapping between the SMN components and their prospective physical deployments is also shown in the Fig.2. For example, the Local Collector could be located at a Distribution Transformer because it would be easy to power the Collector and obtain measurements from other feeder devices. On the other hand, the independent deployment for Local Collector may provide more flexibility to adjust its coverage range. The Central Collector is likely to be deployed closer to the centers. If the utility's coverage region is not very large, only one Central Collector located close to the centers may suffice. Otherwise, Central Collectors may be placed at the Distribution Substation as the second tier data relay. Since the HAN consists of electric appliances which are manufactured by different vendors and has much

flexibility in implementation especially on application layers, the utility operating SMN may leave it open and focus their design on the upper level network. On the other hand, the design of network from the Local Collector to the Central System does not only depend on the communication requirements of SMN because it also includes the electrical devices which serve the power systems other than AMI. Therefore, the main focus of the MAC protocols design for SMN lies on the segment from Smart Meter to Local Collector (NAN).

3.2. Communication Technologies

A large amount of literature discusses the feasibility of several optional bi-directional communication technologies applied on the SMN. We summarize the various options and highlight their pros/cons in Fig.3.

In Power Line Carrier (PLC), the data is transmitted over electricity transmission lines along with electrical power [9]. Its communication performance depends on several factors, such as frequency, propagation distance and existence of transformers because the data signals cannot go through the transformers. PLC has gained a lot of attractions because it uses the existing power lines as signal carrier and no extra cabling fee is needed. Therefore, many countries (e.g. Singapore) adopted it for broadband communication services. However, PLC also suffers from several disadvantages, such as high signal attenuation, high noisy medium and lower scalability, which leads to the termination of deployment in some countries (e.g. US) [8].

ZigBee is a wireless communication technology that consumes low power at the device side [10]. Thanks to its low cost and easy implementation, this technology has already been widely used in the Smart Home network by many AMI vendors, such as Itron and Landis Gyr. They produce smart meters and measuring devices integrated with ZigBee protocol to monitor and control the Home Energy Status. On the other hand, there are still some constraints on ZigBee for its practical application on the SMN. For example, its short range confines this protocol in the application domain of HAN. Furthermore, the processing capabilities and memory size of the ZigBee device are expected to be improved for more advanced functions and communication requirements of the SMN.

Machine Type Communications over Cellular Network allows the Smart Meters and Local Collector to exchange information via low data load communication service, which has been supported by multiple mature cellular network standards, such as LTE [11]. It is the popularity and easy implementation that make this technology become an attractive candidate option. Furthermore, the long range and high data rate provide the utility more flexibility to design and implement the SMN. However, the concern about reliability, security and delay performance makes a barrier for the implementation of this technology in practice, especially under the condition of heavy traffic load.

Finally, WiFi is a communication technology that allows devices to exchange data wirelessly based on IEEE 802.11

Standards. Its popularity, mature development and unlicenced spectrum make it on the top of the candidate technology list. Furthermore, it is also a cost efficient network with dynamic self-healing and distributed control, which makes it easier to be implemented. On the contrary, the capacity, scalability and security issues are the main challenges for its application on the SMN. Therefore, in order to solve these challenges, a new standardization task group IEEE 802.11ah is established and aimed at creating a WiFi based standard to support wireless communication between Smart Meters and Local Collector as one of its primary use cases. According to [14], IEEE 802.11ah compliant devices will utilize Multi-Input Multi-Output, Orthogonal Frequency Division Multiplexing (MIMO-OFDM) at frequencies below 1 GHz, where there is no licensing and regulatory issues. The most discussed channelization for .11ah in the US focusses on the 902 - 928 MHz band, which is currently free. The .11ah Working Group appear to have settled on 1 MHz and 2 MHz as the possible channel bandwidth [13]. In addition to the benefit of free spectrum, the signal transmission below 1 GHz generally suffer less propagation path loss, enabling the network to achieve larger coverage, as verified in the calculations in following section.

Technology	Applications	Benefits	Limitations
Power Line Carrier	WAN, NAN, HAN	No Extra Cabling Fee, High Security	High Noisy Medium, Low Scalability
Messaging over Cellular Network	HAN, NAN, WAN	Mature Development, Long Range	Low Data Rate, Low Robustness, Low Security, Costly Spectrum Fees, Low Scalability
WiFi	HAN, NAN, WAN (with multi-hop)	Mature Development, Free License, High Robustness	Low Security, Low Scalability
ZigBee	HAN	Low Cost	Short Range, Low Security, Low Data Rate

Figure 3. *Comparison among optional Communication Technologies for SMN.*

4. MAC Protocols Design Performance Metrics and Important Research Issues

MAC protocols must be designed to match the differing objectives for the various types of Smart Grid data as well as adapt to the different network topology scenarios. Furthermore, the special features and applications in SMN also address some new challenges to the suitable MAC protocols design. We next outline the broad performance metrics for MAC protocols as they relate to Smart Grid operations, and identify some specific research challenges for SMN MAC protocols.

4.1. Different Data Types

Usually, the Smart Meter traffic may be classified into two different classes: periodical and event-triggered data. The former includes energy consumption information while the latter is largely data from protection devices (relays, reclosers etc. that monitor local fault status) or electric vehicle charging stations. We list a table (Fig.4) of several representative traffic examples with their important properties. Given the different requirements and features of traffic types, different MAC protocols should be designed specially for each traffic type. For example, [28] targets on the event-driven data, while [27] is designed for the periodical reporting data.

Traffic Examples	Delay Requirements	Trigger Type
Outage Alert	Seconds	Event Triggered
Billing Information	Minutes to Hours	Periodical
Demand Response	Seconds or Minutes	Periodical
Real-time Pricing Information	Seconds or Minutes	Periodical
EV Charging Information	Seconds or Minutes	Event Triggered

Figure 4. *Examples of Data with Different Communication Requirements.*

4.2. Delay

Different types of data induce different communication requirements within a SMN. For example, the energy consumption data is delay tolerant. On the other hand, the delay sensitive data - such as those reporting a fault and protection related messages - should have higher priority over others, so as to minimize end-to-end latency. Therefore, how to minimize the SMN latency - such as that of the last hop between the Smart Meter and the Local Collector is a primary concern. In general, several factors impact the delay, such as the choice of the communication technology, the network architecture, and most notably, the MAC protocol. A good MAC protocol can coordinate the uplink transmissions of multiple communication nodes to reduce the collision probability significantly, resulting in lower delay.

Usually, it is convenient to design MAC protocols for one type of traffic. For example, a polling based (taking-turns) protocol such as Point Coordination Function (PCF) defined in 802.11 is well-suited to reporting data with bounded delay guarantees. The total duration of one polling cycle increases only linearly with the number of nodes (in contrast to exponential increase in delay with random access systems as the aggregate load increases) and provides a guaranteed delay bounds. However, the efficiency of such protocols declines rapidly with the number of nodes. On the other hand, random access MAC protocols such as Distributed Coordination Function (DCF) in 802.11 have been designed to provide reasonable efficiencies in terms of throughput at low to moderate loads, but the delays escalate rapidly as the average load increases. In order to solve this challenge, [28] proposes two grouping based DCF MAC protocols: TDMA-DCF and Group Leader DCF-TDMA scheme. These two schemes divide all Smart Meters into several groups to reduce the competitive channel access and are both directed at 802.11 type networks operating at the frequencies below 1GHz, which has been adopted by the IEEE 802.11ah TG.

However, most Smart Grid scenarios comprise of a mix of traffic, e.g. regular traffic and emergency traffic. To serve both types within a DCF framework, the notion of traffic

classes were introduced via Enhanced Distribution Channel Access (EDCA) defined in 802.11e [4], to prioritize low-latency (event-driven) data over non time-critical data applications (such as email). A combination of EDCA and PCF, Hybrid Coordination Function Controlled Channel Access (HCCA) tries to serve multiple traffic types by granting higher priority to some particular kinds of data via polling algorithm, which is centralized controlled by Access Point (AP). On the other hand, the performance against scalability issue of these hybrid MAC protocols under densely populated network still needs to be evaluated.

4.3. Scalability

Scalability, which is the most significant challenge in SMN, requires that MAC protocols continue to perform well as the number of SMs offering data scales. Referring to [16], one Local Collector is required to support a network with up to 6000 SMs when considering MAC performance. According to [15], the urban outdoor path loss model for 900MHz RF in dB is $P_{L,dB}(r) = 8 + 37.6 \times \log_{10} r$, where r denotes the distance between the Smart Meter and Local Collector. Then based on the parameters listed in the table (Fig.5) [14], the received power and noise power at the receiver terminal are given as:

$$P_{RX,dB}(r) = P_{TX,dB} + G_{dB} - P_{L,dB}(r)$$
$$= 0 + 3 - (8 + 37.6 \times \log_{10} r) \quad (1)$$

$$P_N = k \times T \times W = 1.3 \times 10^{-23} J/K \times 290K \times 2MHz \quad (2)$$

Parameter	Value	Parameter	Value
Channel Bandwidth W	2 MHz	Transmitter Power $P_{TX,dB}$	30 dBm
Antenna Gain G_{dB}	3 dB	Temperature T	290 K
Transmission Frequency	900 MHz	Boltzmann's Constant k	1.38×10^{-23} J/K
SM Density ρ	1000 per km^2	Cell Radius R	1200 m
$SINR_{dB,th}$	14 dB	$P_{RX,dB,th}$	-129 dB

Figure 5. Parameter List for Transmission Rate and Hidden Node Calculation.

Referring to the AWGN capacity derivation with BPSK modulation in [18], we may draw a figure of the transmission rate and received power with respect to r. As shown in the Fig.6, the RX received power at 1200m satisfies a reasonable received power threshold (-120dB) and achievable transmission rate at 4500m exceeds 100kbps, the minimal required data rate set in IEEE 802.11ah Use Case [16]. This implies that one Local Collector may need to communicate with individual Smart Meters located 1200m away using a star topology. Given typical SM density ρ (1000-6800 SMs per km²) [21], [22], it is possible to have in excess of 6000 Smart Meters communicating to one Local Collector. Clearly, the design of MAC protocols for such large number of nodes, invites challenges about scalability of any chosen MAC

protocol for SMN as discussed next.

For random access protocols based on carrier sensing - such as DCF (CSMA/CA) in IEEE 802.11 - such a large coverage area corresponding to a single Collector cell will lead to significant hidden nodes. In the traditional random access protocols, such as CSMA/CA, all nodes listen to estimate channel status (busy/idle) based on energy thresholding. If the node observes the channel to be continuously idle for a specific interval, it starts contending for the channel via a random back-off process. However, when the coverage of the Local Collector is enlarged, one Smart Meter (hidden node) may not be able to detect ongoing transmission of other meters due to the degenerated radio channel condition. Then this hidden node may initiate its own transmission because it determines the channel as idle, which leads to a collision.

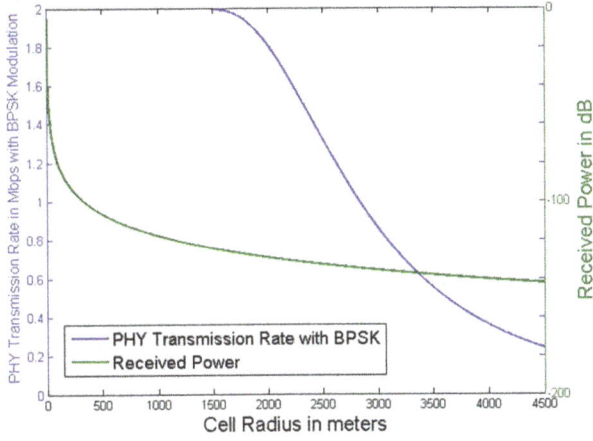

Figure 6. PHY Transmission Rate of the Communication with BPSK Modulation and Received Power at Local Collector with respect to Their Distance.

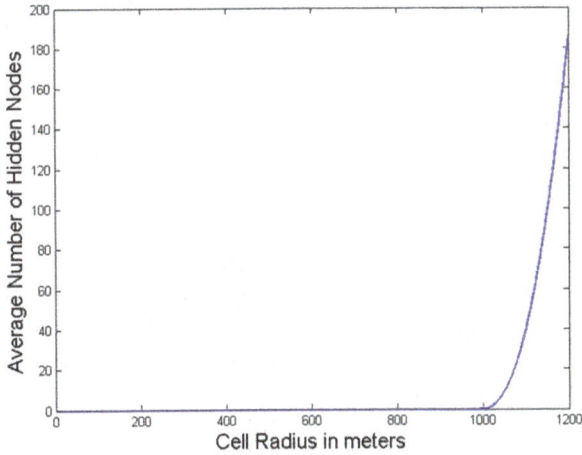

Figure 7. Network Topology for Hidden Node Calculation.

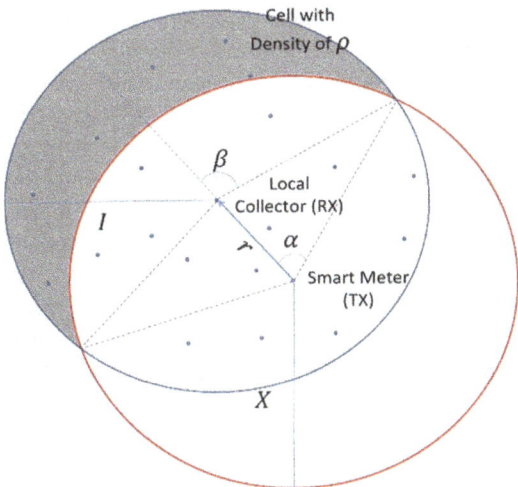

Figure 8. Results for Hidden Node Calculation.

In order to investigate the number of hidden nodes with growing Local Collector coverage, we consider the network topology of a disk with radius of R and uniformly distributed SM density ρ (Fig.7), the distribution of Smart Meter deployment with respect to r is $f(r) = 2r/R^2$, where $0 \leq r \leq R$. Since the down-link communication from Local Collector may cover all the Smart Meters successfully, thus the hidden node only exits in up-link communications where Smart Meter is always the transmitter (TX) and Local Collector is always the receiver (RX). According to [20], the area inside the interference range of the receiver I and outside the carrier sensing range of the transmitter X is defined as the hidden area $A(r)$ (shadowed zone) for a given Smart Meter, which is given as:

$$If\ r \leq X - I, A(r) = 0$$

$$If\ r > X - I, A(r) = \beta I^2 + rX|\sin(\alpha)| - \alpha X^2 \quad (3)$$

where $\alpha = \cos^{-1}(X^2 + r^2 - I^2/2rX)$ and $\beta = \pi - \cos^{-1}(I^2 + r^2 - X^2/2rI)$.

The Carrier Sensing Range X can be calculated based on the equation $P_{TX,dB} + G_{dB} - P_{L,dB}(X) = P_{X,dB,th}$, where $P_{X,dB,th}$ is the carrier sensing threshold for the received power such that the received signal can be detected. Furthermore, the Interference Range I can be derived based on the following equation:

$$I(r) = \min\left[R, \{y|P_{RX,dB}(r) - 10\log_{10}(P_N + P_{RX}(y)) = SINR_{dB,th}\}\right] \quad (4)$$

where $SINR_{dB,th}$ is the threshold for Signal Noise Interference ratio such that received signal can be decoded successfully.

After that, we use $N_{hidden} = \int_0^R \rho A(r)f(r)dr$ to obtain the mean number of hidden nodes inside the network and plot it with respect to cell radius R based on the parameters listed in Fig.5. As shown in the Fig.8, the number of hidden nodes increases dramatically when the coverage of the network is enlarged. Although the DCF in 802.11 proposes the RTS/CTS algorithm to reduce the occurrence probability of hidden nodes event, its effect on such a large area network still needs to be investigated. On the other hand, the increasing number of Smart Meters may also increase the data load in the SMN, which basically results in stronger competition for the medium access. Then the collisions and following retransmissions happen more frequently, which directly aggravate the performance.

In order to solve this scalability issue, the IEEE 802.11ah TG is considering using improved DCF and PCF [14]. The modified DCF with Contention Factor and Prohibition Time is one of the suggestion new options. Before the contention phase, the Local Collector broadcasts a prohibition time T and contention factor $0 < Q < 1$, according to the current network congestion status. After that, each Smart Meter generates a random number r which follows a uniform distribution on the unit interval and compare it to Q. Then the Smart Meter may contend for the channel if $r \leq Q$ and, otherwise, it keeps silent until the prohibition time T passes. In order to further relieve contention congestion, the MAC scheme may also divide all the Smart Meters within a cell into several groups and provide different groups with different parameters Q and T or allow them to contend for the channel group by group. The collision probability is expected to decrease dramatically (compared to traditional DCF applied to all SMs), as a result.

In order to solve the scalability challenge for PCF, IEEE 802.11ah TG proposes a modified PCF scheme, Probe and Pull MAC (PP-MAC) [14]. After partitioning all the SMs into groups, the Local Collector broadcasts a Probe message to a certain group of Smart Meters before the contention free phase. After that, the Smart Meters having data to send reply a short Probe-ACK concurrently with the use of Zadoff-Chu sequences. By assigning these orthogonal sequences to each Smart Meter and multiplying their own messages with their respective sequences, the cross-correlation of the simultaneous short Probe-ACK transmissions is reduced, so that the Local Collector is able to resolve these parallel ACKs and identify the different transmitters. After that, the Collector schedules and only polls the Smart Meters with Probe-ACK, which leads to a shorter polling cycle and a more efficient taking-turns MAC protocol.

4.4. Fairness

This seeks to measure whether each node in the SMN obtains a fair share of system resources. Fairness can be quantified in terms of the access probability to the shared channel by each node - ideally, this should be equal (independent of the node) assuming that all SMs require identical data rates. In general, the notion of Proportional Fairness should be applied, based on different data rate requirements by different nodes. For example, the Least Completed First Send (LCFS) Principle proposed in [27], provide an efficient way to guarantee the fairness of access opportunity of each SM.

4.5. Security

Data security in SMN is an extremely vital issue as it relates to household or customer information (e.g. energy consumption profile) that is considered private. Therefore, it is necessary to encrypt the message to prevent eavesdroppers from intercepting the message. Although cryptographic tools and algorithms are relatively mature, these will result in extra load on the SMN. An open question is whether the known features of SM data may be exploited to develop simplified yet effective cryptographic approaches. Secondly, end-point authentication is also indispensable for SMN; whenever the data collector receives a message of energy consumption report, it has to authenticate the identity of the sender. Specifically, defenses against two common types of attacks will be of high priority. Integrity in data communications between SM and Local Collector may be compromised by a relay or man-in-middle attacker. And such communications may be targeted for disruption via Denial-of-Service (DoS) attacks by saturating the Local Collector with a large number of spurious external communications, so that it cannot respond to the legitimate traffic [12]. Within this context, it is noted that most SM communications are regular as it

reporting actions are typically scheduled. Therefore, we may exploit such features to filter out malicious accesses by an attacker, by identifying anomalous access traffic patterns.

4.6. Expandability

For SMN, the expandability means the ability of this network to accommodate the new communication nodes (Smart Meters) to its existing capacity. It is noted that the deployment of Smart Meters will not occur according to a fixed schedule; for example, whenever a house is built, a newly installed Smart Meter will need to be introduced into the existing NAN covered by the corresponding Local Collector. This introduction procedure, which may include registration, identity authentication, geographical location identification, has to be conducted automatically and is used for the Local Collector to determine the newly installed Smart Meter is ready to work inside its coverage. Therefore, how to realize this introduction procedure should be a part of the MAC protocols design. For example, whenever one newly installed Smart Meter is online, it sends a request to its associated Local Collector to report its own identification and geographical location. After that, the collector registers this new meter and replies it via a message with some necessary setting information. Then how to automatically modify the parameters of current communication systems due to the newly registered Smart Meter still needs to be analyzed.

4.7. Fault Detection

For SMN, the fault can be categorized into two kinds of cases. The first one is data fault, which means the data involved in the message has some errors. These data errors may be caused by monitoring errors or malicious message altering. Fortunately, the data collector may detect such kind of fault by some statistical algorithms, such as comparison between the current data and historical data. This kind of fault and corresponding solutions mainly occur at application layer. What the MAC protocol designers need to consider is another one, communication fault, which means that some Smart Meters cannot communicate with the Local Collector directly. These communication problems may be caused by the malfunction of Smart Meters or the degeneration of wireless communication environment. Since communication fault leads to the loss of message, which is intolerable for SMN, it is required to design a MAC protocol to detect the 'silent node' inside the SMN as quickly as possible. For example, the Local Collector may exploit the idle communication intervals to poll every Smart Meter and expect its feedback. After that, the collector may detect the silent meters by checking the missing feedbacks. Furthermore, how to schedule the poll-feedback actions in detail and improve its efficiency still need to be investigated.

5. Conclusion

This paper highlighted the design consideration for the SMN, the two-way communication network for next generation AMI systems. We listed several important AMI applications and their benefits to the current power grid. The features and requirements of SMN which have much impact on its MAC protocols design were discussed, which includes the SMN architecture and some candidate communication technologies. Most importantly, we highlighted several important MAC performance evaluation metrics, including traffic types, delay, scalability, fairness, security and expandability, and also gave a survey of associated research issues for SMN MAC protocols, including a summary of current status for MAC protocol design effort in IEEE 802.11ah TG.

References

[1] C. Lima, "Smart Grid Logical Communications Reference Architecture," IEEE P2030 SG1 Architecture, Jan. 2010.

[2] AEIC Team, "AEIC/SmartGrid AMI Interoperability Standard Guidelines for ANSI C12.19 Communications", Version 2.0, Nov.19, 2010.

[3] "International Standard Information Technology -- Telecommunications and Information Exchange between Systems -- Local and Metropolitan Area Networks -- Specific Requirements -- Part 11: Wireless LAN Medium Access Control and Physical Layer Specifications", IEEE 802.11 WG, ref.no.ISO/IEC 8802-11:1999(E), IEEE Std.802.11, 1999.

[4] IEEE 802.11 WG, IEEE 802.11e/D5, "Draft Supplement to Standard for Telecommunications and Information Exchange Between Systems -- LAN/MAN Specification Requirements -- Part 11: Wireless Medium Access Control and Physical Layer Specifications: Medium Access Control Enhancements for Quality of Service," Aug. 2003.

[5] T.Khalifa, K.Naik, and A.Nayak, "A Survey of Communication Protocols for Automatic Meter Reading Applications", IEEE Communications Survey and Tutorials, vol.13, No.2, Second Qtr, 2011.

[6] W.Wang, Y.Xu, and M.Khanna, "A Survey of the Communications Architectures in Smart Grid", Computer Networks, vol.55, no.16, 2011.

[7] V.C.Gungor, D.Sahin, T.Kocak, and S.Ergut, "Smart Grid Communications and Networking", Turk Telekon Technology Report, 2011.

[8] V.C. Gungor, D. Sahin, T. Kocak, S. Ergut, C. Buccella, C. Cecati, and G.P. Hancke, "Smart Grid Technologies: Communications Technologies and Standards", IEEE Transactions on Industrial Informatics, vol.7, no.4, 2011.

[9] R.P. Lewis, P. Igic, and Z. Zhongfu, "Assessment of Communications Methods for Smart Electricity Metering in the UK", IEEE PES/IAS Conference of Sustainable Alternative Energy (SAE), 2009.

[10] Y. Peizhong, A. Iwayemi, and C. Zhou, "Deveploing Zigbee Deployment Guideline under WiFi Interference for Smart Grid Applications", IEEE Transactions on Smart Grid, no.1, 2011.

[11] H. Tan, H. Lee, and V. Mok, "Automatic Power Meter Reading System Using GSM Network", IEEE International Conference on Control Application, 2007.

[12] J. Kurose, and K.Ross, "Computer Networking A Top-Down Approach, Fifth Edition", Pearson Education Inc, 2010.

[13] S. Aust, and R. Prasad, "IEEE 802.11ah: Advantages in Standards and Further Challenges for Sub 1 GHz Wi-Fi", 2012 IEEE International Conference on Communications, Jun. 2012.

[14] IEEE 802.11ah TG, "Specification Framework for TGah", IEEE 802.11-11/1137r15, May 2013.

[15] IEEE 802.11ah TG, "TGah Channel Model Proposed Text", IEEE 802.11-11/0968r3, Nov. 2011.

[16] IEEE 802.11ah TG, "Potential Compromise for 802.11ah Use Case Document", IEEE 802.11-11/0457r0, Mar. 2011.

[17] Y. Yang, X. Wang, and X. Cai, "On the Number of Relays for Orthogonalize-and-Forward Relaying," IEEE WCSP 2011, Nanjing, China.

[18] J.G. Proakis, "Digital Communications, Fifth Edition", McGraw Hill Publishing House, 2009.

[19] National Electrical Manufacturers Association, "American National Standard for Utility Industry End Device Data Tables", 2008.

[20] H. Ma, H. Alazemi, and S.Roy, "A Stochastic Model for Optimizing Physical Carrier Sensing and Spatial Reuse in Wireless Ad Hoc Networks", IEEE International Conference on Mobile Adhoc and Sensor Systems, Nov.2005.

[21] Electric Power Research Institute, "Summary of EPRI Test Circuits", www.epri.com.

[22] IEEE 802.11ah TG, "Association ID Management for TGah", IEEE 802.11-11/0088r1, Jan.2011.

[23] Y. Yang, and S. Roy, "PMU Deployment for Three-Phase Optimal State Estimation Performance", IEEE International Conference on Smart Grid Communications 2013, Vancouver, Canada.

[24] Y. Yang, and S. Roy, "PMU Deployment for Optimal State Estimation Performance", IEEE GLOBECOM 2012, Anaheim, CA, USA.

[25] B. Park, D. Hyun, and S. Cho, "Implementation of AMR system using power line communication, IEEE/PES Transmission Distribution Conf. Exhibition, Oct. 2002.

[26] A. Yarali, "Wireless mesh networking technology for commercial and industrial customers", Proceeding Electrical Computing Engineering CCECE, May 1–4, 2008, pp. 000047–000052.

[27] Y. Yang, and S. Roy, "PCF Scheme for Periodic Data Transmission in Smart Metering Network with Cognitive Radio", IEEE GLOBECOM 2015, San Diego, CA, USA.

[28] Y. Yang, and S. Roy, "Grouping Based MAC Protocols for EV Charging Data Transmission in Smart Metering Network, IEEE Journal of Selected Area on Communications, vol 32, no 7, July 2014.

17

Financial Risk Management for Designing Multi-echelon Supply Chain Networks Under Demand Uncertainty

De Gu[1], Jishuai Wang[2]

[1]Key Laboratory of Advanced Process Control for Light Industry (Ministry of Education), Institute of Automation, Jiangnan University, Wuxi, China
[2]Suzhou Institute of Biomedical Engineering and Technology, Chinese Academy of Sciences, Suzhou, China

Email address:
gude@jiangnan.edu.cn (De Gu), wangjs@sibet.ac.cn (Jishuai Wang)

Abstract: This paper presents a methodology to include financial risk management for the design of multiproduct, multi-echelon supply chain networks under uncertainty. The method is in the framework of two-stage stochastic programming. Definitions of financial risk and downside risk are adapted. Using these definitions, financial risk management constraints are introduced and a new two-stage stochastic programming model is established. Case studies illustrate the applicability of such financial risk management. Trade-offs between expected cost and risk are also analyzed.

Keywords: Supply Chain, Financial Risk Management, Downside Risk, Uncertainty

1. Introduction

In today's highly competitive environment, it is important to maximize the total profit by managing supply chains. A variety of academic disciplines have been focused on supply chain for many years. [1] A supply chain begins with procurement of raw materials and ends with finished products shipped to customers. [2] It is an integrative approach used to manage the inter-related flows of products and information among suppliers, manufacturing plants, warehouses, and customers. [3] This paper considers the design of multiproduct, multi-echelon supply chain networks.

Early research on supply chain network design can be accredited to Geoffrion et al. [4] who firstly formulated a mixed integer linear program (MILP) model, the objective is to decide the optimal location of intermediate distribution facilities between plants and customers. Alikens et al. [5] reviewed significant contributions of best warehouses site planning, and extended the model considering site inventory. Vidal et al. [6] presented an extensive review of strategic production-distribution models. Special emphasis was placed on models for global logistics systems and further research in this area. Tsiakis et al. [2] designed a multiproduct, multi-echelon supply chain network in a MILP formulation. The objective was to minimize of the total cost of the network. Assavapokee et al. [7] proposed a MILP model to

solute the problem of designing the infrastructure of the reverse production network to reduce the amount of waste stream. Cafaro et al. [8] presented a mixed-integer nonlinear programming (MINLP) model to optimally determine the structure of shale gas supply chain. Kalaitzidou et al. [9] proposed a model in a MILP framework for the design of supply chain networks by providing flexibility on facilities' location and operation.

But the above researches were all deterministic. In reality, there are many uncertain factors including supplier changes, output fluctuating, and demand uncertainties. [10] These factors will have impactions on the total profit of supply chain, therefore it's necessary to develop methods to address the problem of supply chain network design in the presence of uncertainty. In general, there are two approaches which can be used for dealing the problem of supply chain network design under uncertainty: one is probability based method, and the other is scenario based. Talaei et al. [11] proposed a optimization model to design a multi-product closed-loop green supply chain network, in which demands and costs were supposed to be fuzzy parameters. But in real world, it's hard to obtain the probability distribution of uncertainties, so scenario based method is widely used. [12] Financial risk management is a scenario based method which provides new insights into the trade-offs between risk and profitability of aimed objectives, and has been used in the tactic level of supply chain. [13] You *et al.* [14] considered the risk management for mid-term planning under demand and freight rate uncertainty. Cardoso *et al.* [15] proposed a MILP formulation to integrate different risk measures into planning of closed-loop supply chains. This paper will discuss the financial risk management in the strategic level such that designing of multi-echelon supply chain network.

This article is organized as follows: In section 2, we review the definitions of financial risk and downside risk. In section 3, detailed mathematical formulations of the multi-echelon supply chain network design problem based on the model proposed by Tsiakis *et al.* [2] are presented. Then the problem is formulated as a two-stage stochastic programming model to take into account risk management constraints. Finally, we demonstrate the effectiveness of this novel approach using several examples. Concluding remarks are given in the last section.

2. Financial Risk Management

2.1. Two-Stage Stochastic Programming

Financial risk can be defined as the probability of not meeting a certain target profit (maximization) or cost (minimization) level. For the two-stage stochastic problem (SP), the financial risk can be expressed by the following equations [13]

$$Max \quad E[Profit] = \sum_{s \in S} p_s q_s^T y_s - c^T x \quad (1)$$

$$s.t. \quad Ax = b \quad (2)$$

$$T_s x + W y_s = h_s \quad \forall s \in S \quad (3)$$

$$x \geq 0 \quad x \in X \quad (4)$$

$$y_s \geq 0 \quad s \in S \quad (5)$$

where x is the first-stage decision variables, y_s is the optimal second-stage solution for scenario s with probability p_s. The objective is the second-stage profit minus the first-stage costs. q_s, T_s, and h_s are uncertain parameters.

2.2. Definition of Financial Risk

In the two-stage stochastic programming, if the second-stage profit is $Profit_s(x)$, then financial risk associated with a target profit Ω can be expressed by the following probability

$$Risk(x, \Omega) = \sum_{s \in S} p_s z_s(x, \Omega) \quad (6)$$

where p_s is the probability of scenario s, and z_s is a binary variable

$$z_s(x, \Omega) = \begin{cases} 1 & if \ Profit_s(x) < \Omega \\ 0 & otherwise \end{cases} \quad \forall s \in S \quad (7)$$

The expected value of the profit can be expressed as

$$E[Profit(x)] = \sum_{s \in S} p_s \xi_s \quad (8)$$

The above scenarios ξ_s are sorted in ascending sequence, such that $\xi_{s+1} \geq \xi_s$. If the expected profit is known as $\underline{\xi} < Profit(x) < \overline{\xi}$, then the relationship between expected profit and risk is

$$E[Profit] = \overline{\xi} - \sum_{s \in S} Risk(x, \xi)(\xi_{s+1} - \xi_s) \quad (9)$$

2.3. Downside Risk

If the definition of $\delta_s(x, \Omega)$ is the positive deviation from a profit target Ω for design variable x, such that

$$\delta_s(x, \Omega) = \begin{cases} \Omega - Profit_s(x) & if \ Profit_s(x) < \Omega \\ 0 & otherwise \end{cases} \quad (10)$$

Then for profit with discrete distribution, downside risk can be written as

$$DRisk(x, \Omega) = \sum_{s \in S} p_s \delta_s(x, \Omega) \quad (11)$$

The relationship between expected profit and downside risk can be expressed as

$$E[Profit(x)] = \overline{\xi} - DRisk(x, \overline{\xi}) \quad (12)$$

2.4. Financial Risk Constraints

When the objective of a model is to minimize the total costs, and at the same time minimize the financial risk at every profit level, a weight $\rho_n(\rho_n \geq 0)$ is introduced. By using ρ_n, a multi-objective optimization can be reduced to single objective by imposing a penalty for risk at different target profits Ω_n

$$Min \quad \sum_{s \in S} p_s q_s^T y_s + c^T x + \sum_{s \in S} \sum_{n \in N} p_s \rho_n z_{sn} \quad (13)$$

s.t. $(2) - (5)$

$$q_s^T y_s + c^T x \geq \Omega_n - U_s(1 - z_{sn}) \quad \forall s \in S, n \in N \quad (14)$$

$$q_s^T y_s + c^T x \leq \Omega_n + U_s z_{sn} \quad \forall s \in S, n \in N \quad (15)$$

U_s is a positive number big enough to ensure only one of (14) and (15) is true, z_{sn} is a binary variable to calculate the risk of the objective function.

3. Mathematical Models

3.1. Multi-echelon Supply Chain Networks Design

The problem consists of determining the number, capacity, and location of warehouses and distribution centers to be set as well as the transportation links that need to be established, and the flows and production rates of materials in the network.

3.2. Constraints

This paper is based on the model of Tsiakis *et al.* [2], and financial risk management constraints are applied and analyzed. There are six sorts of constrains. Binary variables and continuous nonnegative variables are included refer to location and logistics.

3.2.1. Network Structure Constraints

$$X_{mk} \leq Y_m \quad \forall m,k \qquad (16)$$

$$X_{kl} \leq Y_k \quad \forall k,l \qquad (17)$$

When warehouse m exists, there are transportations between warehouse m and distribution center k as shown in (16). The relation between distribution center k and customer l is expressed as (17).

3.2.2. Logical Constraints

$$Q_{ijm} \leq Q_{ijm}^{\max} Y_m \quad \forall i,j,m \qquad (18)$$

$$Q_{imk} \leq Q_{imk}^{\max} X_{mk} \quad \forall i,m,k \qquad (19)$$

$$Q_{ikl} \leq Q_{ikl}^{\max} X_{kl} \quad \forall i,k,l \qquad (20)$$

$$\sum_i Q_{imk} \geq Q_{mk}^{\min} X_{mk} \quad \forall m,k \qquad (21)$$

$$\sum_i Q_{ikl} \geq Q_{kl}^{\min} X_{kl} \quad \forall k,l \qquad (22)$$

When distribution center k exists, there are transportations limitations as shown in (18) to (22).

3.2.3. Material Balances Constraints

$$P_{ij} = \sum_m Q_{ijm} \quad \forall i,j \qquad (23)$$

$$\sum_j Q_{ijm} = \sum_k Q_{imk} \quad \forall i,m \qquad (24)$$

$$\sum_m Q_{imk} = \sum_l Q_{ikl} \quad \forall i,k \qquad (25)$$

$$\sum_k Q_{ikl} = Dem_{il} \quad \forall i,l \qquad (26)$$

There is no inventory for each site as shown from (23) to (26).

3.2.4. Resource Constraints

$$P_{ij}^{\min} \leq P_{ij} \leq P_{ij}^{\max} \quad \forall i,j \qquad (27)$$

$$\sum_i \rho_{ije} P_{ij} \leq R_{je} \quad \forall j,e \qquad (28)$$

Resource limitation for each product is written by (27) and (28).

3.2.5. Capacity Constraints of Warehouses and Distribution Centers

$$W_m^{\min} Y_m \leq W_m \leq W_m^{\max} Y_m \quad \forall m \qquad (29)$$

$$D_k^{\min} Y_k \leq D_k \leq D_k^{\max} Y_k \quad \forall k \qquad (30)$$

$$W_m \leq \sum_{i,k} \alpha_{im} Q_{imk} \quad \forall m \qquad (31)$$

$$D_k \geq \sum_{i,l} \beta_{ik} Q_{ikl} \quad \forall k \qquad (32)$$

Each warehouse and distribution center has capacity limitation for certain product as shown from (29) to (32).

3.2.6. Transportation Costs Constraints

Because the transportation cost is a piecewise linear function of the material flow, the transportation cost between plants and warehouses can be written as

$$\sum_{r=1}^{NR_{fjm}} Z_{fjmr} = 1 \quad \forall f,j,m \qquad (33)$$

$$\overline{Q}_{fjm,r-1} Z_{fjmr} \leq Q_{fjmr} \leq \overline{Q}_{fjmr} Z_{fjmr} \quad \forall f,j,m,r=1..NR_{fjm} \qquad (34)$$

$$\sum_{i \in I_f} Q_{ijm} = \sum_{r=1}^{NR_{fjm}} Q_{fjmr} \quad \forall f,j,m \qquad (35)$$

$$C_{fjm} = \sum_{r=1}^{NR} [\overline{C}_{fjm,r-1} Z_{fjmr} + (Q_{fjmr} - \overline{Q}_{fjm,r-1} Z_{fjmr}) \frac{\overline{C}_{fjmr} - \overline{C}_{fjm,r-1}}{\overline{Q}_{fjmr} - \overline{Q}_{fjm,r-1}}] \qquad (36)$$
$$\forall f,j,m$$

The transportation cost between warehouses and distribution centers can be written as

$$\sum_{r=1}^{NR_{fmk}} Z_{jmkr} = 1 \quad \forall f,m,k \qquad (37)$$

$$\overline{Q}_{fmk,r-1} Z_{fmkr} \leq Q_{jmkr} \leq \overline{Q}_{fmkr} Z_{jmkr} \quad \forall f,m,k,r=1..NR_{fmk} \qquad (38)$$

$$\sum_{i \in I_f} Q_{imk} = \sum_{r=1}^{NR_{fmk}} Q_{fmkr} \quad \forall f,m,k \qquad (39)$$

$$C_{fmk} = \sum_{r=1}^{NR} [\overline{C}_{fmk,r-1} Z_{jmkr} + (Q_{fmkr} - \overline{Q}_{fmk,r-1} Z_{fmkr}) \frac{\overline{C}_{fmkr} - \overline{C}_{fmk,r-1}}{\overline{Q}_{fmkr} - \overline{Q}_{fmk,r-1}}] \qquad (40)$$
$$\forall f,m,k$$

The transportation cost between distribution centers and final customers can be written as

$$\sum_{r=1}^{NR_{fkl}} Z_{fklr} = 1 \quad \forall f,k,l \qquad (41)$$

$$\overline{Q}_{fkl,r-1} Z_{fklr} \leq Q_{fklr} \leq \overline{Q}_{fklr} Z_{fklr} \quad \forall f,k,l,r=1..NR_{fkl} \qquad (42)$$

$$\sum_{i \in I_f} Q_{ikl} = \sum_{r=1}^{NR_{fkl}} Q_{fklr} \quad \forall f,k,l \quad (43)$$

$$C_{fkl} = \sum_{r=1}^{NR} [\overline{C}_{fkl,r-1} Z_{fklr} + (Q_{fklr} - \overline{Q}_{fkl,r-1} Z_{fklr}) \frac{\overline{C}_{fklr} - \overline{C}_{fkl,r-1}}{\overline{Q}_{fklr} - \overline{Q}_{fkl,r-1}}] \quad (44)$$
$$\forall f,k,l$$

3.3. Objective Function

Because the demand Dem_{il} is uncertain, this paper takes scenario-based method, we assume that the probability of scenario s occurring is ψ_s ($\sum_{s=1}^{S} \psi_s = 1$), then all variables corresponding to a certain scenario will have one additional index. The objective function can be expressed as

$$\min \quad \sum_m C_m^{WH} W_m + \sum_k C_k^{DH} D_k + \sum_{s=1}^{S} \psi_s (\sum_{i,j} C_{ij}^P P_{ijs} + \sum_{i,m} C_{im}^{WH} (\sum_j Q_{ijms})$$
$$+ \sum_{i,k} C_{ik}^{DH} (\sum_m Q_{imks}) + \sum_{f,j,m} C_{fjms} + \sum_{f,m,k} C_{fmks} + \sum_{f,k,l} C_{fkls}) \quad (45)$$

The first two terms are fixed costs of setting up warehouses and distribution centers which are design variables. The others are control variables related to different scenarios. The third term is manufacturing cost, the fourth and fifth terms are logistic costs of warehouses and distribution centers, the sixth term to the eighth term are the transportation costs of the whole network.

3.4. Financial Risk Management Model

As the objective of the supply chain network design is to minimize the expected cost, once the calculated objective is larger than the expected one, the risk exists. Financial risk management is aim to control the bias between real cost and the expected cost Ω_n. Two additional constraints of risk management should be added to the former model

$$\sum_m C_m^{WH} W_m + \sum_k C_k^{DH} D_k + \sum_{i,j} C_{ij}^{PH} P_{ijs}$$
$$+ \sum_{i,m} C_{im}^{WH} (\sum_j Q_{ijms}) + \sum_{i,k} C_{ik}^{DH} (\sum_m Q_{imks})$$
$$+ \sum_{f,j,m} C_{fjms} + \sum_{f,m,k} C_{fmks} + \sum_{f,k,l} C_{fkls} \geq \Omega_n - U_s(1 - z_{sn}) \quad (46)$$

$$\sum_m C_m^{WH} Y_m + \sum_k C_k^{DH} Y_k + \sum_{i,j} C_{ij}^{PH} P_{ijs}$$
$$+ \sum_{i,m} C_{im}^{WH} (\sum_j Q_{ijms}) + \sum_{i,k} C_{ik}^{DH} (\sum_m Q_{imks})$$
$$+ \sum_{f,j,m} C_{fjms} + \sum_{f,m,k} C_{fmks} + \sum_{f,k,l} C_{fkls} \leq \Omega_n + U_s z_{sn} \quad (47)$$

The objected function is

$$\min \quad \sum_m C_m^{WH} W_m + \sum_k C_k^{DH} D_k +$$
$$\sum_{s=1}^{S} \psi_s (\sum_{i,j} C_{ij}^{PH} P_{ijs} + \sum_{i,m} C_{im}^{WH} (\sum_j Q_{ijms}) + \sum_{i,k} C_{ik}^{DH} (\sum_m Q_{imks})$$
$$+ \sum_{f,j,m} C_{fjms} + \sum_{f,m,k} C_{fmks} + \sum_{f,k,l} C_{fkls}) + \sum_{s=1}^{S} \sum_{n=1}^{NR} \psi_s \rho_n z_{sn} \quad (48)$$

U_s is a positive number big enough to ensure only one of (46) and (47) is true, ρ_n is a nonnegative variable to control the weight of different target profits.

If downside risk is referred as the risk measurement, when Ω is the expected cost, the following constraint should be added

$$\delta_s \geq (\sum_m C_m^{WH} W_m + \sum_k C_k^{DH} D_k) -$$
$$\sum_{s=1}^{S} \psi_s (\sum_{i,j} C_{ij}^{PH} P_{ijs} - (\sum_{i,m} C_{im}^{WH} (\sum_j Q_{ijms})$$
$$+ \sum_{i,k} C_{ik}^{DH} (\sum_m Q_{imks}) + \sum_{f,j,m} C_{fjms} + \sum_{f,m,k} C_{fmks} + \sum_{f,k,l} C_{fkls})) - \Omega \quad (49)$$

The objected function is

$$\min \quad \mu\{\min \quad \sum_m C_m^{WH} W_m + \sum_k C_k^{DH} D_k +$$
$$\sum_{s=1}^{S} \psi_s (\sum_{i,j} C_{ij}^{PH} P_{ijs} + \sum_{i,m} C_{im}^{WH} (\sum_j Q_{ijms}) +$$
$$\sum_{i,k} C_{ik}^{DH} (\sum_m Q_{imks}) + \sum_{f,j,m} C_{fjms} + \sum_{f,m,k} C_{fmks} + \sum_{f,k,l} C_{fkls})$$
$$+ \sum_{s=1}^{S} \sum_{n=1}^{NR} \psi_s \rho_n z_{sn}\} + \sum_{s=1}^{S} \psi_s \delta_s \quad (50)$$

Decisions vary corresponding to different Ω. Decisions are conservative when Ω is smaller, conversely, decisions are aggressive. So it is helpful for decision makers to make selections for their preference.

4. Case Study

We use the second case of Tsiakis et al. [2]. All models were implemented in Lingo 11.0 on an Intel Core 3.50 GHz/4G RAM platform. The case contains three plants, three warehouses, three distribution centers and eighteen customers with fourteen products demands. Besides three given scenarios, we will increase eighteen scenarios, that the number of scenarios is twenty-one.

4.1. Two-Stage Stochastic Model

We first suppose customer demands of added scenarios are normal distribution, mean values are the same to scenario 2 of Tsiakis et al. [2], variances are 5% of mean values. We assume that all twenty-one scenarios have equally probability, i.e., $\psi_s = 1/21$ ($s = 1....21$). Without risk management constraints, the risk under expected cost is shown in Figure 1. For the cost of each scenario, its corresponding value shows the risk level. In Figure 1, the dotted line is the expected cost 1957.046 (k£/week) in the two-stage stochastic programming, but it's risk is 35%, that is to say, there exists 35% probability that the real cost building the supply chain network exceeds the expected value. Therefore financial risk management constraints should be added to help decision makers with their different risk attitudes.

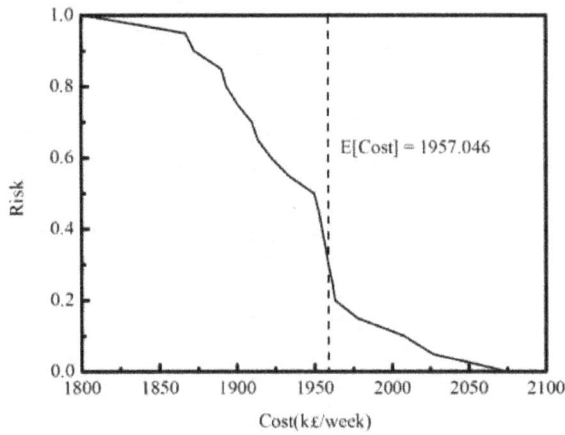

Figure 1. Solution that minimizes the expected total costs.

4.2. Financial Risk Management Analysis

To illustrate the usefulness of the risk management constrains, three different cost targets Ω are used. A set of risks with hypothetical solutions at cost targets with $\rho = 100$ are depicted in Figure 2. The operational decisions of warehouses are shown respectively in Table 1.

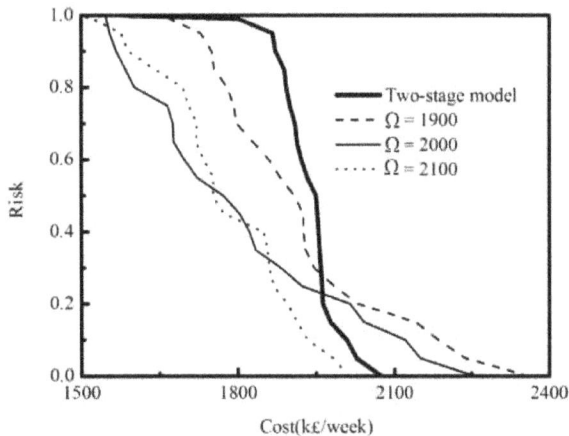

Figure 2. Solution using risk constraints with different values of Ω.

Table 1. Solutions corresponding to different values of Ω.

Ω	Warehouse Throughput (te/week)		
	U.K.	ES	IT
1800	2687	873	788
1900	2436	681	775
2000	2716	952	920

After managing financial risk, risk at each cost will typically result in better performance around the specific target Ω. For example, From Figure 2, at the cost = 1950, the risk of Ω = 1900, 2000, and 2100 are 35%, 25%, and 10% respectively, the risk of the two-stage stochastic model is 50%. It is up to the decision maker to choose Ω and ρ accordingly with his/her risk preference. When $\Omega = 2000$,

there is no risk when expected cost larger than 2250, but the risk is up to 50% when the expected cost is 1800. Therefore downside risk constraints are introduced to measure the risk integral between Ω and expected costs.

4.3. Downside Risk Analysis

Downside risk management can reduce the risk of specific target Ω, as the objective of the two-stage stochastic model is 1957.046, we assume Ω = 1950. The solution is shown in Figure 3. The full line is the risk of the two-stage stochastic model, the dotted line is the risk with downside risk management constraints. Although objectives of both models are approximately equal, risk for 1950 reduces from 50% to 10%. It is helpful for a risk-averse investor having low risk for some conservative profit aspiration level.

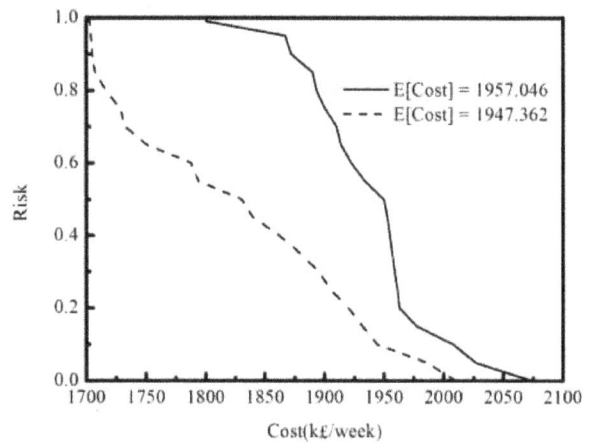

Figure 3. Solution using downside risk management.

5. Conclusion

Supply chain networks design is a hard task because of the intrinsic complexity and interactions with outer cooperators, as well as the considerable uncertainty in product demands. This paper proposes a detailed two-stage stochastic programming formulation with risk constraints which aims to minimize the risk level at certain expected cost. Under demand uncertainty, the trade-off between risk and total cost is analyzed using downside risk as measurement. The result is able to provide a full spectrum of solutions for decision makers to make choices accordingly with their risk preference.

Notation

C —— Cost
D —— Distribution center capacity
Dem —— Product demand
N —— Expected profit selection
NR —— Amount of transportation flows
P —— Production capacity
Q —— Transportation amount
S —— Scenario
U —— Calculated cost
W —— Warehouse capacity
X, Y, Z —— Binary variables

Superscript
min —— Minimum value
max —— Maximum value
DH —— Distribution center
PH —— Plant
WH —— Warehouse

Subscripts
e —— Manufacturing resource
f —— Product family
i —— Product
j —— Plant
k —— Distribution center
l —— Customer
m —— Warehouse
n —— Weight of certain expected cost
r —— Discount range of transportation flow cost
s —— Scenario

Greek letters
α, β, ρ ——Index
δ —— Risk
ξ —— Cost
ψ —— Probability
Ω —— Expected cost

References

[1] D. J. Garcia, F. You, Supply chain design and optimization: Challenges and opportunities. Computers & Chemical Engineering, 2015, 81: 153-70.

[2] P. Tsiakis, N. Shah, C. C. Pantelides, Design of multi-echelon supply chain networks under demand uncertainty. Industrial & Engineering Chemistry Research, 2001, 40 (16): 3585-3604.

[3] T. Santoso, S. Ahmed, M. Goetschalckx, A. Shapiro, A stochastic programming approach for supply chain network design under uncertainty. European Journal of Operational Research, 2005, 167 (1): 96-115.

[4] A. M. Geoffrion, G. W. Graves, Multi-commodity distribution system design by benders decomposition. Management Science, 1974, 20: 822-844.

[5] C. H. Aikens, facility location models for distribution planning. European Journal of Operational Research, 1985, 22 (3): 263-279.

[6] C. J. Vidal, M. Goetschalckx, Strategic production-distribution models: A critical review with emphasis on global supply chain models. European Journal of Operational Research, 1997, 98 (1): 1-18.

[7] T. Assavapokee, W. Wongthatsanekorn, Reverse production system infrastructure design for electronic products in the state of Texas. Computers & Industrial Engineering, 2012, 62(1): 129-140.

[8] D. C. Cafaro, I. E. Grossmann, Strategic planning, design, and development of the shale gas supply chain network. AIChE Journal, 2014, 60(6): 2122-2142.

[9] M. A. Kalaitzidou, P. Longinidis, P. Tsiakis, M. C. Georgiadis, Optimal Design of Multiechelon Supply Chain Networks with Generalized Production and Warehousing Nodes. Industrial & Engineering Chemistry Research, 2014, 53(33): 13125-13138.

[10] I. Heckmann, T. Comes, S. Nickel, A critical review on supply chain risk–Definition, measure and modeling. Omega, 2015, 52: 119-132.

[11] M. Talaei, B. F. Moghaddam, M. S. Pishvaee, A. Bozorgi-Amiri, S. Gholamnejad, A robust fuzzy optimization model for carbon-efficient closed-loop supply chain network design problem: A numerical illustration in electronics industry. Journal of Cleaner Production, 2015.

[12] G. Cairns, P. Goodwin, G. Wright, A decision-analysis-based framework for analysing stakeholder behaviour in scenario planning. European Journal of Operational Research, 2016, 249(3): 1050-1062.

[13] A. Barbaro, M. J. Bagajewicz, Managing financial risk in planning under uncertainty. AIChE Journal, 2004, 50 (5): 963-989.

[14] F. You, J. M. Wassick, I. E. Grossmann, Risk management for a global supply chain planing uncder uncertainty: modes and algorithms. AIChE Journal, 2009, 55: 931-946.

[15] S. R. Cardoso, A. P. Barbosa-Póvoa, S. Relvas. Integrating Financial Risk Measures into the Design and Planning of Closed-loop Supply Chains. Computers & Chemical Engineering, 2016, 85: 105-123.

18

A Study on the Indoor Positioning Method of a Motorcar Detection System Based on CSS (Chirp Spread Spectrum)

Ho-Hung Jung[1, *], Changlong Li[2], Key-Seo Lee[2]

[1]Seoul Metro, Seoul, Korea
[2]School of Robotics, Kwangwoon University, Seoul, Korea

Email address:

metro2line@daum.net (Ho-Hung Jung), kslee@kw.ac.kr (Key-Seo Lee)

Abstract: Accidents that happen on the Urban Railway or in railroad operating institutions can sometimes involve passengers and often occur with night workers who are carrying out system maintenance and management duties. Various safety related systems and technologies are currently being developed in order to prevent accidents from happening, but it is difficult to find a system that is safe and reliable in regards to unforeseen circumstances of duties carried out at night, on and off the tracks, as well as in tunnels. Despite continuous health and safety education, collision accidents between workers and operating vehicles (motorcars) during maintenance and management at night are regularly occurring, threatening precious lives and rendering the loss of property. The purpose of this paper is to prevent accidents between trains, between motorcars, and between trains and people in the Urban Railway and railroad operating institutions through an RT-ToF (Round Trip Time of Flight), appropriate for vehicles and trains, by developing a device that can trace the distance and location by using an RF module.

Keywords: CSS, RT-ToF, Motorcar, Indoor Positioning, RTLS, Positioning

1. Introduction

Accidents that happen on the Urban Railway or in railroad operating institutions can sometimes involve passengers and often occur with night workers who are carrying out system maintenance and management duties. Various safety related systems and technologies are currently being developed in order to prevent accidents from happening, but it is difficult to find a system that is safe and reliable in regards to unforeseen circumstances of duties carried out at night, on and off the tracks, as well as in tunnels. Despite continuous health and safety education, collision accidents between workers and operating vehicles (motorcars) during maintenance and management at night are regularly occurring, threatening precious lives and rendering the loss of property.

In order to prevent such accidents, a system that allows the tracing of a location between vehicles (motorcar ⇔ motorcar) or between a vehicle and a maintenance worker needs to be established through a bilateral detecting system with distance measurement technology along with positioning technology using an RF modulation method.

Object recognition, positioning, action analysis, etc. are in fact key technologies that support an ever- present generation, and as we enter into such a ubiquitous era, the locating system has become quite significant in its position. Nevertheless, existing location services, such as GPS (Global Positioning System), are inappropriate for indoors since their development was designed for the outdoor environment. Consequently, an interest in studies on location systems and the necessities that are compatible for indoors are increasing. Thus, developing a service that allows indoor positioning is essential considering that the service is unsuitable for most consumers who need precise positioning, such as inside a building. Among them, the IEEE 802.15.4a, selected by the UWB (Ultra Wide Band) standard group for low speed, low power, and capacity for wireless positioning, is appropriate for an indoor distance measurement. After considering the processing method of signals that are received from the wireless communication infrastructure and the infrastructure operating technology, etc. when conducting a consecutive indoor and outdoor wireless positioning, the actual method of positioning for the RTLS system, as with GPS and LBS, is classified into three main methods which are, Triangulation, Scene Analysis, and Proximity. Out of these three, locating through triangulation is the most superior method and is based on the technologies RSSI (Received Signal Strength Indication) and ToA (Time of Arrival).

Of the two methods, UWB and CSS, that are PHY standards of low speed positioning WPAN, the IEEE 802.15.4a aims to simultaneously achieve the distance measurement with data communication that's based on low power consumption, which is the purpose of IEEE 802.15.4a PHY. With the current UWB base, distance measurement that requires precision within scores of centimeters is actually possible, and with the CSS base, a precise measurement with a 1~2m margin of error is also possible regardless of the distance. Nevertheless, an identical positional value cannot always be calculated due to the characteristics of wireless communication. Therefore, a position correction technique is proposed in this paper in order to increase the accuracy of positioning through major technologies using the RTLS location calculation method.

Diverse methods of bilateral distance measurement have already been currently proposed but so far no case on implementing the performance in underground tunnels has been reported. In this paper however, the application of a positioning system in Urban Railway sections within tunnels was reviewed by applying Ranging and ToA methods using CSS technology, which uses the RF module to measure distance. This CSS (Chirp Spread Spectrum) technology has much better transmission and static characteristics in comparison to those of Zigbee, a standard technology of IEEE 802.15.4, making it suitable for accurate positioning indoors and within short distances.

The CSS (Chirp Spread Spectrum) method and the RT-ToF (Round Trip-Time of Flight) method have been used to measure the actual locations.

2. Indoor Positioning Method Based on CSS

2.1. CSS (Chirp Spread Spectrum)

CSS technology has been extensively used since 1940 for military radar and submarine sound ranging technology, consuming low power while creating powerful radio waves to multi-paths in order to accommodate communication to a broad area. In March 2007, the Institute of Electrical and Electronics Engineers had adopted the CSS technology as one of the physics class technology for the national standard IEEE 802.15.4a, the WPAN standard for indoor wireless communications (IEEE 802.15.4a, 2007). The CSS measures distance through the triangulation method of TOA.

The CSS-based RTLS, which was standardized in the 2007 IEEE 802.15.4a, had showed an enhanced performance in comparison to the existing RSSI and Zigbee method. The Chirp Spread Spectrum aims for tag positioning in a maximum distance of 300m with margin of error within a 1m radius, through a data-rate of 1Mbps and 10dBm transmission power by using CSS signals through a Bandwidth of 22MHz at 2.4GHz ISM-Band.

CSS Characteristics.

Chirp Pulses are a transmission type of technology that implements communication through Pulse signals, which are interference resistant, after they are generated as radio waves increase or decrease in accordance to time. The interaction formula of the changing radio waves is as follows.

$$S_{cp}(t) = Re\left[exd\left[j\left(\omega_s + \frac{\omega_{BW}}{2T_{cp}}t \right)t + \theta_0 \right] \times \left[u(t) - u\left(t - T_{cp} \right) \right] \right]$$

Each composition is as follows.

T_{cp}: Duration of the Linear Chirp Signal.

ω_s: The starting pulse for the start of the Linear Chirp's sweeping.

ω_{BW}: Sweeping Bandwidth.

u(t): Unit Step Function.

The Chirp signal changes to a consistent inclination while the phase changes at a faster pace as time passes, as the above formula is composed of the duration of the Linear Chirp signal, the start pulse at the start of the Linear Chirp's sweeping, the sweeping bandwidth, and the unit step function.

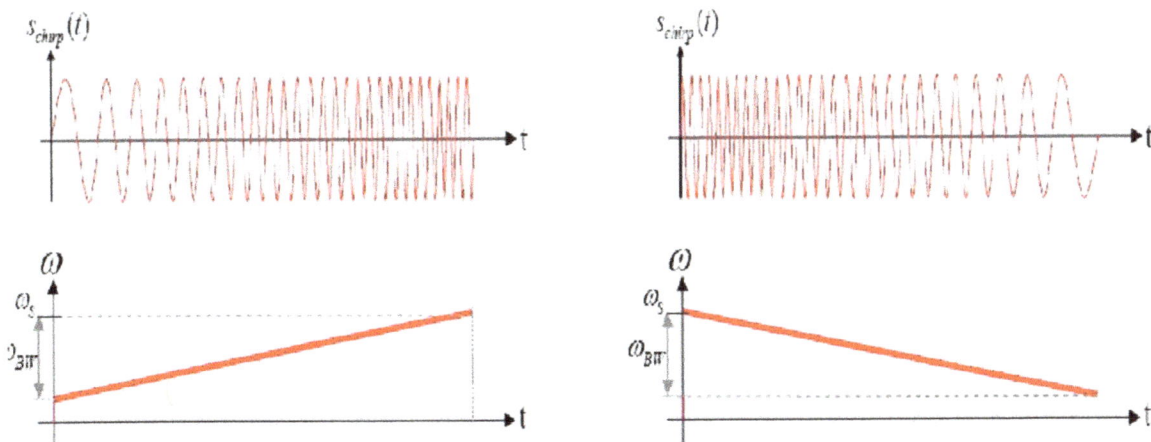

Fig. 1. *Chirp Pulse Signals and Frequency Conversion.*

The CSS method is capable to implement a positioning margin of error within 1m, the target value, even in an indoor environment since the direct waves and multi-path signals can be separated in the receiver by de-correlation, including even the multi-path signals with a path difference within 1m, due to the physical characteristics of the Chirp signal. The advantages of the CSS method is that it allows a long-distance transmission as it is resistant to confusion and multi-path fading since it employs Chirp pulse signals with the compressed power of Chirp pulses that is spread across a broad bandwidth. As well as that the system can be formed with a low power consumption due to Chirp signal generation using DDL that is composed of simple passive devices.

2.2. RT-ToF (Round Trip-Time of Flight)

The RT-ToF (Round Trip-Time of Flight) method was originally proposed to solve the problems of ToF and TDoF, as the name demonstrates ToF is a method using a Round Trip.

The ToF method using RTT concludes the time needed for the signal to travel back and forth between tags.

In other words, supposing that the time taken for the radio wave that is generated from the RFID reader to reach an RFID tag, then for the response to the RFID tag to reach the RFID reader is identical as t_{prop}, and that the time taken for the tag to receive the signal and respond to the reader is t_{tag}, then the time it takes for a trip back and forth between the reader and the tag, t_{RTT}, then results as the following formula.

$$t_{RTT} = 2_{prop} + t_{tag}$$

Arranging this into a formula of t_{prop} is as follows.

$$t_{prop} = \frac{t_{RTT} - t_{tag}}{2}$$

Here, the t_{tag} is the value that is necessary for the RFID tag to process data experimentally, while the t_{RTT} measures each signal's a response from the reader. Since the t_{RTT} needs to be approximately 10μsec in order to provide positioning accuracy of meters, the time information of the RFID reader and μsec standard must be provided.

For this, a high-precision oscillator should be used, which acts to increase the price and the size of the circuit.

Aside from this, the downside is that the time it takes to calculate the transmission time of radio waves takes 2 to 6 times more than the original ToF method, so consequently the amount to be calculated increases slightly. Nevertheless, this is minor detail compared to the limitations of modulating the starter.

Distance Measurement Calculation.

The calculated formula of the distance measurement method using the locating technology, a combination of the above formulas, subsequently sending and receiving radio waves, measuring speed, and referring to the image below, is as follows.

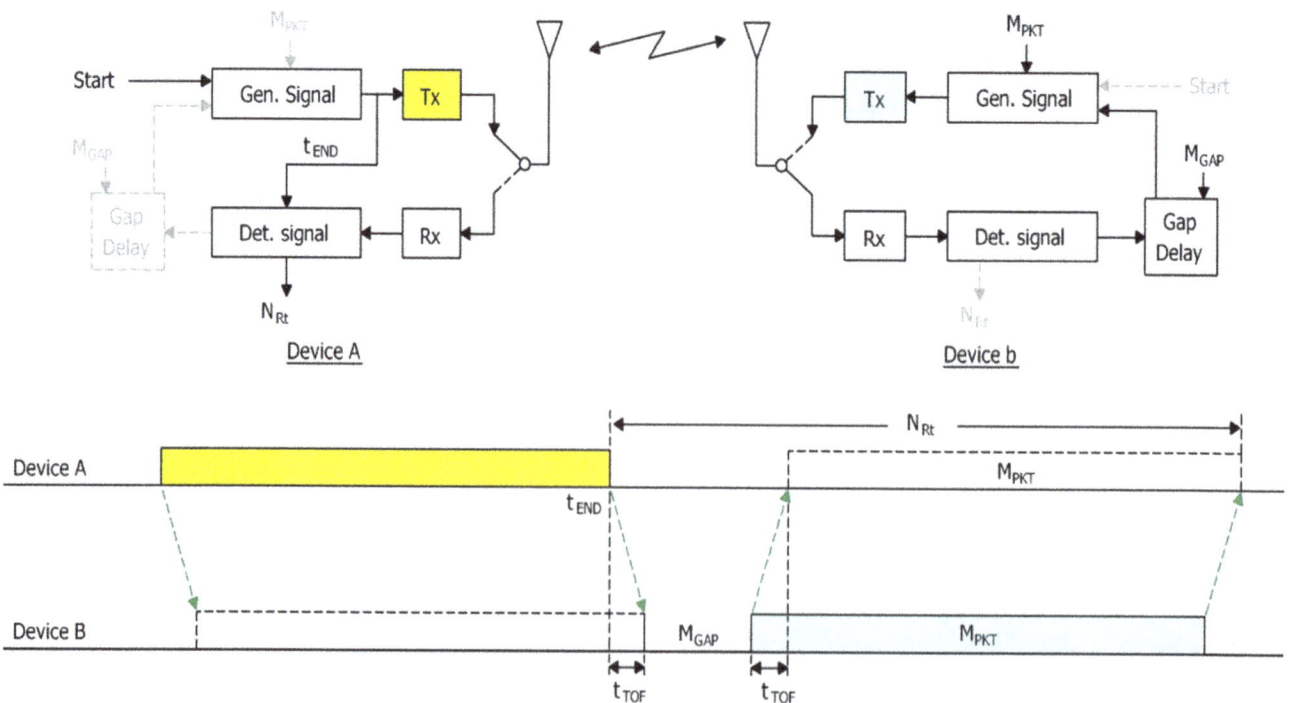

$$\hat{d}_{12} = c \times \hat{t}_{TOF} = c \times (N_{RT} - (M_{GAP} + M_{PKT})) \times T_{CLK}/2$$

Fig. 2. Concept of the RT-ToF Distance Measurement.

2.3. Experiment of Error Correction Filtering Application and Analysis of Results

2.3.1. Error Correction Filtering

The error can only continue to increase if it is difficult to procure LOS between the reader and tag. In particular, the position recognition error increases because of the distance measured through multiple paths, such as the refracted radio waves in a curve section, which cannot be trusted. In order to solve this problem, the error correction filtering was used. The most commonly used algorithm in position recognition systems uses arithmetic mean and the weighted average techniques.

With the arithmetic mean technique, it is problematic to obtain an approximate value since it is simply calculated through the mean when a massive error occurs. Therefore, the error correction filtering was then developed by applying the weighted average technique, which applies the weighted value in accordance to the measured time.

The data filtering method is as follows.

Firstly, when trying to delete a particular range of data, specify the data's minimum and maximum value. Confirm whether the collected data is within the range that is designated by the user, and if the data is outside the range, then return it.

Secondly, there is a specific method that's used when the user wishes to delete any predictable data. Thus, the data transmission number is decreased by deciding whether or not to transmit data after considering the data's margin of error.

The order of the error correction algorithm that was actually implemented is as follows.

1. 10 distance data is measured through 5ms of scan time.
2. Of the 10 distance data, detect and delete the minimum value.
3. Of the 10 distance data, detect and delete the maximum value.
4. Calculate the average number from the remaining eight data: (data sum/8 = avr).
5. Save the 1 time measurement avr data 1.
6. Collect avr data 2 by measuring the distance from the opposing equipment.
7. Calculation is completed through new_avr = (avr data 1 + avr data 2) / 2.
8. New_avr distance data is the valid data, which is automatically saved and displayed on the screen.

Through this process, a more stabilized distance data is calculated.

2.3.2. Underground Tunnel Site Fixed Point Experiment (Straight Line)

By applying an error correction filtering system, the tag was measured at Line 9 in the underground tunnel whereby every 10m for 10~50m while moving. The Reader System was connected with a motorcar antenna and the Tag System connected with a worker antenna. The below images reveal a comparison from the tags location when applied with only

the actual location and the Kit calculation method, with the tag location measured through the data filtering method.

(a) Server System.

(b) Reader Antenna.

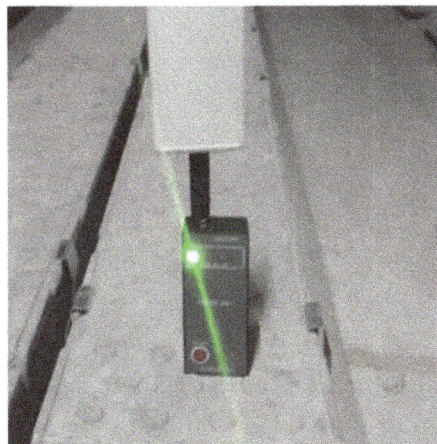

(c) Tag System.

Fig. 3. Site Experiment Photographs.

Distance(m)

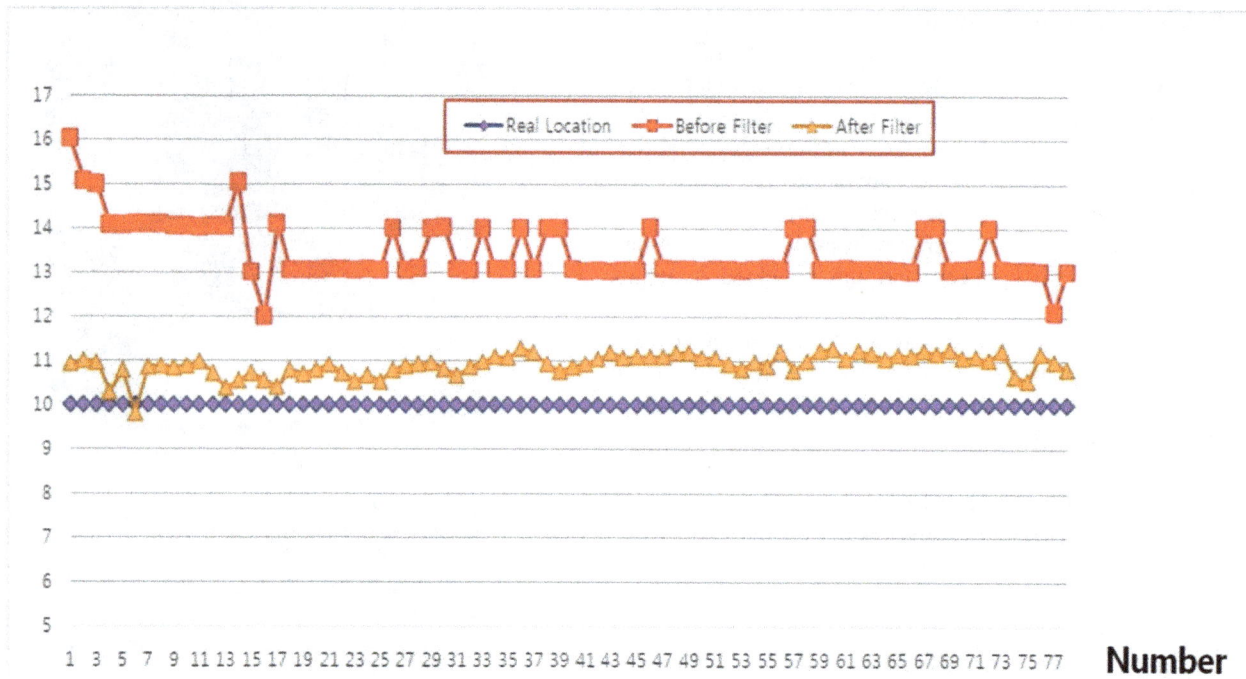

(a) 10m.

Distance(m)

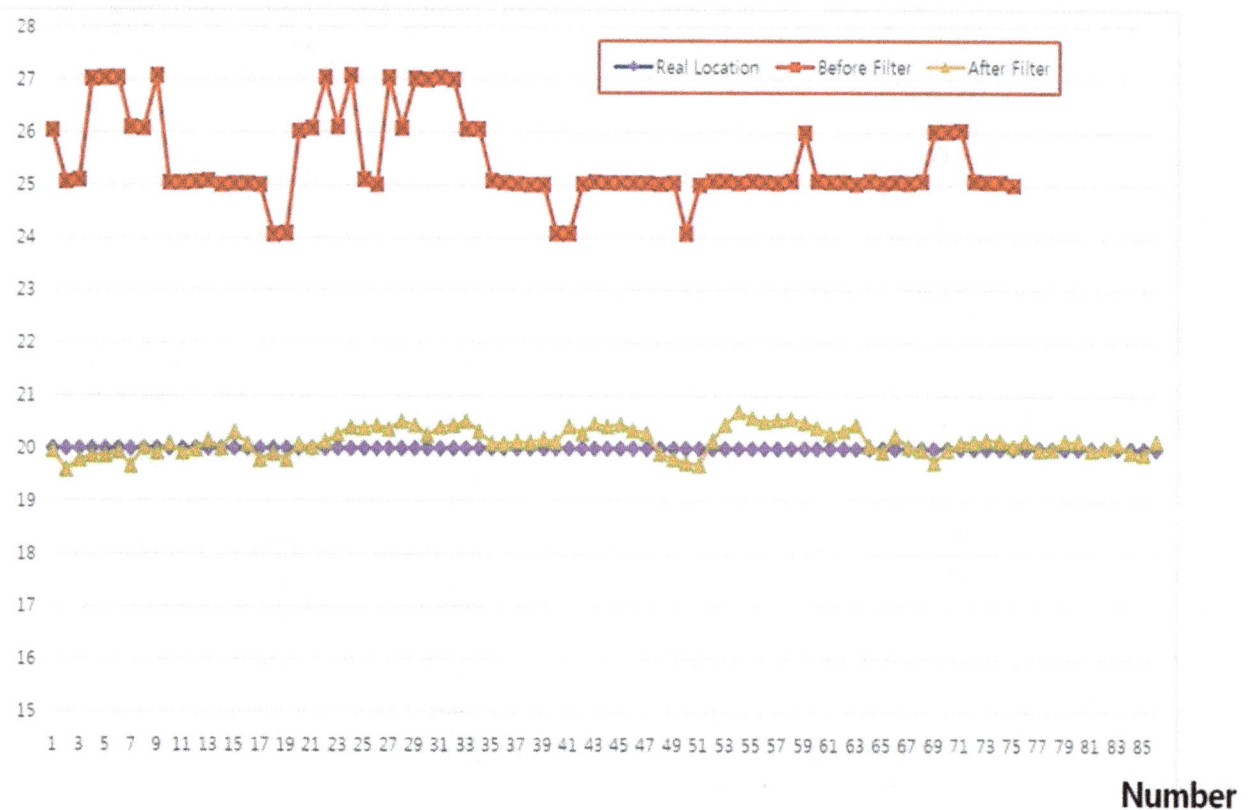

(b) 20m

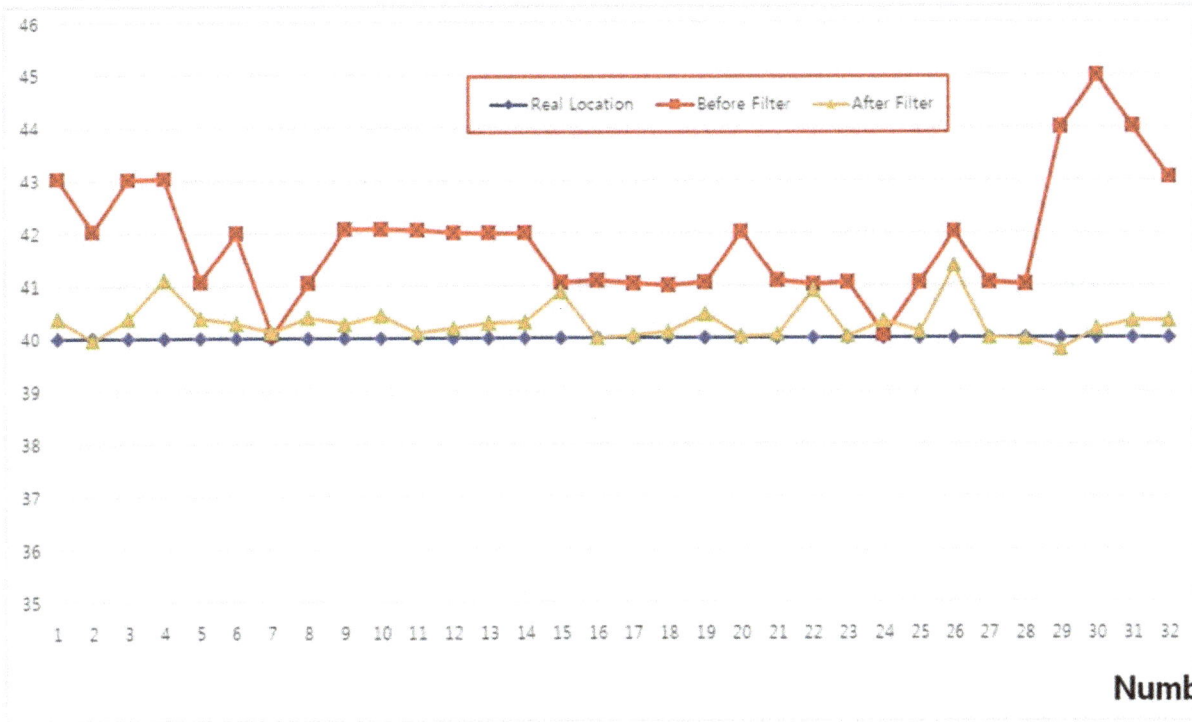

(c) 30m

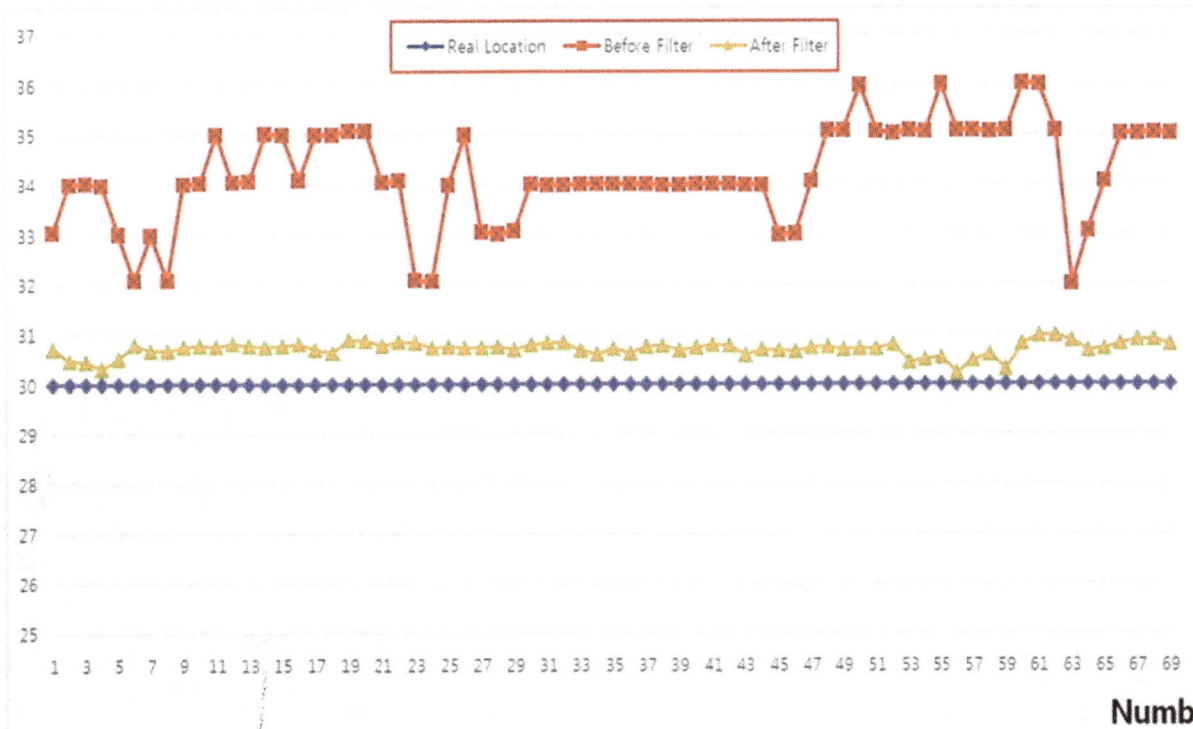

(d) 40m.

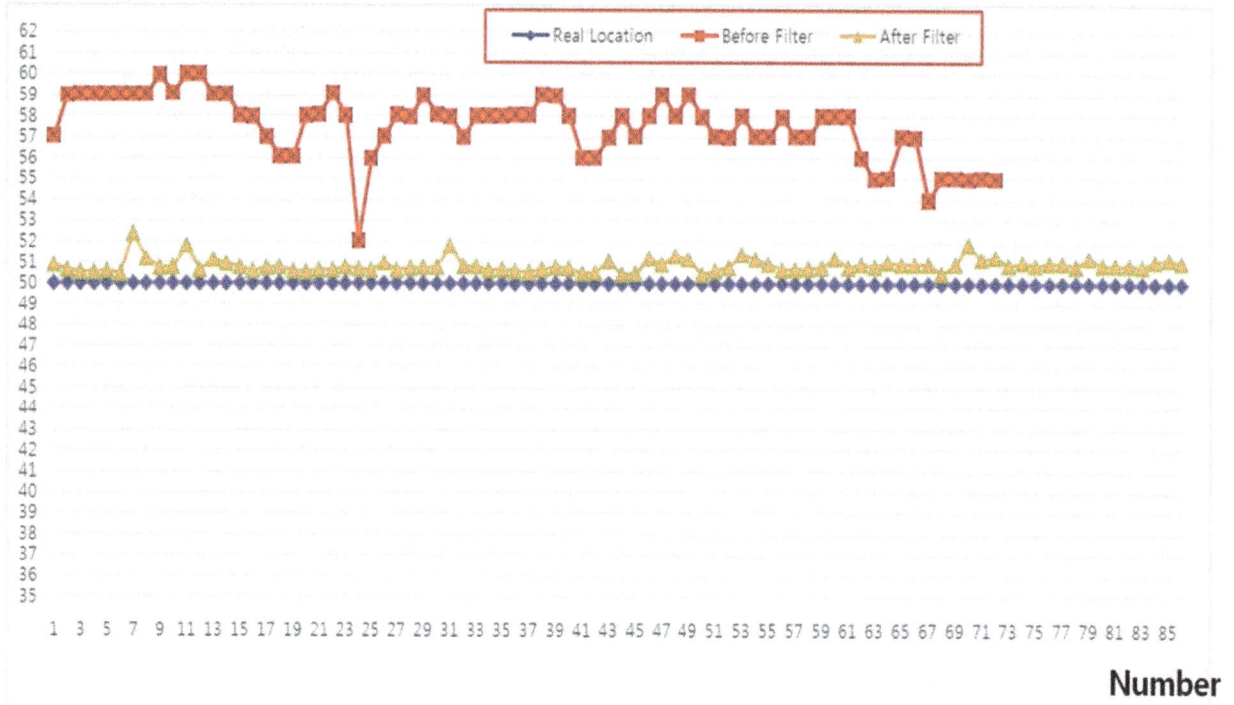

(e) 50m.

Fig. 4. *Measurement distance before/after error correction filtering (10m ~ 50m).*

Fig 4 reveals the comparison of the measurement distance before and after applying the error correction filtering. The blue diamonds are the actual location of the tag, the red squares are the measured location before filtering, and the yellowish green triangles are after the error correction filtering. As displayed in the image, when a high error occurs the error correction filtering is applied, reducing the error to within 2m.

2.4. Establishment of a Bilateral Detection System for Urban Railway Maintenance Vehicles

The bilateral detection system designates the highest address to the exchange frequency that uses CSS to assort internal and external, upper and lower lines and select at the initial operation. When the highest address is designated to the internal (upper) line, the address information on the internal line can be exchanged. In contrast, the external (lower) line can be designated in exactly the same way. The exchange time is analyzed and the distance displayed after the address is designated, and despite the rate of error showing a slight difference according to the speed movement, the goal is to the measure the location ± 1m when in a stationary condition and ± 3m when in a motion condition of 45km/h.

A system capable of preventing safety accidents was proven by providing the information of an opposing motor within a short-distance having calculated the distance through a RT-ToF method radio wave exchange and displaying the distance.

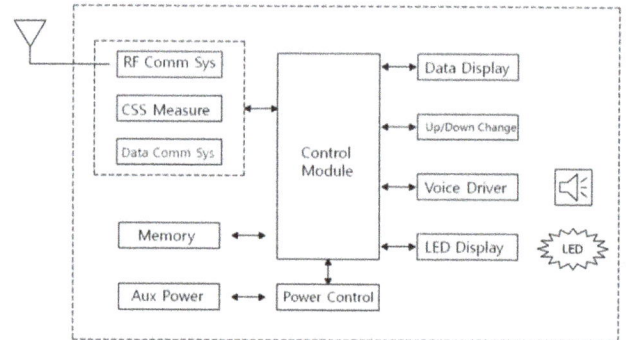

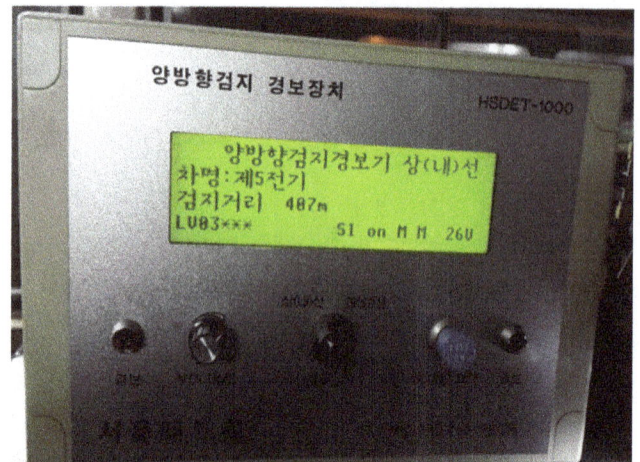

Fig. 5. *Bilateral Scanning Alarm Device for Motorcars.*

3. Conclusion

In order to realize a motorcar's positioning value or a worker's risk information discovered through RTLS technology, the positioning technology of low cost and high efficiency is necessary.

By reviewing various transmission technologies while comparing and analyzing them the CSS method, which is not affected by existing AP antennas through underground tunnels, was selected as the most appropriate positioning technology for underground sites.

The RTLS technology has several methods for measuring location. The methods include RSS (Received Signal Strength), technology, ToA (Time of Arrival) technology, TDoA (Time Difference of Arrival) technology, AoA (Angle of Arrival) technology, and RT-ToF (Round Trip Time of Flight) technology. After analyzing these different methods the RT-ToF technology was selected as the most appropriate for underground sites.

Lastly, in order to improve the error of position recognition, which is said to be the limitation of the position recognition technology, various error correction algorithms were analyzed and applied. The bilateral detection system for the Urban Railway repair and maintenance car was design and established through the selection of a positioning technology for the position recognition system, selection of a position detection technology for transmitting the position recognition system and various site information, and the selection of an error correction filtering for a correct position recognition.

Acknowledgment

The present research has been conducted by the research grant of Kwangwoon University in 2013.

References

[1]   Junhyeong Joo, Hyeonuk Kim, Seonghyeon Yang, "The Design and Implementation of a CSS based Location Aware System", Journal of Korea Institute of Information and Communication Engineering, pp. 1549-1558; 2012.

[2]   IEEE Standard 802.15.4a-2007: MAC and PHY Spec for Low-rate W-PAN, IEEE.

[3]   Daewon Kim (2011) Safety Management System for a Large Construction Project Using Real-time Location System, Doctor's Degree, Korea University.

[4]   Yeongho Kim (2011) Adaptive RTLS Algorithm of Passive RFID Tag, Doctor's Degree, Chonnam National University.

[5]   Uisang Ahn (2013) Implementation of UWB RTLS System, Master's Degree, Kwangwoon University.

[6]   Gyeongguk Lee, "Real Time Location System (RTLS) based on Chirp Signal", Information Processing Society Journal: 2008.

[7]   Orthotron, "Chirp 2410 Transceiver" Manual.

[8]   Ho-Hung Jung, Changlong Li, Yang – Ok Ko, Key-Seo Lee, "Study on Precise Positioning using Hybrid Track Circuit System in Metro" JKIECS Vol. 8 No. 3, pp. 471-477; 2013.

[9]   Deung-Ryeol Yoo, Kyeng-Hwan Hwang, Key-Seo Lee, "A Study on Architecture Design of Output Module for SIL4 Safety Related System", JKIECS Vol. 10. No. 10, pp. 1079-1086, 2015.

[10] Young-Soo Park, Key-Seo Lee, "A Study on Development of Interlocking Inspection System for Electronic Interlocking System", JKSR Vol. 5, No. 2, pp 104-111, 2002.

19

Design Optimization of Amazon Robotics

Jun-tao Li[1], Hong-jian Liu[2]

[1]School of Information, Beijing Wuzi University, Beijing, China
[2]Graduate Department, Beijing Wuzi University, Beijing, China

Email address:

ijuntao@bwu.edu.cn (Jun-tao Li), 522927485@qq.com (Hong-jian Liu)

Abstract: In the era of e-commerce, the logistic distribution center is put in the center role of order picking for the sake of meeting the needs of different customer orders, hence, improving the automation and work capacity of distribution center becomes research priority in the fields of logistics and warehousing. The objective of this article is to-solve the shortcomings of currently traditional distribution center picking system with high automation by introducing a new method of picking provided by Amazon's Kiva system, that is, mobile racking with goods is broke out to sorting table by Kiva mobile robots named AGVs (automated guided vehicles), which could not only reduce the walk time and labor cost, but improve efficiency. This article starts with the constitutive requirements of the Amazon KIVA robotic systems, and then some key problems of Kiva picking system and design optimization about task allocation and path planning of multi-robots are demonstrated respectively in this article. Finally, the content is summarized and the application of robotic system in is simply demonstrated and prospected.

Keywords: Amazon Kiva Robotics, Picking System, Design Optimization

1. Introduction

In the operation procedure of distribution center, picking is considered as the primary field to improve operation efficiency of warehousing as it occupies plentiful funds which accounts for fifty percent of overall distribution center operating costs. As one of the largest multinational E-company, Amazon has created a new logistics model that could be called "shelf to the people" by using Kiva robotic system during the US holiday season last year. In Massachusetts, a city of America, Quiet Logistics is located as a big Operations Center with area of 25,500 square meters [1]. In this distribution center, there are 200 robots to help workers deal with 10thousands to 20 thousand order items per day. With the help of these robot, the overall efficiency of picking system would improve by 3.5 times to 5 times. In this new picking system, robots search for the mobile racking full of products based on the bar code and then move to the front of the review packager directly, then items required on this order were chosen by a review packager to make secondary sorting and packaging review, etc. Amazon employees just standing in a fixed point to complete the whole work, that is to say, the timing cost of pickers spent

walking could be reduced close to zero in Kiva system. Kiva robotic system could not only reducing unnecessary labor costs, but its thin fuselage and highly efficient route planning also increase the storage space by reducing the passage area to improve throughput of the overall system. Thus, the emergence of Kiva robots better meets the needs of E-company to efficiently handle a large number of diversified orders effectively and relieve the order delivery pressure of s that the electricity commercial companies faced with during the holiday season such as "Black Friday" [2].

Compared with the traditional distribution center, Amazon's Kiva system design a new type of multi-robot system for order picking instead of ATS (automatization tridimensional storehouse), long distance belt conveyor, Carousel, etc. Kiva robot is only 76 cm long, 64 cm wide and 41 cm high. Because of its own limitations, the mobile rack is designed very small (5 layers) and each layer could only put 2-3 items. Kiva robot could carry a ton of items in spite of its small size. When order signal is received, the robot was sent to ship the mobile rack with required items to the picking zone [3]. These robots could even find their charging stations to replenish their energy when they are about to run out of energy. These robots could work in an orderly manner without collision, mainly due to the manipulation of the software [2]. Therefore, time cost of the robotic system could significantly be reduced and picking efficiency could be improved by optimizing the allocation of freight tasks and their work and return path through the information system.

At present, There is more attention on the design optimazation of traditional distribution center picking system with high automation [4], but a little research about kiva robot at home and abroad, and kiva robot algorithms has not been public, therefore, this article will focus on the design optimization about task allocation and path planning of multi-robots, which represents some theoretical value.

2. Design of Amazon Robotics

Amazon Robotics is designed for the sake of meeting growing consumer demands in e-commerce by using a better system to provide order fulfillment solutions. As the wholly owned subsidiary of Amazon.com, Amazon Robotics makes it more convenience, more effective, faster in e-Commerce order real-time processing through automation. Most advanced robotic technology is used to achieve highly automated distribution center operations. These methods includes autonomous mobile robots, sophisticated control software, language perception, power management, computer vision, depth sensing, machine learning, object recognition, and semantic understanding of commands.

In particular, a complete robotic picking system (Goods to person or Goods to man, G2P or G2M) consists of three parts [5]:

2.1. The Back end Workstation System

This system uses distributed software / hardware

architecture, firstly, information about items and suppliers is stored in the back end system in each warehouse. This kind of warehouse storage is generally in a random fashion, and SKU is placed in a appropriate position in storage according to the frequency of shipments of each SKU, in the meantime, high-speed real-time storage is realized; secondly, order processing is accomplished by connecting with ERP (Enterprise Resource Planning) and WMS (warehouse management system).

The best route of picking is automatically planned when the order information arrives to autonomous mobile robots and then the most appropriate picking station is automatically searched, in the same time, the order processing sequence is also calculated automatically; afterwards, adequately scheduling and dispatching robot resource becomes a notable feature to make a plurality of robots running at the same time work in an orderly space and try to avoid the queues, cross-collision and other conditions, while the charging time is scheduled according to assignments of each robot.

2.2. The High-Speed Mobile Communication System

The high-speed mobile communication system is used to guarantee stable wireless network in the warehouse and also to avoid the "offline hour" during the picking work.

2.3. Autonomous Mobile Robots

It is generally believed that the height form factor is an important factor to consider in the design of robots. Robots is considered that it would run more stable if it was designed shortly, as a result, Kiva robots are 40cm high. Kiva's autonomous mobile robots apparatus comprising:

Table 1. *Design of Amazon AGV.*

Designation	Function	Weakness
Lifting Device	When the autonomous mobile robots reaches to the bottom of the mobile rack, it could lift the mobile rack off the ground through the screw device and send it to the picker.	The speed is limited due to its high center of gravity of the moving objects, so the robot vehicle need to pay special attention to slow up the suspension when it start or stop.
Close Sensing System	The sensing system, or collision detection system contains infrared sensor systems and collision- avoidance system that could quickly detect other objects including human beings appeared around the robot, and also can quickly stop the robotic car in order to avoid a collision, which would make the whole system collapses, and minimize any risk to its employees.	The collision detection system could only stop the automatic guided vehicle to avoid collision rather than choosing another route.
Navigation System	Kiva robots are equipped with two cameras（as shown in Fig. 2）, one is designed to read the bar code at the bottom of mobile rack in order to help robotic cars to make information identification, and another one is designed on the bottom of the robotic car for the sake of helping the car to read a bar code on the road to identifying the route of running.	It is inconvenient that robots could only move in the warehouse according to the fixed trajectory for Amazon warehouse needs frequently moving shelf.
Charging System	Automated guided vehicle electronic device is powered by four composition lead-acid batteries. When the car battery is low, the car will automatically go to the charging station for charging.	Robots maybe lined up for the charging which would decrease the efficiency of the whole system.
Traveling System	In the bottom of both sides of the trolley mounted rubber wheels, its movement is controlled by two independent DC motors	Amazon chose the more simple differential scheme structure, other driven wheel balance is needed in addition to the drive wheels, otherwise the vehicle will be down.
Control System	Control system is responsible for guiding the robotic car make path selection, walking and trolley loading and unloading operations and other functions.by connecting with other systems, robotic cars could operate as effectively as possible	

Figure 1. Navigation System of Amazon Robotics.

Figure 2. Kiva robots are equipped with cameras.

3. Key Issues in Kiva Robot Operation

The most advanced robot design of Amazon's Kiva picking system is that it can run accurately by following the identification code on the ground in distribution center, so as to avoid mistakes caused by collision between each other. At the same time, robotic cars are designed that there would be no more than 10 a robot fails at the same time in each layer of the warehouse; therefore, the working accuracy of the Amazon robots can reach to 99.99% [6], which could greatly reduce the probability of human error. When kiva robot receiving orders of picking, they will line up under setting rules and came to picking area and then pickers would remove the goods. Finally, Kiva robots follow the path back to the storage area to wait for the next picking work.

Kiva system perfect make the connection between the new and advanced artificial intelligence technology and logistics automation technology, and it also make the integrated use of computer systems, robotic car, the shuttle shelves and workstations and other basic tools to achieve automation of picking process in distribution center. And specifically, the Kiva robot operation mode is summarized as follows:

(1) Receive and process orders

These Kiva robots are both fast and quiet in positioning the aimed mobile rack where contains the items in the picking order after it receiving the picking order information, and the picking orders are assigned to different order picking stations. After these robots received the wireless transmission of

digital commands from central computer, by scanning a bar code label on the ground through the sliding shelf to the ground, then 1.2m wide, 340kg heavy shelve is lifted [7]. Amazon's picking robots are equipped with the advanced visual system that can use bar codes to track every product on the shelf, at the time fetch the appropriate shelves when order pickers coming to.

(2) Complete tasks and get back

When the picking is finished, the information of finishing tasks would be sent to the system and then the system determines the new storage location of mobile rack based on the current overall situation and sent the results to the robots to select return routine, and then the robotic car will return back corresponding to the storage location. This position is selected is the result of re-calculation under normal circumstances as a reason why it is different from the previous shelf position.

(3) Be released or charged.

The robot will be released and continue to accept new tasks after it placed mobile rack back to the storage area [8]. Meanwhile, during the system operation time, the robot remaining charge will also be real-time monitored and low battery robotic car will be guided to a charging station. When charging is completed, robots continue to return to the system operation.

4. Design Optimization of Kiva Robot

Kiva robot complete the order task in four steps which is shown in Figure 2.

Step 1. Kiva robots receive orders, and go to the position of the mobile rack under selection.

Step 2. Kiva robots transport the mobile rack from the original location to the site into a queue and to prepare to be selected.

Step 3. Orders are satisfied by picked, packaged and re-checked by Pickers

Step 4. The mobile rack which has been picked would be sent back to a now storage location by Kiva robots.

As can be seen from figure 3, design optimization of Amazon's kiva robot could become a reality in the following three aspects:

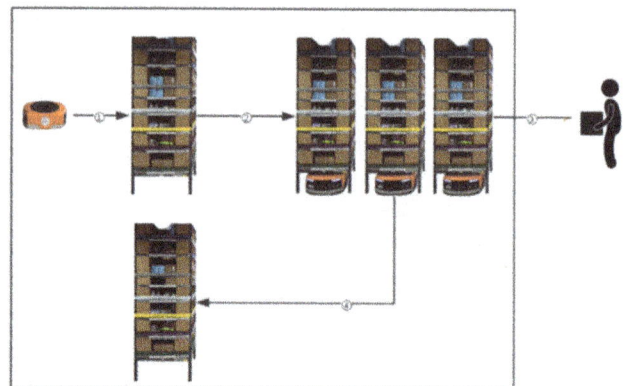

Figure 3. Four steps of Kiva robot complete the order task.

4.1. Task Allocation of Multi-Robots

In distribution center applied Kiva system, automatic mobile robots should be designed aimed to avoid robot in spare time to ensure the robot to perform tasks in the full time [9]. These robots in a time window could have three different kind of states: ① performing tasks; ② be released into the next round when task is finished; ③ Waiting for performing task. In this optimization, the robots only have the first two states in a perfect working environment that are performing tasks and be released into the next round. Therefore, the optimization of multi-robots task allocation can effectively reduce time cost in step one, and then improve the working efficiency of the entire system.

In Kiva Systems, mobile racks are arranged in the area of storage in a regular way to prepare to be available for robots to find by its definite position, coordinates, and bar code [10]. Task allocation of multi-robots is optimized by comprehensively considering arrival sequence and position of robots under the mobile rack as well as the status of each robot, and robotic cars match the nearly mobile rack in task in the system:

Assume that within a window of time for the process of order picking, during this time, the task store shelves in order of arrival of the waiting shelf, waiting for the robotic car to come to work. Mobile rack that to be processed are considered as set S1 {M1, M2, M3, M4, M5...}, while robot at rest in the current time as set S2 {R1, R2, R3, R4, R5...}. The current positions of each Mobile rack that to be processed and robotic car at rest in S1 and S2 are known. Consider arrival sequence and position of robots under the mobile rack as well as the status of each robot to make the task allocation. the specific method to Determine the task allocation is to consider the Manhattan Distance between not allocated mobile rack with all current free AGVs, and the nearest AGV whose Manhattan Distance is smallest would be chosen to suit for the not allocated mobile rack and then update the set S1 and set S2 in this system, and then the next task sequentially selecting a free trolley shelf in accordance with the method described above.

For example, assuming orders in a time window has been processed, the task mobile rack to be picking are lined up in the order, that is M1, M2, M3, M4, M5…., From the first one to the last, M1 come to the first to be selected the robotic car in free. Assuming that there are 5 automated robotic vehicle, R1, R2, R3, R4, R5, the nearest car is chosen to perform this task shelves by comparison task Manhattan distance between M1 and robotic vehicles. If the R2 is calculated to be the nearest one in Manhattan Distance, it would be matched with the task of M1, that is to say, task allocation is completed between M1 and R2. When there are more than one robotic cars have the smallest Manhattan Distance with M1, there will be a short time judgment of the car number and the smallest number would be selected by the controlling system to perform the current task. The number of elements in set S2 will be updated and then mobile rack continue to be allocated. All the mobile racks are assigned task after end of work.

This type of task allocation model gives full consideration about the sequence of mobile rack in task, that is priority of shelf and it also consider the current form for all robotic vehicle in the system. The task allocation is finished based on the nearest Manhattan Distance between robotic cars and mobile racks which make the picking system performs tasks with a shorter time and smaller walking path.

4.2. Path Planning of Multi-Robots

After completing picking task, the robot needs to carry the mobile rack to shelf storage area. In the process of return, new storage location is produced with following different path of return. By considering and rank the frequency the mobile racks are chosen, the higher frequency mobile rack would be placed in a closer position to the platform while the low frequency mobile rack would be placed in a further location that away from the site, and picking tasks are finished through the information systems by real-time monitoring and real-time adjustments.

The design optimization of the Path planning of multi-robots would not only to further enhance the Kiva robot's walking efficiency in the operation process, but also to provide a lot of room for improvement for the path optimization in the next process [11].

5. Summary and Prospect

In the robotics ecosystem, Amazon Robotic comes up with the new automated picking solutions. In this picking system Kiva robots can automatically arrive to mobile rack and lifted it to the picking station according to the order needs and inventory information. Pickers are prompted for which items in the order should be picked from the mobile rack; all of his work is reaching for items without walking, which greatly enhance the level of automation of distribution centers. At present, Google, Apple and other technology giants have set foot in the field of robotics [12]. Currently, China's e-commercial enterprises have also made efforts in robotic innovation to follow and bring cutting edge technology into the logistic field. For instance, About a dozen of Geek + robots are put into use in Tmall (Tianjin) warehouse smart picking area of 2000 square meters in year 2015 during the holiday season. It is believed that there are greater possibilities of optimization and a better application foreground.

Acknowledgements

This study is supported by intelligent logistics system Beijing Key Laboratory (NO: BZ0211) and Beijing Intelligent Logistics System Collaborative Innovation Center.

References

[1]   News & Trends. "Amazon is equipped with robots to meet holiday season". World vision, vol. 23, pp. 36-37, 2014.

[2]   Industry dynamics. "Amazon use robots handle orders". Journal of Robot Technique and Application, vol 1, pp58-60, 2015.

[3]   X. M. Zhang. "The order picking optimization and algorithm research based on Kiva system," Master Thesis, Beijing: Beijing University of Posts and Telecommunications, 2015.

[4]   D'Andrea R, Wuman P. Future challenges of coordinating hundreds of autonomous vehicles in distribution facilities//Techonologies for Practical Robot Applications, 2008. TePRA2008. IEEE International Conference on. IEEE, pp 80-83, 2008.

[5]   J. P. Wu. "Analysis and prospect of Amazon kiva robot application," Journal of Logistics & Material Handling, vol. 10, pp 159-162, 2015.

[6]   M. Zamirian, A. V. Kamyad, M. H. Farahi. "A novel algorithm for solving optimal path planning problems based on parametrization method and fuzzy aggregation," Journal of Physics Letters A. 2009 (38).

[7]   P. Klaus. Logistics research: a 50 years' march of ideas. Logistics Research. 2009.

[8]   J. Enright, P. R. Wurman. "Optimization and Coordinated Autonomy in Mobile Fulfillment Systems. Automated Action Planning for Autonomous Mobile Robots," 2011.

[9]   S. Qi. "Study on path planning of AGV system," Master Thesis, Zhejiang: Zhejiang University, 2012.

[10] R. Peter. Wurman, Raffaello D'Andrea, and Mick Mountz. "Coordinating hundreds of cooperative, autonomous vehicles in warehouses," Journal of AI Magazine, vol 29, no. 1, 2008.

[11] Ruey-Jer "Bryan" Jean, R. R Sinkovics, D. Kim. "The impact of technological, organizational and environmental characteristivs on electronic collaboration and relationship performance in international customer-supplier relationships," Journal of Information and Management, 2014.

[12] T. Frank, C. Q. Zhang. "To the robot era of apple, amazon and Google," Journal of World Science, vol 3, pp 53-54, 2014.

20

Evolutionary Model for Virus Propagation on Networks

Arnold Adimabua Ojugo[1], Fidelis Obukowho Aghware[2], Rume Elizabeth Yoro[3],
Mary Oluwatoyin Yerokun[4], Andrew Okonji Eboka[4], Christiana Nneamaka Anujeonye[4],
Fidelia Ngozi Efozia[5]

[1]Dept. of Math/Computer, Federal University of Petroleum Resources Effurun, Delta State, Nigeria
[2]Dept. of Computer Science Education, College of Education, Agbor, Delta State, Nigeria
[3]Dept. of Computer Sci., Delta State Polytechnic, Ogwashi-Uku, Delta State, Nigeria
[4]Dept. of Computer Sci. Education, Federal College of Education (Technical), Asaba, Delta State, Nigeria
[5]Prototype Engineering Development Institute, Fed. Ministry of Science Technology, Osun State, Nigeria

Email address:
arnoldojugo@yahoo.com (A. A. Ojugo), ojugo_arnold@yahoo.com (A. A. Ojugo), aghwarefo@yahoo.com (F. O. Aghware),
rumerisky@yahoo.com (R. E. Yoro), an_drey2k@yahoo.com (A. O. Eboka), agapenexus@hotmail.co.uk (M. O. Yerokun),
anujeonyechristy@gmail.com (C. N. Anujeonye), fenngo31@yahoo.com (F. N. Efozia)

Abstract: The significant research activity into the logarithmic analysis of complex networks will yield engines that will minimize virus propagation over networks. This task of virus propagation is a recurring subject and design of complex models will yield solutions used in a number of events not limited to and include its propagation, network immunization, resource management, capacity service distribution, dataflow, adoption of viral marketing amongst others. Machine learning, stochastic models are successfully employed to predict virus propagation and its effects on networks. This study employs SI-models for independent cascade and the dynamic models with Enron dataset (of e-mail addresses) and presents comparative result using varied machine models. It samples 25,000 e-mails of Enron dataset with Entropy and Information Gain computed to address issues of blocking, targeting and extent of virus spread on graphs. Study addressed the problem of the expected spread immunization and the expected epidemic spread minimization; but not the epidemic threshold (for space constraint).

Keywords: Stochastic, Immunize, Network, Vertices, SIS, SIR, Search Space, Solution, Models

1. Introduction

Networks are dynamic and their normal operation is continually threatened by unethical users referred to as hackers. They employ use of harmful and malicious programs called malware to wreak havoc to users of networks. Today, Internet has become a high target for the spread of such malwares – as hackers do damage globally, much more easily and faster. Thus, its early detection is imperative to minimize damage caused by it (Desai, 2008).

Malwares are known to attach copies of itself, alters the behaviour as well as modifies attributes of its host machine's files without user's knowledge (Szor, 2005). Malwares also can sometimes, modify their codes as they infect, to include an evolved copy (Dawkins, 1989). Malwares are grouped into simple, encrypted, polymorphics and metamorphic viruses. Malware are considered malicious software if they

consist of codes, scripts, active contents and other software – designed to disrupt or deny operations, gather information that tends to loss of privacy or exploitation, gain unauthorized access to system resources, and other abusive behaviour (Singhal and Raul, 2012). Thus, software codes are considered a malware based on the perceived intentions of its creator rather than any particular feats.

Antivirus (AVs) is designed to detect, prevent and remove malwares such as viruses, worms, Trojans, spyware and adware. AVs detection mechanism are broadly grouped into: (a) signature-based scans for signature, and to evade it – virus makers create new virus strings that can alter their structure while keeping its functionality via code obfuscation, and (b) code emulation creates sandbox so that files are executed within it while scanning for virus. If virus is detected, it is no threat as it is running in controlled environment that limit damage to host machine (Singhal and

Raul, 2012).

AVs can often impair system performance as any incorrect decision may lead to security breach since it runs at the operating system's kernel. If an AV uses heuristics, its success depends on a right balance between positives and negatives. Today, virus may no longer be executables but macros, which present security risk and antivirus heavily relies on signature-detection. Some viruses evade detection effective (Filiolel, 2005) via code obfuscation and encryption methods. Studies show the best AVs can never yield a perfect detection since all scanners can yield a false positive result and identify benign files as malware (Bishop, 2005).

The idea is to model a network using graph theory with emails as nodes. It implements SI (susceptible-infect) design for virus propagation on graph with a view of immunization problem that helps to deal with extent, targeting and blocking of virus propagation on a network. SIR model with independent cascade is used as it aid inoculated nodes stay completely invulnerable to viral attacks.

1.1. Network Topologies

Networks are used for spreading of data – making it easier for users to disseminate useful data as well as viruses. The problem of virus propagation has been a recurring subject and ongoing research notes that every harmful data spread over such networks are considered as malware or viruses as can be interchangeably used; while the process of impeding the spread of such harmful data (malware) over such social network is referred to as network immunization. This aims to prevent the spread of such malwares, protect such networks from virus attacks and control data and sensitive information leakages – while at same time noting that our resources such as vaccination, AVs and influences are quite costly and limited in their capability to discover such malware. With such AVs and vaccinations, users aim to achieve the best effect; while still allocating the least resources possible (Ojugo et al 2013).

Hackers (or adversary) wreak more havoc being aware of the propagation model used to avert such attacks. In simplest form, a social network is seen as a complex graph. Thus, the propagation model has as input a graph G = (V,E), state vector $S_v^{(t)}$ for each node vertex v ∈ V at t, and an internal parameter vector P. Based on the states of all interacting nodes, it outputs a new state vector $S_v^{(t+1)}$ for each node at t+1. Models are applied to synthetic data with graph types (Mitchell, 1997; Giakkoupis et al, 2010; Kermach and McKendrick, 1927; Pastor-Satorras and Vespignani, 2002) as:

a. Scale-Free Networks: Probability that node x in network is of degree k is proportional to $k^{-\gamma}$ with γ > 1. Scale free graph are modeled as by Barabasi and Albert (1999). It inserts nodes sequentially with each node linked to an existing one chosen with a probability proportional to its current degree in a tree-fashion with grandparent-parent-children-grandchildren structure and it builds graph with exponent γ = 3 denoted with G_{sf}. Each node in the graph can be autonomous but must be connected to an existing one. Thus, two nodes are connected together on the graph via physical link

between two corresponding autonomous systems. Such is referred to as Autonomous Systems. Another scale free graph consists of undirected edges between nodes, also termed Co-Author graphs.

b. Small World Networks are those with small characteristics path length L (the average shortest path between any pair of vertices) and large clustering coefficient C (the average fraction of pairs of neighbours of a node also connected to each other). We generate small-world graphs using the generating model proposed in Watts (1999). G_{swL} denotes small-world graphs with path length feat; while G_{swC} to denote those of large clustering coefficient.

Graphs G_{swL} are influenced by α, which intuitively determines the probability of two nodes being connected given a number of their common neighbours. It controls to what extent the graph has small and densely connected components. As α nears infinity, G_{swL} becomes a random graph. Conversely, graphs of G_{swC} are influenced by q, which determines the probability of an edge in the lattice being rewired to connect to a random node in the graph. Thus, initialized on a ring lattice, each node is of degree k. Small values of q entails G has high clustering coefficient and large average path length; while large values q creates random graphs. For q-values close to 0.01, the generated graphs are small-world graphs. Note that G_{swL}, G_{swC} and G_{sf} are quite distinct graphs.

1.2. SI-Models for Epidemic Spread

Two major models are: Susceptible-Infect-Remove (SIR) and Susceptible-Infect-Susceptible (SIS). In SIR, a node may be in any of these states: (a) susceptible: if the node has no virus but will become infected if it is exposed to it, (b) infected: if the node has the virus and can pass it on to others, and (c) removed: if the node had the virus but has been recovered or virus dies. The node is permanently immunized and can no longer participate in propagation, and a particular node cannot be infected twice. Conversely SIS, a node can be cured but not immunized. Thus, it can be infected again. Such node switches between susceptible and immunized.

Giakkoupis et al (2010) and Lahiri and Cebrian (2010) A graph holds these definitions as true:

a. Network is directed or undirected graph for propagation of virus. A node is represented as v ∈ V; and edge (u,v) ∈ E represents interactions between two individuals or nodes in the system. It also assume that the graph is drawn from a specific family (algorithm consider all possible graphs). For G = (V,E) as a dynamic network, E is set of edges that are time-stamped, $(u,v)_t$ ∈ E are interactions at t ∈ Z^+. In a typical SI setting, set of nodes are initialized as *activated*. The propagation process proceeds in discrete time-steps such that at each time-step, an activated vertex may come in contact with *inactive* vertices. This continues till a stop criterion is satisfied or there are no more inactive vertices.

b. The virus propagation model that determines how the virus is spread on the network.

c. Immunization model aims to minimize viruses spread and an immunized node cannot transfer or receive a virus. It is conceptually removed from graph. Cost of immunization model is, number of nodes immunized.

d. Adversary with knowledge of the propagation model, plants d copies of virus in G, to maximize speed of spread is denoted as F_d. An adaptive adversary is one who has knowledge of choices made by immunization algorithm; while a randomized adversary places copies of virus, uniformly at random on the network.

1.3. Independent Cascade Model

It is a discrete-time special case SIR model in which at t = 0, an adversary inserts d copies of virus to some nodes on graph. If node x is infected the first time at t, it has single chance to infect any neighbours y currently uninfected. Probability that x succeeds with y is P_{xy}. If x succeeds, y is infected at t+1; Else, x tries again in the future (even if y gets infected by another neighbour). This process continues and stops after n-steps if no more infections are possible. It requires a nodes stay infected exactly once and it is the independent cascade model following Kempe et al (2003). Graph of size M, has M_d subset of nodes and d copies of virus placed on the network. With propagation complete, $S(M_d,G)$ is expected number of infected nodes. Expectation exceeds all random choices made by propagation model. Eq. 1 is maximum expected number of infected nodes and maximum exceeds all possible initial virus placements.

$$S_d(G) = \max_{M_d} S(M_j, G) \qquad (1)$$

$A_d = arg \max_{M_d} S(M_d, G)$ corresponds to choices made by an adaptive adversary. $S_d(G)$ is epidemic spread in G. A similar definition of epidemic spread of randomize adversary is Eq. 2, that defines expected epidemic spread as where the expectation takes over all possible positions of the d viruses placed on the network and given by:

$$S'_d(G) = E_{M_d}[S(M_d, G)] \qquad (2)$$

1.4. Dynamic Propagation Model

In SIS, viruses are seen as dynamic birth-death process that evolves overtime. It continues to either propagate or eventually die. An infected node x spreads virus to node y in time t with infection rate of $\frac{\beta}{\delta}$ and probability β. At same time, an infected node may recover with probability δ. With adjacency matrix T, $\lambda_1(T)$ is largest eigen-value of T. The condition $\frac{\beta}{\delta} < \frac{1}{\lambda_1(T)}$ holds true as epidemic threshold and is sufficient for quick recovery, easily proven (Ganesh et al, 2005; Wang et al, 2003).

2. Statement of Problem

While networks are an effective way to spread data, they also help with spread and propagation of malwares and viruses. These have significant implication on the network as it can destroy user data and/or become a means to retrieve useful, confidential data from unsuspecting users. It thus becomes imperative to deal with means that help user curb the spread of viruses on networks.

3. Network Immunization Problem: Proposed Framework / Design

Typical challenges in SI propagation model are:

1. Extent: With specific subset of initially activated vertices in network and propagation model used, how many vertices are expected to be activated after a specific time/period?

2. Targeting: Which vertices are targeted as initiators by an adversary to result in max extent of spread (Cohen et al, 2003)? This is a hard NP to solve optimally, regardless of propagation model used (Kempe et al, 2003)

3. Blocking: Which vertices are targeted for immunization to minimize the expected number of activated vertices (Singhal and Raul, 2012; Dezso and Barabasi, 2002)?

The immunization problem is thus defined as thus:

Problem 1 – Spread Immunization: Given graph G, a number of d initial copies of viruses and number k. We immunize k nodes in G such that expected spread $S_d(G')$ in the immunized graph is minimized. The role of the adversary is played by the influence-maximization model of Kempe et al (2003), whose proof is omitted due to space constraint.

Problem 2 – Expected Epidemic Spread Minimization: Given graph G, a number of d initial copies of viruses and a number k. We immunize k nodes in G such that the expected epidemic spread $S'_d(G')$ in the immunized graph is minimized. As a hard NP-complete task that attempts to immunize G with random strategy for influence spread and closely related to the sum-of-squares partition task as studied in Aspnes et al (2005).

Problem 3 – Threshold Maximization: Given G, a number of d copies of viruses and an infection rate of $\frac{\beta}{\delta}$, we immunize the minimum number of k nodes in G so that $\frac{\beta}{\delta} < \frac{1}{\lambda_1(T)}$ holds true. Thus, the epidemic spread $S'_d(G')$ in the graph is minimal. The task attempts to immunize G with influence spread while seeking the minimal number of nodes that can be immunized.

4. Experimental Framework

Machine learning as a branch of artificial intelligence is a scientific discipline that deals with development and design of algorithms that allows machines (computers) to evolve its behaviour based on empirical data such as sensors data and databases. A learner takes advantage of data to capture its characteristics of interest of their underlying and unknown probability distribution. Such data may illustrate relationships between observed variables. Major focus on machine learning is to automatically learn to recognize complex patterns and make intelligent decisions from it (Singhal and Raul, 2012).

4.1. Dataset

The Enron e-mail dataset is one of the largest, e-mail dataset available, representing a dynamic social network. Each node in network is an e-mail address with a directed time-stamped edge as e-mail sent between two addresses. Lahiri and Cebrain (2010) obtained all e-mail headers of 1,326,771 time-stamped e-mails from 84,716 addresses and 215,841 unique timestamps as non-uniformly covering a period of approximately 4years.

We sampled a subset of 25000 addresses representing about 30% for the graphs G_{sf}, G_{swL} and G_{swC} families. In all cases, we used p = 0.25, q = 0.009 and α = 6 to generate the graphs. These result in models' graph having low average path length and high clustering coefficient. There exists the relationship between parameters (p, q and α) and the clustering coefficient as studied in (Watts, 1999). α starts with value 1 till it reaches 6. The clustering coefficient drop as α increase and for small values of q, high clustering coefficient is observed while clustering coefficient drops as q tends to 1.

4.2. Genetic Algorithm (GA)

GA is inspired by Darwinian evolution (survival of fittest) and consists of a pool of solutions chosen for natural selection to a specific task. Each potential solution is an individual for which an optimal is found via four operators namely: initialize, select, crossover and mutation (Coello et al, 2004). Individuals with genes close to its solution's optimal is said to be fit, and the fitness function determines how close an individual is to the optimal solution.

Theorem 1: With G, adjacency matrix T and infection rate $\frac{\beta}{\delta}$. $\frac{\beta}{\delta} < \frac{1}{\lambda_1(T)}$ is true, if expected time for virus to die is logarithmic. This is a function of the number of nodes in the graph against an adversary. Many interesting families of graphs holds too that $\frac{\beta}{\delta} > \frac{1}{\lambda_1(T)}$ is recovery rate – so that expected time at which virus dies out is exponential, known as Epidemic threshold.

GA achieves its fitness function as it finds solution to the network. Its dynamic, non-linear model can be made linear so as to resolve it analytically. The dynamic nature of graph as social network makes them impossible to resolve analytically using non-linearity (if considered as a multiple copies model). Let v^t be an n-dimensional vector of states at t-steps and v_d^t is number of virus copies at node x at t-steps. Initialized at t = 0, v_d^0 is number of d copies planted by an adversary. At t+1, the model evolves for (all) nodes x,y,z in the network, and for each v_d^t copies of virus planted at node x, virus is propagated to node y with probability β. Virus dies with probability 1 − δ, and if Δ = βT + diag(1 − δ,..., 1 − δ) is true, v^t is the expected state of system at time t. Then, model is completely linear if $\Delta v^t = \Delta v^{t+1}$ proven as in (Giakkoupis et al, 2010; Kempe et al, 2003; Kleinberg, 2007 and Hethcotee, 1989).

For GA, operators are (Lahiri and Cebrian, 2010):
a. Initialize/Select: For edge (u,v) at time t, let its corresponding state string be coded as $S_u^{(t)}$ and $S_v^{(t)}$ vectors respectively, which are interactions between the nodes. Thus, we select $S_u^{(t+1)} = S_u^{(t)}$ and $S_v^{(t+1)} = S_v^{(t)}$.
b. A crossover point C is randomly and uniformly selected from the interval [1,β]. Two new states strings or vectors are created by swapping the tails $S_u^{(t)} and S_v^{(t)}$ - where tail is defined by all positions including and after the index C. let these two new vector states strings be denoted as st_1 and st_2 respectively.
c. Objective score of each new state vector is then evaluated according to the fitness function f(x). if any of them have a greater fitness value that either of their parent node, the corresponding parent nodes state vector string is replaced by its offspring for the next iteration, achieved via:

$$S_u^{(t+1)} = \text{argmax}_{x \in \{S_u^{(t)}, S_u^{(t+1)}, st_1, st_2\}} f(x) \qquad (1)$$

$$S_v^{(t+1)} = \text{argmax}_{x \in \{S_v^{(t)}, S_v^{(t+1)}, st_1, st_2\}} f(x) \qquad (2)$$

In the case of ties in fitness score between original and a new string vector, its original string vector is retained – as the offspring cannot outperform its parent. This model is close to GA with spatially distributed population GASDM (Min et al, 2006; Payne and Eppstein, 2006) – except that the GA's selection operator is replaced with real social network data that dictates the sequence of mating operation. The propagation in GASDM occurs as states vector and are modified using crossover. After which, they are subsequently adopted based on fitness value. Major missing components to add meaning to this mapping is the choice of its fitness function, f(x).

Study proposes that the objective/fitness function be achieved via Information Gain.

4.3. Decision Tree / IDA

It uses hill-climbing to search a space for optima. Once a peak is found, it restarts with another randomly chosen starting point (as such peak may not be the only one that exists). Its merit is simplicity with functions with too many maxima. Each random trial done in isolation helps immunize the nodes and overall shape of the domain is transparent to an adversary – because, as random search progresses, it continues to allocate its trials evenly over the space and evaluates as many points in the both regions found with low- and high-fitness values. Its choice is in selecting feats and attributes in graph to test is via information gain at each step while it grows the graph. The algorithm as Mitchell (1997) and Ojugo et al, (2012) is thus:

1. DT (Examples, Target_Attribute, Attributes)
2. //Data Attributes are feats to be tested. Target_Attribute are
3. //values predicted. Return is decision to correctly detect Example
4. Create a Root node of Graph
5. If Examples are positive, Return single_node Root with label = +
6. If Examples are negative, Return single_node Root with label = -

7. If Attribute is empty, Return single_node Root, with label = most
8. common value of Target_Attribute in Examples
9. Otherwise Begin
   a. A ← attribute from attributes that best* classifies Examples
   b. The decision attribute for Root ← A
   c. For each possible value v_i, of A,
   d. Add new branch to G below Root, corresponding to A = v_i
   e. Let Examples v_i be subset of Examples with value v_i for A
   f. IF Examples v_i = empty
   g. THEN add leaf to new branch with label = most common
   h. value of Target_Attribute in Examples
   i. Else below this new branch, add the subtree
10. IDA(Examples v_i, Traget_Attributes, Attributes − {A})
11. End
12. Return Root

Random Forest Algorithm as a decision tree predictor in which each individual is trained on partially, independently sample set of instances selected from the complete training dataset. Predicted output of a classified instance is the most frequent class output of the individual trees (Szor, 2005; McGraw and Morrisett, 2002; Mitchell, 1997).

Bayesian Belief Model describes probability distribution of a set of nodes on G by specifying a set of conditional independent assumptions along with a set of conditional probabilities. Thus, allows stating conditional assumptions that applies to a subset of nodes on the network by providing an intermediate and more tractable solution unlike Naïve Bayes that applies to each instance that assumptions of each graph attribute values are conditionally independent of the target value. Thus, the assumptions is that given target value of an instance, the probability of observing the interactions between nodes in the graph is the product of their probabilities from the individual attributes (Szor, 2005; Alpaydin, 2010; Mitchell, 1997, Harrington, 2012).

Entropy characterizes impurity of an arbitrary collection of nodes on G, which contains both activated (infected) and inactive (uninfected) node. The Entropy is a Boolean classification given by:

$$Entropy\ (E) \equiv -p_\oplus log_2\, p_\oplus - p_\ominus log_2\, p_\ominus \quad (3)$$

Sample consists of n=25000 e-mail address from which we have normal/infected nodes to form G. Normal (inactive/p+) = 20000, infected (activated/p-) nodes where adversary plants viruses p- = 5000. To compute Entropy, we have:

$$Entropy\ (E) \equiv -\frac{20000}{25000} log_2 \frac{20000}{25000} - \frac{5000}{25000} log_2 \frac{5000}{25000}$$

$$E \equiv [-(0.8)log_2\,(0.8)] - [(0.2)log_2\,(0.2)] = 0.0775 + 0.1398$$
$$= 0.22$$

Information Gain is the expected reduction in entropy caused by partitioning the network according to its attributes (infected and uninfected) nodes. IG is info about target function value, given the value of another attribute A. IG of attribute (A) is given by Eq. 4. The Values(A) is set of all possible values of Attribute A, E_v is E subset of attributes A with value v. Our second is the expected entropy after partitioning with attribute A (sum of all entropies of each subset E_v weighted by fraction of $\frac{E_v}{E}$ of E_v).

$$Gain(E, A) \equiv Entropy(E) - \sum_{v \in Values(A)} \frac{|E_v|}{|E|} Entropy(E_v) \quad (4)$$

$$\equiv Entropy - \sum_{v \in \{inactive, immunized\}} \frac{|E_v|}{|E|} Entropy(E_v)$$

$$Gain(E, A) \equiv 0.220 - \left\{ \frac{8000}{25000} * 0.811 \right\} - \left\{ \frac{23000}{25000} * 0.921 \right\}$$
$$= 0.220 - \{- 0.587\} \equiv 0.220 + 0.58$$
$$= 0.807$$

Thus, we choose only top 80% of nodes that are most likely to be infected. IG is updated as below:

$$Gain(E, A) \equiv Gain(E, A) \pm \left[\frac{\sum_{i=0}^{n} Gain(X_i)}{n} \right]$$

4.4. Result Findings and Discussion

After training/testing, model results discovered that with the same amount of seed nodes (that is, viruses planted in the same number of nodes in this case, 5000nodes, on a network), the extent of the network that is blocked from virus attack is 22%; while 81% of the nodes are targeted before a complete network immunization is performed. However:

a. GA took 21seconds to find the solution after 98 iterations (best). CGANN was run 15-times and it found optima each time. Its convergence time that varied between 21seconds and 4 minutes and depends on how close the initial population is to the solution as well as on mutation applied to the individuals in the pool. The model is able to immunized 90% of the nodes before the virus eventually dies out.
b. IDA (at best) took 18seconds after 321 iterations. It was run 25 times and solution found each time on a range between 4seconds and 3minutes. In addition to the facts, the model is able to immunized 94% of the nodes before the virus eventually dies out
c. RFA arrived at solution 2.112seconds after 401 iterations. In addition to the facts as stated earlier on its extent and targeting, the model is able to immunized 97% of the nodes before the virus eventually dies out.

4.5. Rationale for Choice of Algorithms

The comparisons are as follows:
• Stochastic Model: are mostly inspired by evolution laws and biological population cum behaviors. They are heuristics that search a domain space for optimal solution to a task. They use hill-climbing method that are flexible, adaptive to changing states and suited for real-time application. GA guarantees high global convergence to

optimal point for multimodal tasks. It initializes with a random population, allocates increasing trials to regions of space found with high fitness and finds optimal in time. Its demerit is that they are not good with linear systems in that if the optimal is in a small region surrounded by regions of low fitness – the function becomes difficult to optimize.

- Gradient/Greedy Search: A number of different methods for optimizing well-behaved continuous functions have been developed which rely on using information about the gradient of the function to guide the direction of search. If the derivative of the function cannot be computed, because it is discontinuous, for example, these methods often fail. Such methods are generally referred to as *hill-climbing*. They can perform well on functions with only one peak (*unimodal* functions). But on functions with many peaks, (multimodal functions), they suffer from the problem that the first peak found will be climbed, and this may not be the highest peak. Having reached the top of a local maximum, no further progress can be made.

- Iterative Search is a combined random and gradient search that also employs an *iterated hill-climbing* search. Once one peak has been located, the hill-climb is started again, but with another, randomly chosen, starting point. This technique has the advantage of simplicity, and can perform well if the function does not have too many local maxima. However, since each random trial is carried out in isolation, no overall picture of the shape of the domain is obtained. As random search progresses, it continues to allocate its trials evenly over the search space. This means that it will still evaluate just as many points in regions found to be of low fitness as in regions found to be of high fitness.

5. Conclusion

Models have been successfully used today to determine epidemic spread of viruses. Many studies recently on the mathematical epidemiology is focusing on the analytic epidemic thresholds for varying propagation models and different families of network – seeking insight into the nature of such epidemic existence, its threshold and to unveil if such epidemic will continue to spread or eventually die out (Bougna et al, 2003; Barthelemy et al, 2005; Barabasi and Albert, 1999). Models serve as educational, predictive tools to compile knowledge about a task. They also serve as a new language to communicate hypotheses, investigate parameters crucial in estimation and help us gain better insight to a problem domain. Thus, their growth, development, sensitivity and failure analysis helps reflect on the theories and functioning of nature systems.

References

[1] Alpaydin, E., (2010). *Introduction to Machine Learning*, McGraw Hill publications, ISBN: 0070428077, New Jersey.

[2] Aspnes, J., Chang, K and Yampolskiy, A., (2005). *Inoculation strategies for victims of viruses and the sum-of-squares partition problem*. In *SODA*.

[3] Barabasi, A.L and Albert, R., (1999). *Emergence of scaling in random networks*. Science, 286, p23.

[4] Barthelemy, M., Barrat, A., Pastor-Satorras, R and Vespignani, A. (2005). *Dynamical patterns of epidemic outbreaks in complex heterogeneous networks*. Journal of Theoretical Biology, p54.

[5] Boguna, M., Pastor-Satorras, R and Vespignani, A., (2003). *Epidemic spreading in complex networks with degree correlations*. Statistical Mechanics of Complex Networks, p36.

[6] Cohen, R., Havlin, S and Ben-Avraham, D., (2003). *Efficient immunization strategies for computer networks and populations*. Phys Rev Letters, p232.

[7] Dezso, Z and Barabasi, A.L., (2002). *Halting viruses in scale-free networks*. Phys. Rev. E 66, p67.

[8] Filiol, E., (2005). *Computer Viruses: from Theory to Applications*, Springer, ISBN 10: 2287-23939-1.

[9] Ganesh, A., Massouli, L and Towsley, D., (2005). *The effect of network topology on the spread of epidemics*. In *IEEE INFOCOM*.

[10] Harrington, P., (2012). *Machine Learning in action*, Manning publications, ISBN: 9781617290183, NY.

[11] Kempe, D., Kleinberg, J and Tardos, E., (2003). *Maximizing the spread of influence through a social network*. In *SIGKDD*.

[12] Kermack, W and McKendrick, A., (1927). *A contribution to the mathematical theory of epidemics*. Proceedings Royal Society London.

[13] Mitchell, T.M., (1997). *Machine Learning*, McGraw Hill publications, ISBN: 0070428077, New Jersey.

[14] Newman, M.E., (2003). *The structure and function of complex networks*. SIAM Reviews, 45(2), p167.

[15] Ojugo, A., Eboka, A., Okonta, E., Yoro, R and Aghware, F., (2012). *GA rule-based intrusion detection system*, J. of Computing and Information Systems, 3(8), p1182.

[16] Ojugo, A.A., and Yoro, R., (2013a). *Computational intelligence in stochastic solution for Toroidal Queen task*, Progress in Intelligence Computing Applications, 2(1), 10.4156/pica.vol2.issue1.4, p46.

[17] Ojugo, A.A., Emudianughe, J., Yoro, R.E., Okonta, E.O and Eboka, A.O., (2013b). *Hybrid artificial neural network gravitational search algorithm for rainfall*, Progress in Intelligence Computing and Applications, 2(1), 10.4156/pica.vol2.issue1.2, p22.

[18] Pastor-Satorras, R and Vespignani, A., (2002). *Epidemics and immunization in scale-free networks*. Handbook of Graphs and Networks: From the Genome to the Internet.

[19] Singhal, P and Raul, N., (2012). *Malware detection module using machine learning algorithm to assist centralized security in Enterprise networks*, Int. J. Network Security and Applications, 4(1), doi: 10.5121/ijnsa.2012.4106, p61.

[20] Szor, P., (2005). *The Art of Computer Virus Research and Defense*, Addison Wesley Symantec Press. ISBN-10: 0321304543, New Jersey.

[21] Wang, Y., Chakrabarti, D., Wang, C and Faloutsos, C., (2003). *Epidemic spreading in real networks: An eigenvalue viewpoint.* In *SRDS*.

[22] Watts, D.J., (1999). *Networks, dynamics and the small world phenomenon.* American Journal of Sociology, 105, p234-245.

21

Electric Power Remote Monitor Anomaly Detection with a Density-Based Data Stream Clustering Algorithm

Liyue Chen[1], Tao Tao[1], Lizhong Zhang[1], Bing Lu[1], Zhongling Hang[2]

[1]Electric power dispatching control center, State Grid Zhejiang Electric Power Company, Hangzhou, China
[2]Department of Automation, Shanghai Jiao Tong University, Shanghai, China

Email address:
chenly@zj.sgcc.com.cn (L. Chen), taotao980925@qq.com (T. Tao), zlz951@163.com (L. Zhang), lubing007@126.com (B. Lu),
daba@sjtu.edu.cn (Z. Hang)

Abstract: Nowadays data streams are more and more involved in the real industry. In this paper, the authors apply the data stream clustering to the electric power remote anomaly detection and propose a new data stream clustering algorithm based on density and grid (density-based data stream clustering algorithm, DBClustream). The double-frame analysis model is used in the proposed algorithm. In the online component, the authors optimize the initialization of the parameters for the K-means algorithm with a method based on density and grid and use the kernels to represent the micro clusters as the result of the online component. In the offline component, the time fading weight and the dynamic threshold to optimize the performance of the DENCLUE algorithm are proposed. To evaluate the performance of the proposed algorithm, both the evaluation of the anomaly detection and the evaluation of the data stream clustering are adopted. As the experiment result demonstrates, compared with the others algorithms, DBClustream can resolve the multi-density data stream and keep the high detection rate as well as the low false positive rate.

Keywords: Anomaly Detection, Density Based, Clustering Algorithm, Data Stream

1. Introduction

Because of the rapid development of the smart grid, more and more unmanned substations with automation equipment and telecommunication systems come out. And the requirement of the safe daily operations for the centralized monitoring of the substations becomes increasingly high. Any problem in the automated remote system might put the substation in danger and lead to a large accident in the power grid [1]. Currently, the electric remote monitoring relies on prior knowledge and post-hoc analysis, which is hysteretic and cannot satisfy the increasingly complex needs of the power grid.

By monitoring and analyzing the information of the communication between the substations and the control center, the anomalies and the problems can be found as soon as possible, which brings great convenience for the latter exception handling and reduces the possible economic loss. However, a huge amount of the data and the unknown anomalies raise the high requirement to the anomaly detection

methods. Compared with other industry fields, the data of the electric power remote monitor system is mainly characterized by the following features [2] [3]:

1) Discrete: the data acquiring devices periodically collect the information of the substations and send it to the control center;
2) Non-spherical distribution: the data of remote monitor system presents an irregular non-spherical distribution with a number of dense centers;
3) Complexity: the complexity of anomaly detection is proportional to the number of substations and the length of the time to analyze.

Therefore, electric power remote monitor anomaly detection is very difficult.

In this paper, a density-based data stream clustering algorithm (DBClustream) is proposed. It adopts CluStream clustering framework, which has two components. In the online component, DBClustream uses the density-based improved K-means to form the micro-clusters, and uses the density-based method to optimize the initialization of the parameters. In the offline component, DBClustream applies the improved DENCLUE algorithm with fading window model to the core points of the micro-clusters. With the windows model and the dynamic threshold, the proposed algorithm has the abilities to analyze the multi density data stream with the non-spherical distribution.

The paper is organized as follows. In section 2, the related work is briefly discussed. In section 3, the basic conceptions and definitions are presented. In section 4, the proposed algorithm is explained in details. The experimental results of the proposed algorithm are shown in the section 5. In the end, the section 6 is the conclusion of the paper with some directions of future works.

2. Related Works

In the recent twenty years, anomaly detection has become a hot topic in many industries, such as network security and production process. And the reliable testing standards and the viable methods are summed up. Among the viable methods, clustering algorithm is one of the most popular algorithms that have received attention in many fields. The typical clustering algorithms can be divided into the following categories: partitioning method, hierarchy method, density-based method, grid-based method and model-based method.

K-means clustering algorithm is the most frequently used partitioning algorithm. All the points in the dataset are assigned to k groups with k cluster centers. DBSCAN (Density-Based Spatial Clustering of Application with Noise) algorithm and DENCLUE (DENsity-based CLUstEring) algorithm represent two main directions of density-based method. The basic idea of DBSCAN is that for the point in a cluster, the points counted in its neighbor can't be less than a minimum of user setting. [4] As the improved DBSCAN algorithm, IDBSCAN [5] cuts down the execution time and LD-BSCA [6] reduces the number of the input parameters. And the basic idea of DENCLUE is to model the overall point density analytically as the sum of influence functions of the data points. CLIQUE (CLustering In QUEst) algorithm is a combination of the density-based method and the grid-based method. The basic idea of CLIQUE is to distinguish the sparse area from the crowded area in the data space and to form the clusters with the crowded area.

During the research on the data stream, Aggarwal proposed CluStream clustering algorithm, which is a double-frame analysis model and includes online and offline stream processing. [7] The online part uses K-means algorithm to form a number of micro clusters and the offline part uses the improved K-means algorithm to realize the macro clustering and cluster evolution. The double-frame analysis model is adopted in the later data stream algorithm research.

Among existing data stream clustering algorithms, Den-Stream [8], DDenStream [9], D-Stream [10] and MR-Stream are algorithms based on density based clustering. All of them can detect arbitrary shape clusters as well as handling noise. However, the quality of these algorithms is decreased in multi-density data where different regions have various densities [11]. All the algorithms focus either on the

quality of the clustering or the anomaly detection and barely consider both.

3. Basic Conceptions

Definition 1: *Data point weight:* The initial weight of the data point is 1. And the weight of the data point x in the time t_n is defined based on the weight in the time t_0 :

$$w(x, t_n) = w(x, t_0) \times f(t_n - t_0) \qquad (1)$$

The function f is the fading function.

Definition 2: *Time fading function:* The weight of data points or micro clusters is decreased exponentially over time via the time fading function as follows:

$$f(t_n - t_0) = \begin{cases} 2^{-\lambda(t_n - t_0)}, 0 < t_n < (t_0 + t_{threshold}) \\ 0, t_n > (t_0 + t_{threshold}) \end{cases} (\lambda > 0) \qquad (2)$$

Definition 3: *Grid weight:* For a grid at the time t_n , the grid g_i weight is defined as the sum of the weight of the data points that are in the grid:

$$w(g_i, t_n) = \sum_{x \in g_i} w(x, t_n) \qquad (3)$$

Definition 4: *Dense grid:* At the time t_n, when the grids are sorted by the grid weight, the grids that are in the top one-tenth are dense grids.

Definition 5: *Kernel weight:* For a kernel at the time t_n , the weight of the kernel k_i is defined as the sum of the weight of the data points in the cluster c_i hat the kernel represents:

$$w(k_i, t_n) = \sum_{x \in C_i} w(x, t_n) \qquad (4)$$

4. Density-Based Data Stream Clustering Algorithm

DBClustream has an online component and an offline component. In the online component, DBClustream uses the improved K-means algorithm to get the micro-clusters. The initial cluster centers are chosen based on grid and density, making the algorithm more stable and more accurate. In the offline component, the micro-clusters are represented by their core points. DBClustream generates the final clusters with the improved DENCLUE algorithm.

4.1. Online Phase of DBClustream

Table 1. Notations Used In Online Phase.

n	The total number of data points
m	The total number of grids
m_d	The total number of dense grids
k	The total number of initial cluster centers
l	The number of selected features
ε	The threshold on the change of the cluster centers

In this section, we discuss how to use the density-based improved K-means to form the micro-clusters. The used symbols are listed in Table 1.

As the definitions, the relations of parameters are as follows:

$$m_d = m / 10 \qquad (5)$$

$$k = m_d \qquad (6)$$

As the result of the online phase, the micro-clusters should be able to minimize the computational complexity of the offline phase under the basic premise that the diversity of the data points is retained. So we set the value of k as follows:

$$k \approx \sqrt{4 \times 1 \times \frac{n}{10}} \qquad (7)$$

With (5) and (6), we can get the initial value of the parameter m.

The improved K-means algorithm has two steps. In the first step, the data points are mapped into m grids. We take the core points of the m_d dense grids as the k initial cluster centers. In the second step, with the k initial cluster centers, the K-means algorithm is adopted to get the micro-clusters. The pseudo code of the density-based improved K-means algorithm is shown in Table 2.

Table 2. Pseudo Code of Improved K-means Algorithm.

Improved K-means(DS, ε)
1: Input: data stream
2: Output: micro clusters MCs
3: Calculate the number of the micro clusters k with (7)
4: The initial kernels of the micro clusters is K[]
5: K[] = GetInitial Kernels (DS, k)
6: Do
7: ∆d = 0
8: while not end of stream do
9: Read data point x from Data Stream
10: Calculate the distance from x to all the points in K[]
11: MCs（K[i]）= MCs（K[i]）∪x （K[i] is the nearest kernel from x)
12: end while
13: for micro cluster mc in MCs do
14: Calculate the new kernel of mc
15: ∆d+ = distance between new and old kernel of mc
16: Replace the old kernel with the new one
17: end for
18: Until ∆d < ε
19: return MCs
GetInitial Kernels (DS, k)
20: Get the maximum and the minimum value for each selected features in the data space
21: Divide the data space uniformly into m grids
22: Map the data points into the grids
23: Sort the grids by the number of their data points
24: Get the dense grids and their center points
25: return the center points of the dense grids

4.2. Offline Phase of DBClustream

In the offline phase, the improved DENCLUE algorithm on the micro-clusters is applied to get the final result. DENCLUE algorithm is mainly based on the following ideas:

1. the density influence of each data point to the other points in its neighborhood is described by a mathematical function;
2. the global density of the data space can be modeled as the sum of the influence of all data points;
3. the final clusters are obtained by determining the local maximum of the global density of the data space.

In the improved DENCLUE algorithm, the following three changes are proposed:

1. the weight of the kernel is considered in the density influence of the kernel;
2. as none of the kernel is noise point, there is no threshold to eliminate the noise point;
3. a dynamic threshold is applied to merge two adjacent density attractors.

The pseudo code of the improved DENCLUE algorithm is shown in Table 3.

Definition 7: *Kernel density function:* At the time t_n, the density function of the kernel x_i is modeled by the Gaussian influence function and related to the weight of the cluster c that the kernel x_i represents.

$$d(x, x_i, t_n) = w(c, t_n) * e^{-\frac{(x - x_i)^2}{2h^2}} \qquad (8)$$

Definition 8: *Global density function:* At the time t_n, the global density function is defined as the sum of the kernels density functions. Given k kernels described by the vectors $D = \{x_1, ... x_k\}$, the global density function is as follows:

$$d(x, t_n) = \sum_{i=1}^{N} d(x, x_i, t_n) \qquad (9)$$

Definition 9: *Density attractor:* A density attractor is defined as the local maximum of the global density function.

Definition 10: *Dynamic threshold:* The dynamic threshold is the minimum density for two adjacent density attractors to merge. If we use a global threshold in the data space, the low density clusters cannot be revealed in the high density ones. The dynamic threshold is defined as follows:

$$\xi = \theta \times \min(d(x_1, t), d(x_2, t)) \qquad (10)$$

where θ is the dynamic parameter between 1 and 0; x_1 and x_2 are two adjacent density attractors.

Definition 11: *Density reachable:* If the minimum point between two adjacent density attractors is bigger than the dynamic threshold, the two adjacent density attractors are density-reachable and we need to merge the two clusters that the two density attractors represent.

Table 3. *Pseudo Code of Improved DENCLUE Algorithm.*

ImprovedDENCLUE (WKs, θ, ε)
1:    Input: Weighted Kernels WKs
2:    Output: Clusters C
3:    Initialize the set of attractor points A as an empty set
4:    For data point x in WKs do
5:    $x^* =$ FindAttractor (x, WKs,ε)
6:    $A = A \cup \left\{ x^* \right\}$
7:    Add the data point x to the set of points $R\left(x^*\right)$ attracted to x^*
8:    end for
9:    Find all the maximal subsets of attractor points $C \subseteq A$, such that any pair of attractors in C is density-reachable from each other
10:    for c in C do
11:    for x^* in c do $C = C \cup R\left(x*\right)$ end for
12:    end for
13:    return C
FindAttractor(x, WKs, ε)
14:    $t = 0; x_t = x$
15:    Do
16:    Get $B_d\left(x_t, r\right)$ as the set of all points in WKs that lie within a l-dimensional ball of radius r centered at x_t
17:    $x_{t+1} = \dfrac{\sum_{x_i \in B_d(x_t,r)} d(x_t, x_i, t_n) x_i}{\sum_{x_i \in B_d(x_t,r)} d(x_t, x_i, t_n)}$; $t = t+1$
18:    until $\left| x_t - x_{t-1} \right| < \varepsilon$
19:    return x_t

5. Experimental Results

In this paper, the experiment is conducted on a PC with Intel Core Dou i7 2 GHz Processors and 16 GB DDR RAM running Windows 7 operating system. And the DBClustream algorithm is implemented in Matlab. We choose KDD CUP 99 dataset to evaluate the performance of DBClustream. The KDD CUP 99 dataset consists of TCP connection records from nine weeks of LAN net-work traffic by MIT. To assess the clustering quality, we use the most widely used parameter, the cluster purity, which is defined as the average percentage of the dominant class label in each cluster. At the same time, we also adopt the detection rate and the false positive rate to assess the performance of the anomaly detection.

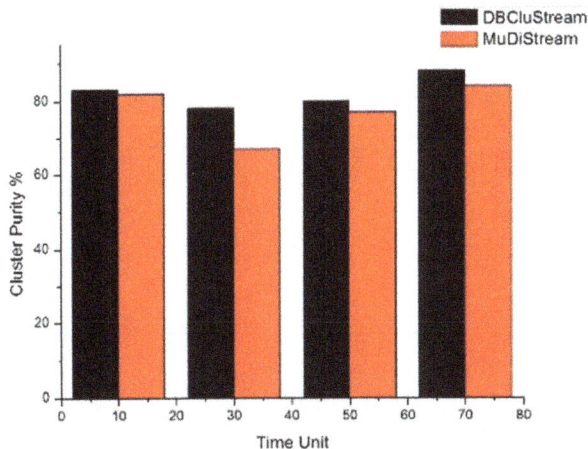

Figure 1. *Comparison of Cluster purity for KDD99 dataset, stream speed = 1000.*

We use the cluster purity to compare the clustering quality of DBClustream and MuDiStream. Figure.1 and Figure.2 show the comparison results of the cluster purity. It can be seen that DBClustream has a very good clustering quality, and is more stable and better than MuDiStream.

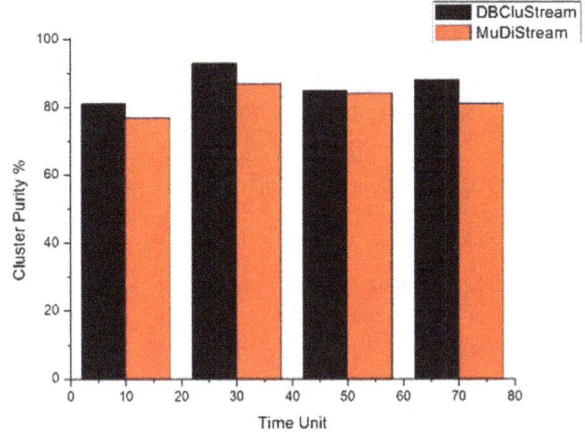

Figure 2. *Comparison of Cluster purity for KDD99 dataset, stream speed = 2000.*

Table 4. *Comparison of detection performance for KDD99 dataset.*

	Detection rate (%)	False positive rate (%)
CURE[12]	81.09~85.10	3.47~5.49
Aprior[13]	87.2~87.5	8.1~17.4
DENCLUE	83.2~89.7	5.3~10
DBClustream	85.0~90.2	2.1~4.7

To assess the anomaly detection performance of DBClustream, we compare the detection rate and the false positive rate of DBClustream with other detection algorithms. From Table 4, we can see that compared with the other algorithms, DBClustream keeps the high detection rate and the low false positive rate at the same time. In the electric power remote monitor system, the low false positive rate is as important as the high detection rate, both of which can cut the unnecessary cost for the smart grid.

6. Conclusions

Considering the feature of the electric power remote monitor system, a density-based data stream clustering algorithm (DBClustream) is described for the electric power remote monitor anomaly detection in this paper. The double-frame analysis model is used in DBClustream algorithm. In order to manage multi density data stream, the improved initialization of the parameters of K-means algorithm with a density-based method, the time fading weight and the dynamic threshold methods are applied in DBClustream algorithm. The experimental results show that the proposed algorithm is more effective on the high detection rate and the low false positive rate than other well-known algorithm. In future, we are going to apply proposed algorithm to the real electric power remote monitor system and improve the performance in effectiveness and efficiency.

References

[1] YANG Huan-hong, YE Hai-ming. "Analysis and Monitoring of Electric Power Tele-control Channel Fault," Journal of Shanghai University of Electric Power, vol. 25, no. 4, pp. 321-324, 2009.

[2] XUE Fei, "The software Design and Implementation of IEC60780-5-104 Protocol," M.S. the-sis, School of Control and Computer Engineering, North China Electric Power University, Beijing, China, 2012.

[3] WANG Jing, "Research of Online Intelligent Alarm," M.S. thesis, School of Electrical Sys-tem and Automation, North China Electric Power University, Beijing, China, 2008.

[4] Zhiwei SUN, Zheng ZHAO. "A Fast Clustering Algorithm Based on Grid and Density," Electrical and Computer Engineering, pp.2276-2279, May 2005.

[5] B. Borah, D. K. Bhattacharyya. "An Improved Sampling-Based DBSCAN for Large Spatial Databases," Int. Conf. on Intelligent Sensing, pp. 92–96, 2004.

[6] G. Wei, H. Wu. "LD-BSCA: A Local Density Based Spatial Clustering Algorithm," in IEEE Symposium on Computational Intelligence and Data Mining. IEEE Computer Society, 1999, pp. 291–298.

[7] Aggarwal C, Han J. "A Framework for Clustering Evolving Data Streams," Proceedings of the 29th VLDB Conference, pp. 81-92, 2003.

[8] F. CAO, M. ESTER. "Density-based clustering over an evolving data stream with noise," Proceedings of the 2006 SIAM International Conference on Data Mining, pp.328-339, 2006.

[9] M. Kumar, A. Sharma. "Mining of Data Stream Using DDen Stream Clustering Algorithm," IEEE International Conference in MOOC, pp. 315-320, 2013, doi: 10.1109/MITE.2013.6756357.

[10] Y. Chen, L. Tu. "Density-based clustering for real-time stream data," Proceedings of the 13th ACM SIGKDD international conference on Knowledge discovery and data mining, pp. 133–142, 2007.

[11] Amineh Amini, Hadi Saboohi. "A Multi Density-based Clustering Algorithm for Data Stream with Noise," IEEE on Data Mining Workshops, pp. 1105-1112, 2013, doi: 10.1109/ICDMW.2013.170.

[12] ZHOU Ya-jian, XU Chen. "Unsupervised Anomaly Detection Method Based on Improved CURE Clustering Algorithm," Journal on Communications, vol. 31, no. 7, pp. 18-23, 2010.

[13] CUI Guan-xun, LI Liang. "Research on an Intrusion Detection System Based on the Im-proved Apriori Algorithm," Computer Engineering & Science, vol. 33, no. 4, pp. 40-44, 2011.

22

Improvement Research PLC Automatic Control System Based on Small and Medium Logistics Classification

Zhengzheng Cong, Cong Li, Yifeng Shao, Zhize Zhou, Chunmeng Liang

Department of Mechanical Engineering, Guilin University of Aerospace Technology, Guilin, China

Email address:

Congzz@guat.edu.cn (Zhengzheng Cong), licong@guat.edu.cn (Cong Li), 369468271@qq.com (Yifeng Shao), 2507663523@qq.com (Zhize Zhou), 7608782202@qq.com (Chunmeng Liang)

Abstract: PLC automatic control classification system is based on commitment to increase express courier company and distribution center for the delivery of processing power, making it more efficient for the shipment have free processing capabilities. Based on the scientific method, prior to the courier company handling capacity estimates, and the company in shipment processing equipment capital investment to enhance the product in the industry's competitiveness. It is used of the existing two-dimensional code scanning recognition technology. Through the automatic identification of microcontroller programming combined ratio for the sensor to achieve classification. It works to reduce logistics costs in the classification of goods while improving the efficiency and accuracy of their goods classification.

Keywords: Intelligent Classification, High Precision, Continuous Work

1. Study Background

Based on all of today's domestic shipments by province and country are required for centralized delivery sorting process conditions, resulting in a shipment of more complex and requires a large amount and classification and timely comparison. Due to the rapid development of the courier company, leading to the sorting process is now also fully express stuck in backward manual processing stages. Thus, it affects the efficiency of logistics and accuracy under these conditions the corresponding article classifier ready. Swift classification treasure is mainly used in the classifying of goods, such as express delivery companies to provinces or countries classified items classified as other fields. Mainly used to automate classification of goods, to reduce costs and improve the accuracy and efficiency of classification of goods. Using infrared scanning technology to items brought two-dimensional code identification, and the two-dimensional code information passed SCM console. SCM through incoming digital into analog control amount, and then by the microcontroller output control of the state sensor. Control then transferring articles through the system. After collected the respective collector

items, the collector sensor status changes, the items falls into the corresponding collector. Thus, it makes goods classification.

In order to achieve efficient and accurate classification of goods as the goal. Swift is expected to classification efficiency classification Per 1200 per hour, while the artificial classification of up to 160 bags per hour and cannot continue to work more than eight hours, the specifications for the mass of less than 25kg, volume in the 80*80*100 of items.

2. The Research Objectives

This work is the key technology of the microcontroller can be programmed in advance, and then use the MCU receives sensor status signal and infrared scanner to approach address information, after the microcontroller logic processing, and then send control commands to control the motor to rotate and turn the dial to collect cells from the MCU The opening and closing. Applied to the microcontroller, angle sensors, two-dimensional code scanner, infrared sensors and other equipment. Relatively conventional automatic sorting system, which works great collection device using rotary machines reduce the space occupied, with a high degree of intelligence, an area of small, low one-time cost, classification accuracy and efficiency advantages.

3. Assessment Indicators

It works with the current emerging two-dimensional code technology and MCU technology and sensor technology the perfect combination to form a complete set of goods classification system. PLC automatic control system design for improved innovation goods classification based collector, produced specifically for the goods classification to achieve automatic classification system design..The basic working principle is as follows: The microcontroller programming, scanner input address information items, then SCM issued a directive, the article collector on behalf of photo sensors that address open, when the article passes this time sensitive sensor is blocked. The transport board separate items falling from the transport board, falling into the article corresponding address slot, and then close the transport board, run-down, in the end portion, the process is repeated again.

4. Based on PLC Automatic Control System

China have also appeared suitable for small and medium sorter express courier company. But it's only sorting less than 5kg of shipment, which for a little bulky courier cannot handle. However, companies developed this product also did not enter mass production stage. Current situation of the courier company working in classified areas, found that most of the small and medium sized courier companies have not achieved classified intelligence resulting in high costs, low classification efficiency, accuracy cannot be guaranteed. The main reason for this phenomenon is the capital weak small and medium companies, we cannot afford large-scale sorting systems high time investment. But the courier classification developments must be intelligent. Therefore, it is suitable for small and medium enterprises to improve the characteristics of a high level of intelligence, high efficiency, high precision, low cost classification system. Bold design conceived PLC automatic control of classification system based on: continuous, a large number of classification courier, using automatic assembly line work, and not subject to time, weather and human limitations can reach more than 100 hours of continuous work, and the number per unit time classification pieces more can be reached 900-1200 per hour, manual classification efficiency above 5 times. It was unable to continue high-intensity artificial classification 8 hours.

5. The Content of the Research Projects, Research Objectives, and Key Scientific Problems to Be Solved

The main research work items classification, such as parcel delivery company to classify and other fields. Articles

are intended to achieve automatic classification, in order to improve the situation. Use two-dimensional code scanning technology for goods carried by the two-dimensional code recognition, and the two-dimensional code before the contents of incoming console and the program will be compiled and implemented through the MCU conversion logic functions than the right, and then by the microcontroller output sensor control state, the sensor status changes within the collector, and ultimately for goods classification. Technical key point in the realization of the main items of data within two-dimensional code to achieve the objects of the efficient and accurate classification. Targets will be achieved one day (10 hours) Category 2000, specifications for the mass of less than 20kg, volume in the 80*80*100 of items. Classification efficiency is expected 900-1200 per hour continuous operation for 100 hours and a maximum of 160 artificial classification per hour and cannot continue to work more than eight hours. Products are not suitable for classification slim, long, overweight and classification of fragile items. Technology is the key to this work is that the microcontroller can be programmed, and then use MCU receives status signals and infrared sensor scanner scanned address information, after the microcontroller logic processing, they are send control commands to control motor rotation and then dial the collection grid by the microcontroller.

Works by the transportation system, QR code scanning system, computer network of single-chip microcomputer processing system, sensor devices and chassis. Transportation system is the important part of the conveyor belt and transmission wheel and the radial transmission device to collect one cell, it is mainly used for early express transportation and later will express delivery to the designated collection bag. QR code scanning system is the main component of QR code scanning guns, is used to identify within its code address information and the information transmitted to the computer. Single chip computer network system for receiving and processing signal and sent to transfer control instruction to realize automatic classification effect. Scanning system will address information transmitted to MCU. It commands to control rotate to the corresponding Angle control the radial transmission device to collect one cell will be Courier to the corresponding collection bags.

6. Research Programs to Be Undertaken and Feasibility Analysis

Design concept of the automatic control of PLC-based classification system, which not only has continuous work long hours, a lot of free delivery, the use of automatic assembly line work and so on. Due to stick with a bar code or two-dimensional code scanning input, express categorization process basically intelligent. Based on design ideas, this work consists of the following components: products from the transport system, two-dimensional code scanning system, computer network microcontroller processing systems, sensor devices and a rack.

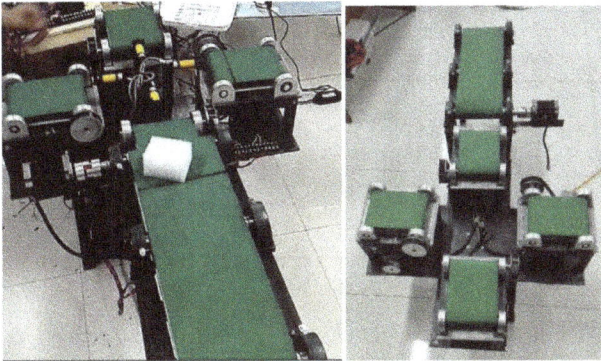

Figure 1. Accessories: computer console, motor, angle sensors, goods collection bag.

The main components of the transportation system and the radial transfer means collected within the cells, mainly for the early delivery of express transport and the latter will transfer to the designated collection bag. Dimensional code scanning system is a major component of the two-dimensional code scanner, for the two-dimensional code identifying the address information and transmits the information to the computer. Computer network microcontroller processing system for receiving, processing and transmission of the classification signal transmission control instructions to automate the classification results. Scanning system will address information is transmitted to the microcontroller. MCU sends commands to control turntable rotation radial transfer means collected within the cells control the collection bag when the courier transferred corresponding to the appropriate angle. The key technology is the process of operation, based on the input and output decency control PLC automatic control system programming and control accuracy as well as infrared scanning identification.

Simulation procedure is as follows: The workers will express a two-dimensional code face up on the first conveyor belt, when the courier run to the second conveyor at the infrared sensor, the signal is transmitted to a computer console, MCU sends control command control first conveyor stops 3S. In this case the courier is still on the second conveyor run, two-dimensional code and two-dimensional code scanner to scan information on the courier and the information is transmitted to the computer console. When the items are transported to the turntable collection grid, single-chip issue control instructions to the computer console before transmission based on a two-dimensional code address information, and precompiled SCM process, control the angle sensor to collect cells to be detected on the corresponding angle. If the turntable is rotated to the appropriate angle to collect cells such as 90 degrees, the computer console, this information is collected MCU issue control instructions to control the inner diameter corresponding to the transfer apparatus to collect cells will express came along and fall into the corresponding radial turntable collection bags, items at this time classification process is completed.

The production process:

First stage is according to the investigation team found that express company now as the low degree of intelligence, low efficiency, high accuracy cannot be guaranteed, artificial cost of these shortcomings. Accordingly we propose using the single chip microcomputer technology, Angle sensors, QR code scanner, infrared sensor technology, and computer network auxiliary processing information collection, each component of a feedback control instruction to complete intelligent classification.

Team in the second stage after a series of discussions on argument to start 3d model of the product, and on this basis to modify the structure of the product, such as design team before spiral collection system and the automatic warehouse system because of the need to apply to the stack to implement more difficult, so later became now rotating disc collecting system in combination with Angle sensor technology and realize express simple and effective classification.

Third stage team began to design parts according to the established 3d model of the machining drawing and purchase the required materials products and commissioning and correction.

Work instructions: Workers will have a parcel with QR code on the side as the first of the conveyor belt, when delivery to the second transmission with an infrared sensor, the signal is transmitted to the computer console, SCM control command control the first conveyor belt to stop 3s. At this point, the Courier is still in the second run on the conveyor belt, and QR code scanner scans the information on the delivery and will transfer the information to the computer console. When goods are transported to the turntable from SCM according to the information transmitted to a computer console QR code address before and compiled microcomputer program control instruction, beforehand control from the Angle of the sensors need to detect the corresponding point of view. When the wheel from rotation to the corresponding Angle such as 90 degrees, computer console to collect information about the single chip microcomputer control instruction control radial transmission device inside the corresponding collection. When express came along the rotary radial and fall into the corresponding collection bags, goods classification process is complete at this time.

7. Conclusions

Works applied to the microcontroller, angle sensors, two-dimensional code scanner, infrared sensors and other equipment. Relatively conventional automatic sorting system, which works great collection device using rotary machines reduce the space occupied. Combined with a microcontroller and sensor technology to automate the classification of a courier 3s, improves the classification efficiency. Technological innovation, new product so that the current two-dimensional code technology and MCU technology and sensor technology is the perfect combination to mix into a complete article classification system.

Thanks to the national college students' innovative entrepreneurial project funding support.

References

[1]   CHENG Lei 1, XU Yu-xian 2, JIN Wei-xing 2 (1.Suzhou Institute of Industrial Technology, Suzhou Jiangsu 215104, China; 2. Suzhou Xinya Electronic Communication Co., Ltd., Suzhou Jiangsu 215132, China); Design of Connector Automatic Assembly Machine Based on the PLC [J]; Equipment Manufacturing Technology; 2012-07.

[2]   WU Yi-zhong~1 GONG Yun-bo~1 ZHANG Guo-quan~(2, 3) (1.National CAD Support Software Engineering Research Center, Huazhong University of Science and Technology, Wuhan 430074, China; 2.Wuhan Polytechnic University, Wuhan 430023, China; 3. Wuhan Rentian Packaging Technology Co., Ltd., Wuhan 430205, China); Research on Key Technologies of Automatic Production Line of Industrial Explosive Secondary Packaging [A]; [C]; 2007.

[3]   Gao De, Xie Hao (Engineering college, Jiamusi university, Jiamusi 154007, China); Develop and Research of Packing machinery MCAI [J]; PACKAGING ENGINEERING; 1999-04.

[4]   Zhou Gao-fei; Research and Development of JQ 700 Type Box-bridge Erecting Machine [J]; Chongqing Architecture; 2008-05.

[5]   L. Wang, Adaptive Fuzzy Systems and Control: Design and Stability Analysis, Prentice-Hall, Englewood Cliff, NJ, USA, 1994.

[6]   R. M. Sanner and J.-J. E. Slotine, "Gaussian networks for direct adaptive control," IEEE Transactions on Neural Networks, vol. 3, no. 6, pp. 837–863, 1992.

[7]   J. T. Spooner and K. M. Passino, "Stable adaptive control using fuzzy systems and neural networks," IEEE Transactions on Fuzzy Systems, vol. 4, no. 3, pp. 339–359, 1996.

[8]   M. Liu, "Decentralized control of robot manipulators: nonlinear and adaptive approaches," IEEE Transactions on Automatic Control, vol. 44, no. 2, pp. 357–363, 1999.

[9]   Y.-G. Leu, T.-T. Lee, and W.-Y. Wang, "Observer-based adaptive fuzzy-neural control for unknown nonlinear dynamical systems," IEEE Transactions on Systems, Man, and Cybernetics, vol. 29, no. 5, pp. 583–591, 1999.

[10] J. T. Wen and D. S. Bayard, "New class of control laws for robotic manipulators—part 1: non-adaptive case," International Journal of Control, vol. 47, no. 5, pp. 1361–1385, 1988.

23

Qualifying Articles of Persian Wikipedia Encyclopedia Through J48 Algorithm, ANFIS and Subtractive Clustering

Seyedtaha Seyedsadr[1], Mohammadali Afsharkazemi[2], Hashem Nikoomaram[3]

[1]Department of Management, Electronic Branch, Islamic Azad University, Tehran, Iran
[2]Department of Management, Tehran Central Branch, Islamic Azad University, Tehran, Iran
[3]Department of Management and Economics, Sciences and Research Branch, Islamic Azad University, Tehran, Iran

Email address:
Sts.sadr@srbiau.ac.ir (S. Seyedsadr), m.ali.akazemi@gmail.com (M. Afsharkazemi), nikoomaram.hashem@gmail.com (H. Nikoomaram)

Abstract: Since Wikipedia encyclopedia is one of the most popular web sites on the internet, providing accurate information is of abundant importance. In this research, the effective variables on quality of Persian articles are identified and a system is, then, designed for judging articles in three quality levels: high quality, cleanup needed, and deletion. First, the variables relating to the articles included in the list of featured articles, good articles, cleanup needed, and deletion articles are collected. Then, two methods are used for the analysis of data: First, a decision tree explains the relationships among the collected variables as rules that are implemented by adaptive neuro fuzzy interference system. Second, the data are implemented by subtractive clustering algorithm and the error of both methods is, finally, measured and compared. The results indicate that the average daily hits, total views, page length, total number of edits, total number of authors, and number of templates used are directly related to quality of Persian articles while the number of recent number of authors is inversely related to quality of articles.

Keywords: Wikipedia Encyclopedia, Quality of Articles, J48 Decision Tree, ANFIS, Subtractive Clustering Algorithm

1. Introduction

Persian Wikipedia is a web-based free online encyclopedia that was founded in December 2003 [1]. This encyclopedia included 373512 articles, 330083 users, and 28 administrators in October 31, 2013 which won the 20th place among the various encyclopedia languages [2]. Free access, constant updates, extensive coverage, and diversity have turned Wikipedia into one of the most unmatched social websites in the cyberspace [3]. Increasing the number of members of a group may usually reduce the effectiveness of teamwork; however, the researchers have assessed wisdom of crowds and collective action as the success factors of Wikipedia [4, 5, 6]. The users residing in the Europe and North America have mostly contributed in writing the articles of Persian Wikipedia and the Iranian users' contribution is only about 45% [1]. Most people who have contributed in the development of Wikipedia are anonymous [1, 7]. Users with different motives have helped development of this encyclopedia. The new users help development of this encyclopedia to satisfy their curiosity and the old users do so with such motivations as providing

information and producing content. Sense of usefulness, sense of finding identity in the society, feedbacks received from the user community, and achieving fame are considered among the most important factors of continuation of contribution of the Internet users with the development of the Persian Wikipedia. Financial incentives have not had any role in collaboration of the users with the Persian Wikipedia [1]. Internet users through having access to the pages of Wikipedia without any limitation and make some changes in those pages except in cases where a particular page is restricted to be edited by the ordinary users by the site administrators [8]. Many people use the site information every day, develop new articles, and edit the older articles. Hence, one of the most important challenges of the site is to deliver an optimum quality level of the articles. People with different ages, cultures, viewpoints, and opinions can express their views on different issues in this encyclopedia regardless of what their specialty is; thus, Wikipedia has prescribed instructions and guidelines for improving the quality of this encyclopedia [9, 10]. Generally, the quality of articles in the site of Wikipedia is determined using opinion polls [11]. Further, the associated research on determining the quality of articles of Wikipedia encyclopedia and using decision tree, ANFIS, and subtractive clustering algorithm in the predictions are examined.

2. Related Work

2.1. Determining the Quality of Wikipedia Articles

Walraven et al. [12] would demonstrate that students and university students spend most of their time searching for information in the cyberspace while they spent less time for assessment and evaluation of the contents. Conducting a research on method of study of a number of students of an American college, Lim [13] suggests that students often search the basic information of the intended subject in the Wikipedia and this encyclopedia directs them towards the use of more specialized sources, in this study, more effort is recommended to improve the quality, precision, and accuracy of the Wikipedia information as well. Lucassen and Schraagen [14] described in a research carried out on the views of the 149 people from different parts of the world that the users do not highly trust on the Wikipedia articles. Wikipedia has been mainly criticized for the low information quality. Stvilia et al. [15] have assessed the information quality criteria in the English Wikipedia by studying the manner of selection of the featured articles and demonstrated that information quality could not be measured with a simple model. Geiger and Halfaker [16] examined the impact of the Wikipedia software robots to prevent vandalism and control the quality of this encyclopedia. Rowley and Johnson [17] have demonstrated that multiplicity of authorship, references, proper structure and grammar would increase the trust of addressees as users while the comments representing the concerns such as citation needed have a negative impact on their trust. Noč and Zumer [18] examined the general situation and the quality of featured articles of the Slovenia Wikipedia based on reviewing the references and sources and concluded that the quality of the featured articles were generally higher than other articles. Kittur and Kraut [4] have suggested a method for improving the quality of articles through coordinating the editors and substantiated the articles that had more editors were of higher quality than the other articles. Improving quality of the articles is achieved, when the editors use coordination techniques. Ram and Liu[3] have examined the effect of the variables of collaboration patterns, number of unique editors, number of edits, article age, article length, and number of unique administrators on quality of articles and indicated that the articles written by professional authors have often higher quality than other articles. Lih [19] measured the quality of Wikipedia articles by the independent variables of linking, number of edits, number of authors, article size, and other metadata from Wikipedia. Priedhorsky et al. [5] have examined the impact of editing and editors on the Wikipedia value. Stein and Hess [20] have demonstrated that there is a significant relationship between the featured articles and increasing number of authors. In a research conducted on the featured articles of the English Wikipedia, Wilkinson and

Huberman [6] have substantiated that there is a direct relationship between increasing the number of editions, the number of authors, contribution in editing the articles, and higher quality. Wöhner and Peters [21] have separated featured and good articles (high quality) from deletion articles (low quality) through examination of the two variables of transient contribution and persistent contribution and calculation of the lifecycle of articles. Saengthongpattana and Soonthornphisaj [22] have developed a system for separating the featured articles from the ordinary ones using fuzzy logic and K-mean algorithm through study of specifications of the featured articles in the Thai Wikipedia such as the number of images, links, main headings, footnotes, etc. Most of the researches conducted on the encyclopedia articles quality have focused on identifying and selection of the featured articles, Anderka et al.[23] have examined the cleanup needed articles and the tags for improving quality of the encyclopedia articles. Xiao et al.[24] separated the featured articles from the start class by C4.5 algorithm using statistical properties and data of the articles such as page length, number of authors, number of editors, number of links, and number of images. Chai et al. [25] presented a model for the assessment and measuring the quality of the posts sent by users at the low, medium and high levels.

2.2. Using the J48 Decision Tree and Fuzzy Algorithm in Predictions

Saravanan et al. [26] have presented a method for determining the status of inaccessible gears in a device. They processed the vibrating signals extracted from the gears and via selecting, the best statistical features of audio signals by J48 decision tree and then implemented the resulting rules using fuzzy logic toolbox of MATLAB software. Therefore, the developed system could detect and distinguish to the defected gears from the sound gears by the audio signal. Omid [27] developed a system using the J48 decision tree and fuzzy algorithm that recognizes the natural open pistachio from the closed pistachio. Jalili and Mahmoudi [28] have developed a system using the J48 decision tree and fuzzy algorithm that separates the two types of Iranian pistachios-Akbari and Kaleghouchi. Noorallah et al.[7] have judged the quality of 226 German Wikipedia articles in a research and assessed low quality articles with (0) and high quality ones with (1). Then, the statistical properties of the articles such as page length, number of authors, number of editors, etc., are classified by J48 decision tree of Weka software. The resulting rules are implemented by adaptive neuro-fuzzy inference system; thus, the researcher develops a system that separates high quality articles from the low quality articles like the human brain.

2.3. Subtractive Clustering Method

Clustering data of a collection aims at a brief display of the behavior of the dominant system in that collection. A set of input and output data of the system are collected. Indeed, the cluster centers specify the desired system behavior. Therefore, each of the cluster centers can be regarded as a base of a rule which is used to describe the intended system behavior [29]. Fuzzy C-Mean clustering Method has numerous applications in unsupervised classification. One of the problems of this method is estimation of the initial quantity of cluster centers [30]. Like many non-linear optimization problems, quality of the FCM solution depends on the initial number and quantity of the cluster centers [29]. Yager and Filev and Yager [30] in 1992 developed a quick method for approximate clustering and estimation of the cluster centers by mountain method. Chiu [29] in 1994 examined and modelled the relationship between five independent variables of population, number of residential units, number of automobile owners, average household income, number of servants and the dependent variable of number of trips by subtractive clustering algorithm in Delaware, New Castle County [29]. A six-dimension space with three rules was covered by SC method and Chiu indicated that SC method reduced computational complexity to a great extent [29, 31]. The greater the radius of clusters, the less the number and centers of clusters will be using SC algorithm; thereby the number of rules will be reduced. The less the radius of clusters, the greater the number of clusters and so the greater the number of rules will be achieved. One of the advantages of SC method is that it is not needed to estimate the clusters [32].Yuan et al. [33] have predicted the quality of software in a research via using SC method. Wei et al.[34] have developed a model for prediction of the stock market of Taiwan by SC method. Malhotra and Sharma [35] have examined and modelled the relationship between nine independent variables such as the number of words in web pages, page size, the number of tables, graphs in the page, etc. and the dependent variable of quality of web pages using SC method. In another research, Afshoon et al. [36] have examined the quality of educational websites by SC method.

2.4. Summing Up

Considering the increasing number of contents produced by the users in the cyberspace, it is difficult to access high quality content. Determining the quality of the produced content is an inevitable requirement [25]. The earlier studies have revealed the necessity of conducting a research on quality of the Wikipedia articles in all languages including Persian. Most of the researches conducted on quality of the Wikipedia articles have assessed the articles of this encyclopedia in two high and low quality levels e. g. [3, 4, 6, 7, 19, 20, 21, 24]. cleanup needed articles cover a part of the Wikipedia articles in which some of the encyclopedia criteria are not observed [37]. These articles cover the intended subject reasonably; however, more references may be needed or the edition instruction of the Wikipedia may not be completely adhered to. As suggested by Anderka et al., this group of articles is less analyzed by the researchers [23, 37]. This research aims at answering to this question, "Do the seven variables (1) average visits of an article per day, (2) total number of visits of an article (view), (3) page length, (4) total number of edits, (5) total number of distinct authors, (6) recent number of distinct authors (rec

authors), and (7) number of transcluded templates in an article affect the quality?", it also aims at designing and implementation of a system that can judge the Persian Wikipedia articles in three quality levels of high, cleanup, and deletion like users and administrators. Figure 1 shows the effective variables on quality of articles of the Persian Wikipedia. Considering the earlier research and potentials of the Persian Wikipedia, the variables of the available articles are collected and saved in the list of featured articles, good articles, cleanup needed articles and articles for deletion. Two methods are used to analyze the obtained data. In the first method, the collected data are classified using the J48 decision tree of Weka software and the relationships between the dependent and independent variables are extracted in the form of rules with if-then structure. The resulting rules are, then implemented by adaptive neuro fuzzy inference system (ANFIS). The designed system separates the Persian articles of the Wikipedia in three high quality, cleanup, and deletion levels, and finally the designed system error is measured. In the second method, the variables data are processed, clustered, and implemented without using the decision tree by subtractive clustering algorithm. The designed system error is measured and compared with the first method. Figure 2 shows the framework of method 1 and 2. Further, the dependent and independent variables and the tools applied in the research are introduced. In section 4, the J48 decision tree is used for processing data and ANFIS and SC algorithm are used to implement the model. The obtained results are analyzed in section 5.

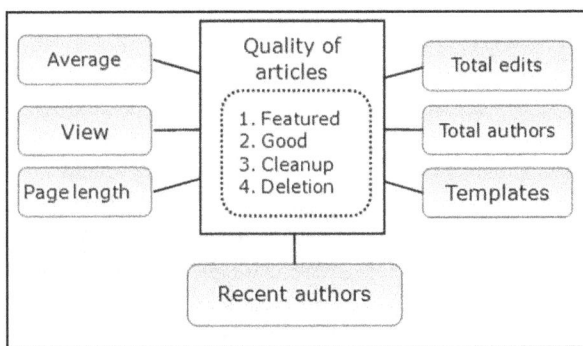

Figure 1. Research variables model.

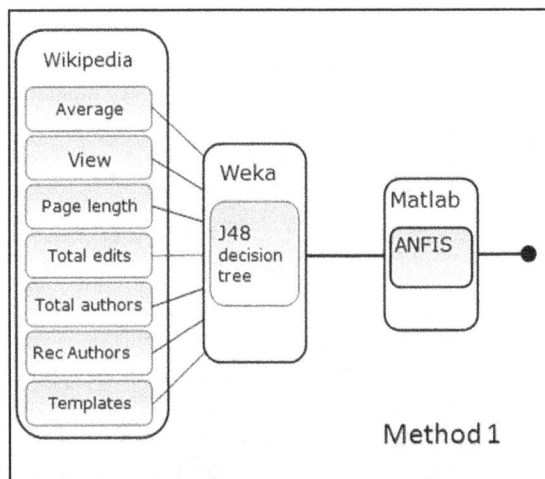

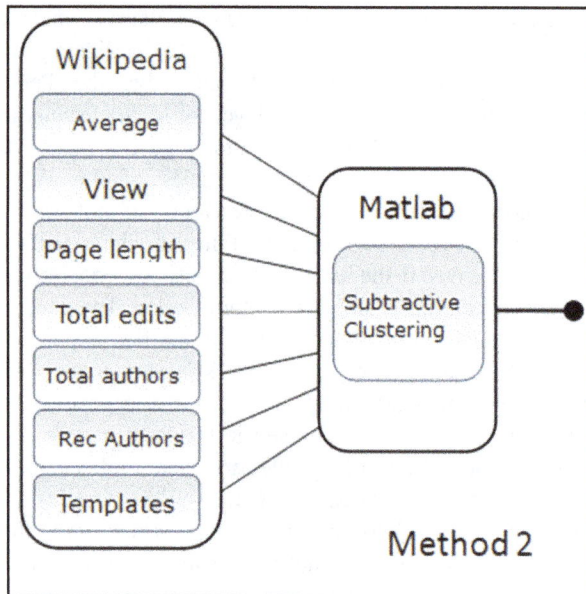

Figure 2. Methods framework.

3. Variables and Tools

The research data are the information related to the articles in the Persian Wikipedia that were extracted from the database of the Wikipedia site in October and November 2013. The purposive sampling method has been used in this research [38].

3.1. Dependent Variables

3.1.1. Articles for Deletion

The Wikipedia site deletes the articles which do not meet the criteria of the encyclopedia contents. Generally, the articles that ignore the copyright or cannot be attributed to reliable sources, new words, violation of biography of those who are alive or advertising are included in the cases of deletion. A feedback form is mainly used to remove articles from the Wikipedia site and users can discuss regarding deletion of articles. The discussion will last at least 7 days. Then, the article will be deleted if the users are agreed to do so [39]. In this research, 75 articles have been examined under the title of deletion articles variable.

3.1.2. Cleanup Needed Articles

Cleanup includes correcting spelling, grammatical, and typographical mistakes, inappropriate tone, and failure to use proper references. Contribution of all users is welcomed in edition or amendment of these articles. Finally, the changes made by a user on an article are discussed and assessed by other users [37]. In the present research, 52 articles in this list have been reviewed under the title of cleanup articles variable.

3.1.3. Good Articles

Good articles of the Persian Wikipedia are the articles with a good prose that are provable, have reliable sources, appropriate coverage of the content, unbiased attitude, and appropriate and relevant image. These articles are first nominated for the opinion poll of other users and are, then, selected as a good

article in case of agreement of users. good articles are the complete articles though they have some problems, too [40, 41]. Sixty-eight good articles have been reviewed in the present research under the title of good articles variable.

3.1.4. Featured Articles

Featured articles are the professional, unique, and comprehensive articles that can be considered as the final source for information of the Wikipedia [42]. The featured articles shall be already selected as good comprehensive articles, with the volume of more than 5 KB; they shall have valid sources and must avoid addressing the marginal unnecessary issues, the Wikipedia instruction must be adhered to in these articles and they shall not be needed to be corrected. The articles are first nominated for the opinion poll of users and are selected as the featured article in case of agreement of users [43]. 89 featured articles under the title of featured articles variable have been analyzed in this research.

3.2. Independent Variables

(1) Average.

A numerical quantification obtained from division of the number of times users visit a particular article into the number of elapsed days since the release of that paper (average=view/days). Specifying the value of this variable for each article is possible using the Toolserver tools [44, 45].

(2) Views.

A numerical quantification, which displays the number of times users visit a specific article. Specifying the value of this variable for each article is possible using the Toolserver tools [44, 45].

(3) Page length.

The space occupied by the article on the encyclopaedia server, and its unit of measurement is bytes.

(4) Total number of edits.

The number of times that the article is written and edited [46].

(5) Total number of authors.

The number of people who have edited the article [46].

(6) Recent number of distinct authors (within the last 30 days).

The number of users who have written or edited the article within the last thirty days or month [46].

(7) Transcluded templates.

It is referred to the codes saved in the template namespace and is used to display the tables at the top and bottom of the article or the flags, etc. templates have also served as function in the programming languages [46].

3.3. Tools

3.3.1. Decision Tree

Decision tree is a unique method of presenting a system, which defines the intended system properly and simplifies the associated decisions [26, 27].

3.3.2. Fuzzy Inference Systems

Fuzzy inference is a process that converts the input into output by the defined rules. The basic structure of the fuzzy inference systems is composed of three sections. section one includes rules that is in the if-then form; section two is a database through which the Membership Functions (MFs) are defined, and finally section three includes the inference mechanism that is achieved with the help of the rules and use of the available data to attain a reasonable output.

Fuzzy inference systems have displayed a successful performance in the areas of data classification, decision analysis, and expert systems [47].

- Takagi-Sugeno Inference systems model.

One rule has been shown in the Sugeno fuzzy model in Eq. 1.

(Eq.1) If x is A and y is B then $z=f(x, y)$

A and B are fuzzy sets, $f(x, y)$ is a function in the conclusion section of the rule. When $f(x, y)$ is a first-order polynomial, the resulting fuzzy inference system is called first-order Sugeno fuzzy model. In addition, if $f(x, y)$ is equal to a fixed amount, the resulting fuzzy inference system is called zero-order Sugeno fuzzy model [48].

- ANFIS.

ANFIS is an acronym made from the first letters of Adaptive Neuro Fuzzy Inference System [31]. Qualitative aspects of human knowledge can be modelled by ANFIS [34]. ANFIS is a hybrid system that takes advantage of the potential benefits of Fuzzy Inference System (FIS) and Artificial Neural Network. Dynamic systems do not usually have linear behavior. Dynamic systems can be modelled with a linear behavior using. Sugeno fuzzy systems can be developed through ANFIS toolbox and tested after training [32].

- Subtractive Clustering method.

Clustering aims at categorizing a very large data set and providing a simple representation of the system behavior [31]. In case no information is available regarding the number of data set clusters, using SC algorithm is a quick. method for estimating the number of clusters and determining their centers [31]. The clustered data can be used to create a fuzzy inference system with the minimum rules and maximum efficiency [29]. ANFIS toolbox supports SC method [31]. In the next part, the variables are processed, classified, and clustered using version 3.6.8 of Weka and version 2, 1, 1 of fuzzy toolbox of MATLAB 8.3.

4. Methods and Results

4.1. Integration of Featured and Good Articles

First, the variables relating to 89 featured articles and 68 good articles are entered Weka in a CSV file format and are classified by the J48 decision tree. Figure 3 indicates that the statistical properties of the featured and good articles are so similar that cannot be separated. Given the similarity of specifications of the good and featured articles of the Persian Wikipedia, both of these articles have been examined under the title of a new dependent variable called high quality in this research.

Given the similarity of specifications of the good and featured articles of the Persian Wikipedia, both of these articles have been examined under the title of a new dependent variable called high quality in this research.

```
=== Summary ===

Correctly Classified Instances        89        56.6879 %
Incorrectly Classified Instances      68        43.3121 %
Kappa statistic                        0
Mean absolute error                    0.4911
Root mean squared error                0.4956
Relative absolute error               99.9748 %
Root relative squared error          100.0001 %
Total Number of Instances            157

=== Confusion Matrix ===

  a  b   <-- classified as
 89  0 |  a = Featured
 68  0 |  b = good
```

Figure 3. *Validation and confusion matrix results for good and featured articles.*

```
Correctly Classified Instances       254        89.4366 %
Incorrectly Classified Instances      30        10.5634 %
Kappa statistic                        0.8203
Mean absolute error                    0.0746
Root mean squared error                0.2469
Relative absolute error               18.889  %
Root relative squared error           55.6181 %
Total Number of Instances            284

=== Confusion Matrix ===

  a   b   c   <-- classified as
 71   4   0 |  a = Deletion
  6  35  11 |  b = Cleanup
  0   9 148 |  c = High quality
```

Figure 4. *Validation and confusion matrix results for high quality, cleanup and deletion articles.*

4.2. Method 1

4.2.1. Classification with the Decision Tree

Variables average, view, page length, number of edits, number of authors, recent authors, and templates are entered Weka in a CSV file format and are classified by the J48 decision tree. Figure 5 shows the decision tree [49].

As seen in figure 4., accuracy=89.4% and kappa statistic = 0.82 that is an acceptable result [50].

Confusion Matrix is shown in figure 4. ab and aa elements of CM matrix reveal that 71 articles have been classified correctly and 4 articles have been classified incorrectly from among 75 deletion articles. ca and ac elements reveal that none of the deletion and high quality articles have been classified incorrectly. cc and cb elements indicate that 148 articles have been classified correctly and 9 articles have been classified incorrectly from among 157 high quality articles. bc, bb, and ba indicate that 35 articles have been classified correctly and 17 articles have been classified incorrectly from among 52 cleanup articles. Given the results of CM matrix, it is concluded that deletion and high quality articles have been separated without any error [50]. The results obtained from J48 decision tree are as follows, see figure 5:

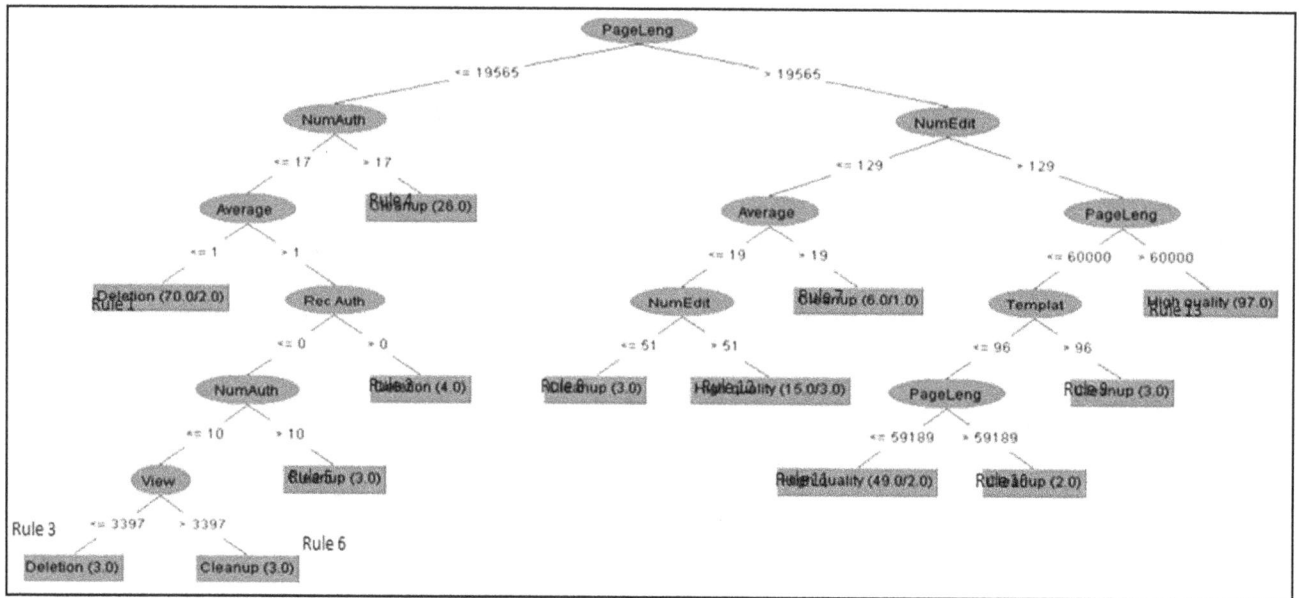

Figure 5. *Decision tree for J48 algorithm.*

4.2.2. Rules

Rule 1.

If (Average=<1) and (NumAuth=<17) and (PageLeng=<19565) then article is Deletion.

Rule 2.

If (Average>1) and (NumAuth=<17) and (PageLeng=<19565) and (RecAuth>0) then article is Deletion.

Rule 3.

If (Average>1) and (View=<3397) and (NumAuth=<10) and (PageLeng=<19565) and (RecAuth=<0) then article is Deletion.

Rule 4.

If (PageLeng=<19565) and (NumAuth>17) then article is Cleanup.

Rule5.

If (Average>1) and (PageLeng=<19565) and (NumAuth>10) and (RecAuth=<0) then article is Cleanup.

Rule 6.

If (Average>1) and (View>3397) and (PageLeng=<19565) and (NumAuth=<10) and (RecAuth=<0) then article is Cleanup.

Rule 7.

If (Average>19) and (PageLeng>19565) and (NumEdit=<129) then article is Cleanup.

Rule 8.

If (Average=<19) and (PageLeng>19565) and (NumEdit=<51) then article is Cleanup.

Rule 9.

If (PageLeng=<60000) and (NumEdit>129) and (Templat>96) then article is Cleanup.

Rule 10.

If (PageLeng>59189) and (NumEdit>129) and (Templat=<96) then article is Cleanup.

Rule 11.

If (PageLeng=<59189) and (NumEdit>129) and (Templat=<96) then article is High quality

Rule 12.

If (Average=<19) and (PageLeng>19565) and (NumEdit>51) then article is High quality.

Rule 13.

If (PageLeng>60000) and (NumEdit>129) then article is High quality.

4.2.3. Implementing with ANFIS

• ANFIS inputs.

The structure of Sugeno rules is the same as the rules obtained by the J48 decision tree. Seven variables are defined as ANFIS inputs [7, 26, 27]. Given the simplicity of relationships and calculations, input Membership Functions are taken into account as trapezoidal [26, 27]. Three MFs have been defined for variable average, two MFs have been defined for variable view, four MFs have been defined for variable Page length, three MFs have been defined for variable number of edits, three MFs have been defined for variable Number of Authors, two MFs have been defined for variable Recent Authors, and two MFs have been defined for variable templates.

• ANFIS output.

Thirteen MFs are allocated to the output, of which 3 MFs are allocated to deletion articles, 7 MFs are allocated to Cleanup articles, and 3 MFs are allocated to high quality articles. Each rule has been specified with a numeric value. The numeric values between -1 and +1 are used to specify the output MFs in a sense that the MFs with the values close to +1 indicate high quality articles, the MFs with the values close to 0 indicate Cleanup articles, and the MFs with the values close to -1 indicate deletion articles. Figure 6, shows the structure of designed system.

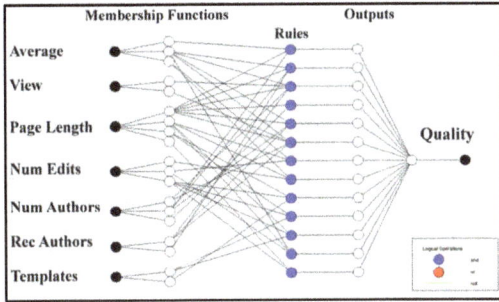

Figure 6. System structure in method 1 (using ANFIS and J48 decision tree).

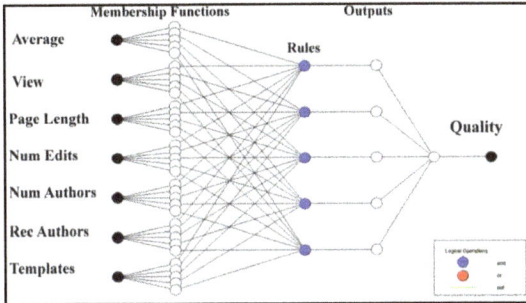

Figure 7. System structure in method 2 (using SC algorithm).

4.2.4. Model Validation

After designing the system, 189 articles (66% of data) have entered the system under the title of training data and 95 articles (33% of data) have entered under the title of checking data [31]. Figure 8, shows the output of the designed system when epoch = 0. The output of designed system with (○), and checking data with (-) are shown, respectively. Root Mean Square Error for training data and checking data are shown in (1) below.

(1) Epoch=0, Training RMSE=0.34, Checking RMSE=0.33.

As seen in figure 9, the error is decreased by increasing epoch. Considering figure 9, when Epochs>600, reduction of the error of checking data is so insignificant that has actually no remarkable impact on reduction of the system error; therefore, the system training is suspended in this stage and the RMS Error values of training and checking data are shown in (2) and the system output is shown in figure 10.

(2) Epochs=600, Training RMSE =0.271, Checking RMSE= 0.279.

Comparison of figures 8 and 10 shows that high quality and cleanup needed articles are more concentrated on the range of 0 and 1 while deletion articles are more scattered around -1. However, the system error is reduced as shown in (2).

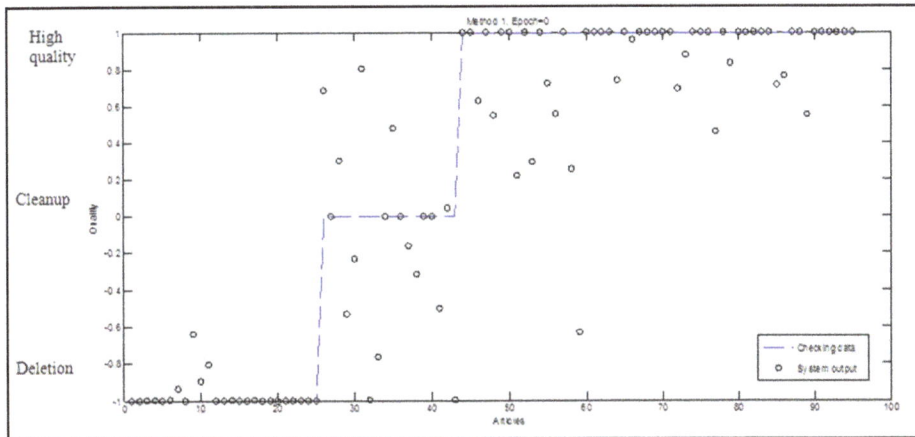

Figure 8. The model output and checking data are shown as circles and solid blue line, when epoch=0, in method 1.

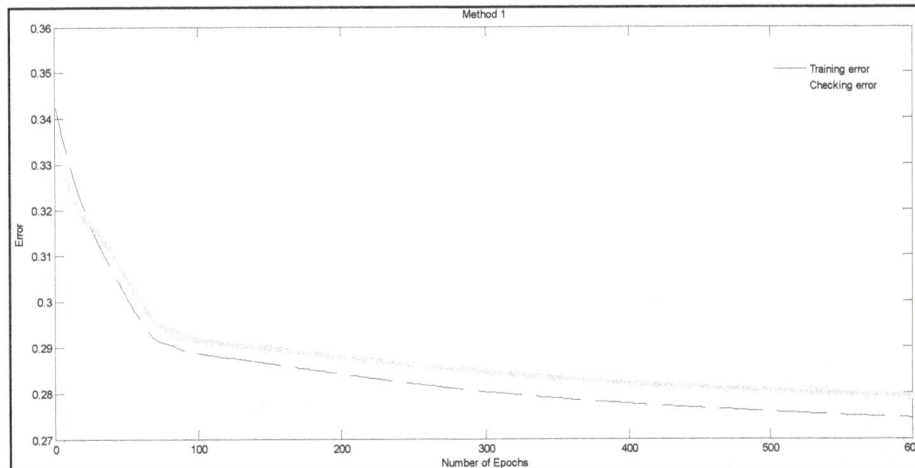

Figure 9. RMS Error values of training and checking data.

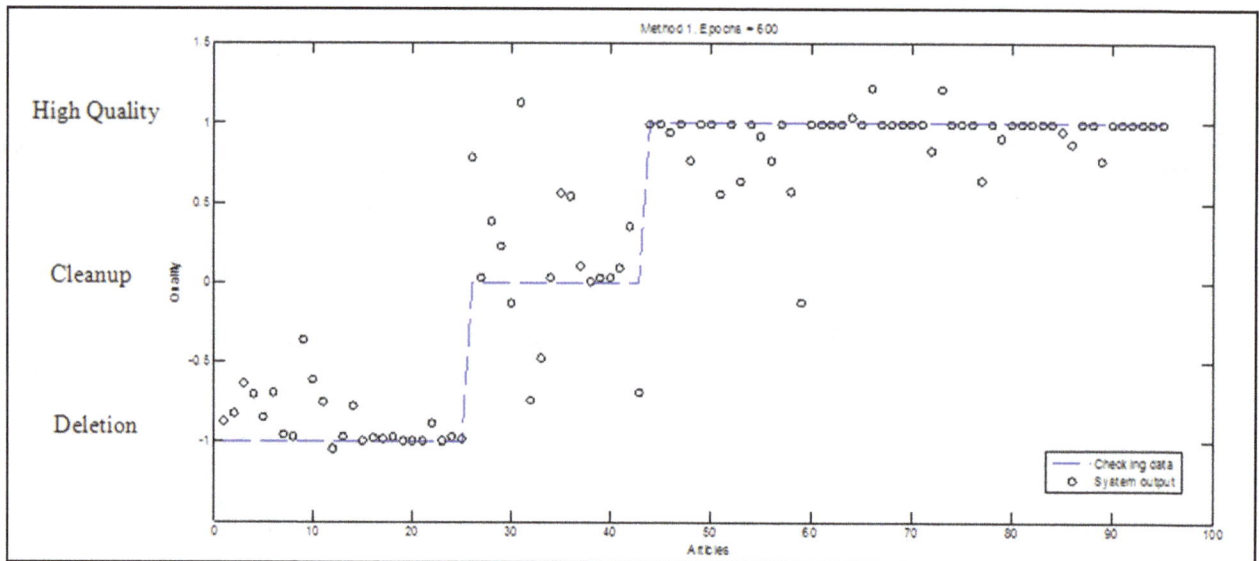

Figure 10. The model output and checking data are shown as circles and solid blue line, when epoch=600, in method 1.

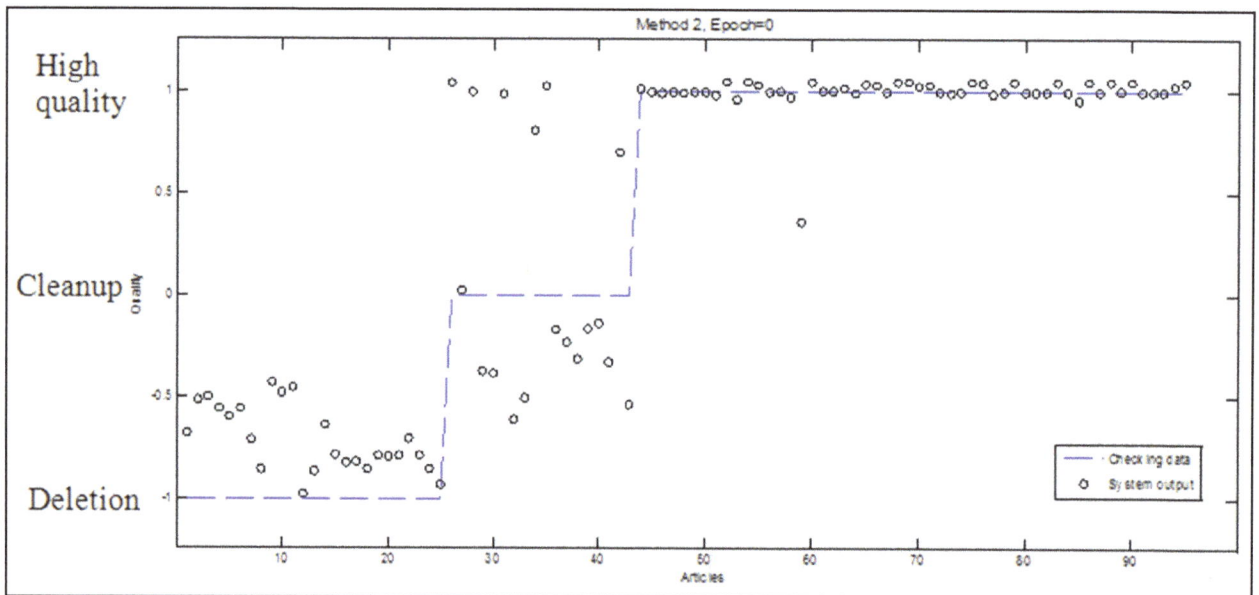

Figure 11. The model output and checking data are shown as circles and solid blue line, when epoch=0, in method 2.

4.3. Method 2

• Implementation with Subtractive Clustering

The membership function used in this method is Gaussian. The independent variables of average, view, page length, and number of edits, number of authors, recent authors, and templates enter the Workspace of MATLAB as Wikidatain and the quality of Wikipedia articles enter as Wikidataout. Instead of using of the titles of deletion, cleanup, and high quality articles, the numerical values of -1, 0, 1 are used, respectively.

(3) [C55, S55]=Subclust ([wikidatain, wikidataout], 0.55).

The command seen in (3) has allocated 5 MFs to each input variables and 5 MFs to the output variables, of which 1 MF is allocated to deletion articles, 1 MF is allocated to cleanup articles and 3 MFs are allocated to high quality articles. Value 0.55 indicates the radius of each cluster [31]. Figure 7, shows the designed system.

After designing the system, 189 articles (66% of data) have entered the system as training data and 95 articles (33% of data) have entered as check data. When epoch=0, RMS Error for training data and checking data are shown in (4). Figure 11, shows the output of the system, when epoch = 0.

(4) Epoch = 0, Training RMSE = 0.31, Checking RMSE= 0.32.

When epochs=600, RMSE for training data and checking data are shown in (5). Figure 12, shows the output of the system.

(5) Epoch = 600, Training RMSE = 0.27, Checking RMSE= 0.29.

Figure 13, indicates that the articles quality is improved by increasing average, view, page length, number of edits, number of authors, and templates variables. Figure 14, indicates that the variable recent authors is reduced by improving the articles quality.

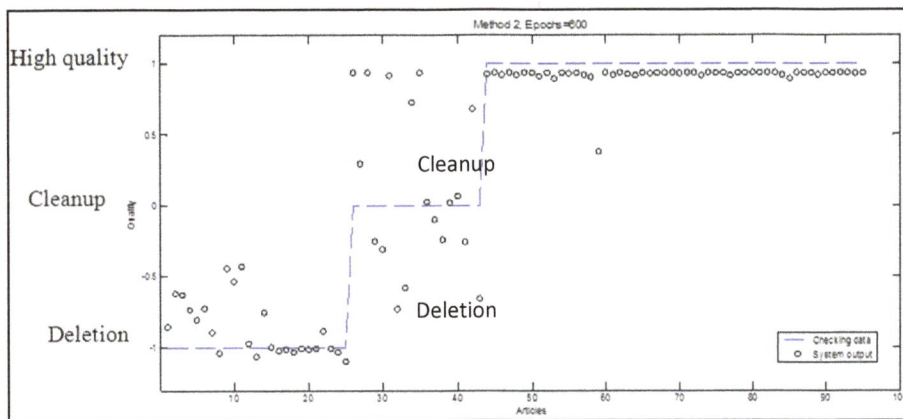

Figure 12. The model output and checking data are shown as circles and solid blue line, when epoch=600, in method 2.

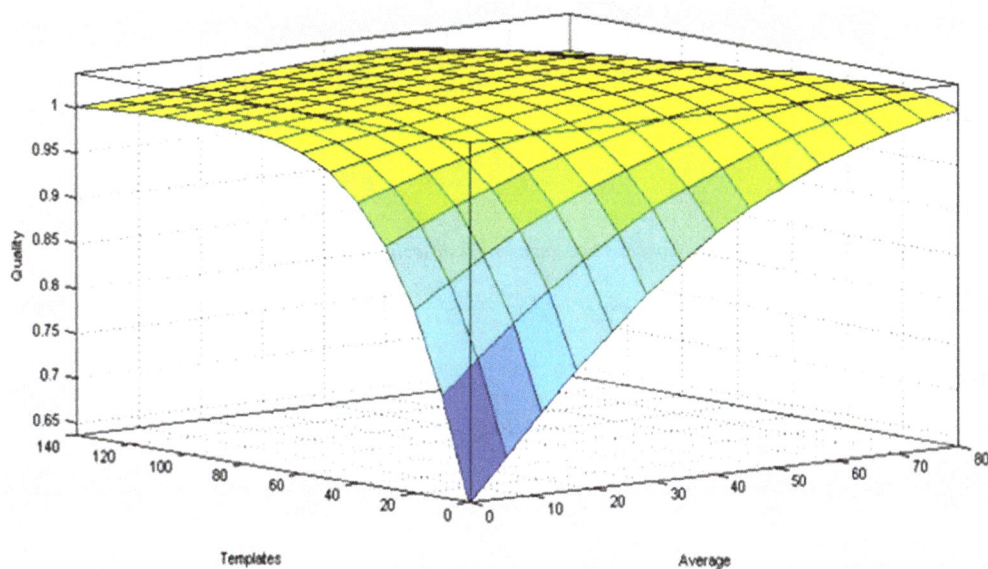

Figure 13A. The increase in Average and Templates variables cause an increase in articles quality.

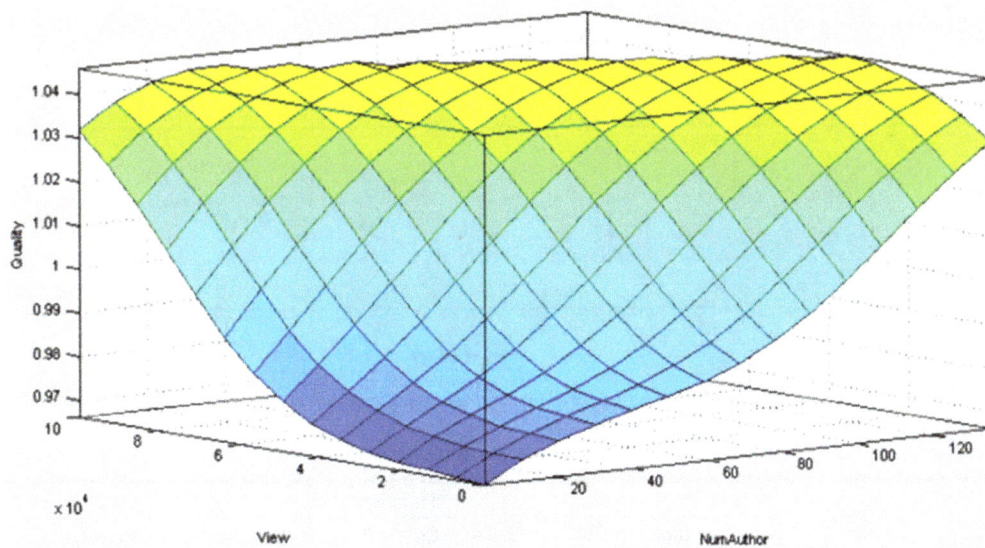

Figure 13B. The increase in view and Num Authors variables cause an increase in articles quality.

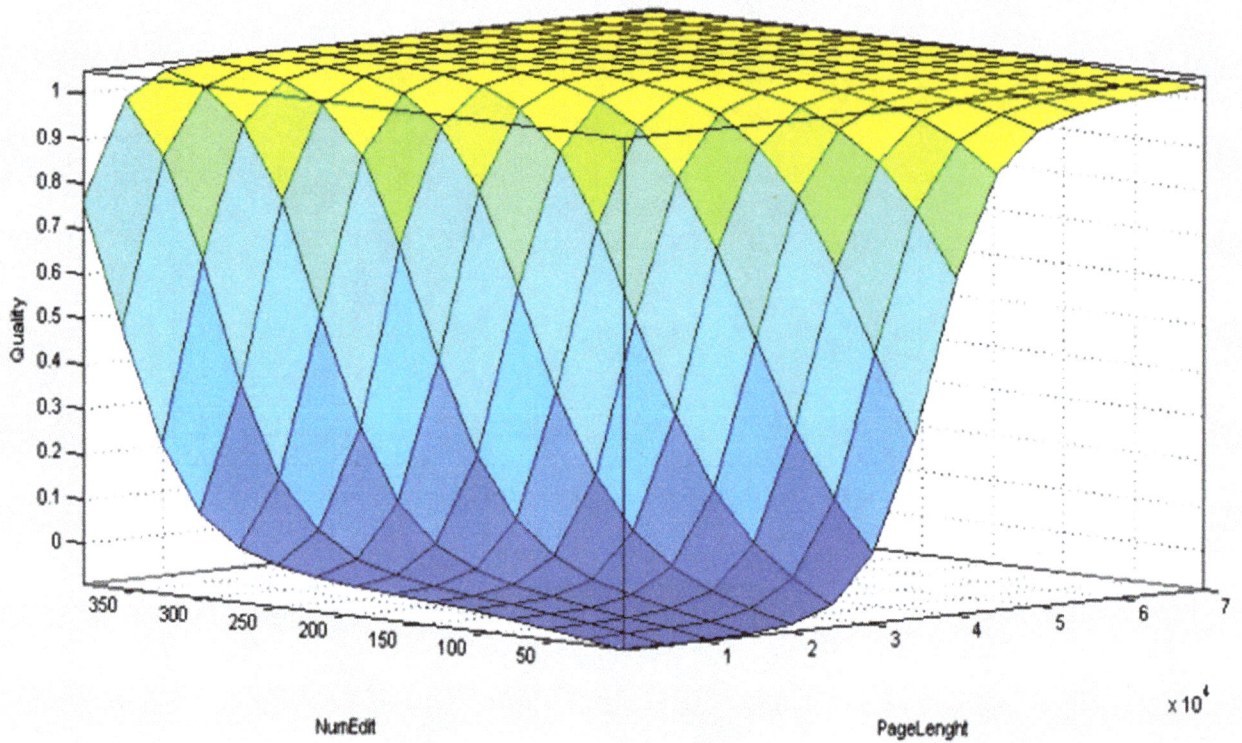

Figure 13C. *The increase in Num Edits and Page length variables cause an increase in articles quality.*

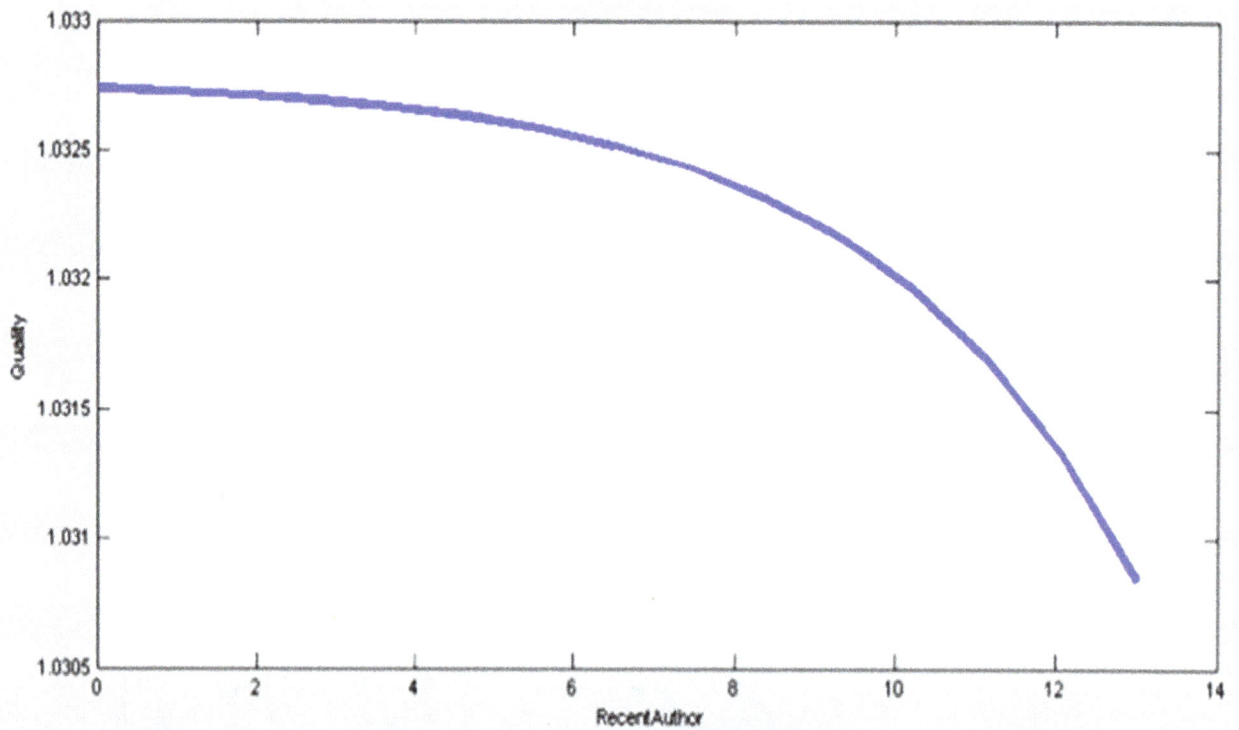

Figure 14. *Increasing the number of Recent Authors variable can lead to decrease the quality of articles.*

Table 1. *Compared two methods used in this research.*

Method	Rules	MFs		Type Of MF	Train Err		Chk Err	
		In	out		Epoch=0	Ep=600	Ep=0	Ep=600
(I)Decision tree & ANFIS	13	19	13	Trap	0.34	0.27	0.33	0.28
(II) Subtractive Clustering	5	35	5	Gaussian	0.31	0.27	0.32	0.29

5. Summery and Conclusion

In this study, the variables related to the articles listed in featured, good, cleanup and deletion Persian Wikipedia were collected. As it was noted in Section 4.1, due to the similarity of selected featured and good articles, these two variables have been studied and integrated under high quality articles. The collected data were entered into the Weka software, and then J48 decision tree has mined 13 rules of data relationships. At this part, the results had shown that 89.7% of the articles were classified correctly by decision tree. Confusion Matrix demonstrated that the decision tree separated high quality and deletion articles without error (shown in Fig. 4). Obtained rules to implement by ANFIS were entered into fuzzy toolbox of Matlab, linear relationship between inputs and outputs was established by ANFIS. In part 4.3, the variables for classification and implementation via Subtractive Clustering method were entered into Matlab software. 5 clusters were allocated to each variable, finally, articles were classified in three levels, including high quality, cleanup needed and deletion. As observed in the Figure 13, that the increased independent variables would cause increased articles quality. However, the figure 14, would indicate that increasing the number of recent authors can lead to decrease the quality of articles. Since when a paper was located as the group of cleanup or deletion articles, authors and editors try more to eliminate the article from those lists, Whereas in the case of articles selected as good and featured articles and their sublime position within encyclopedia recognized, subsequently, the authors of these articles have not been increased in the short term.

Table 1 has compared two methods used in this research, in Subtractive Clustering method, the number of rules reduced, while Membership functions and the error are increased in comparison to the first method. Consequently, the research query is answered and the objectives are fulfilled. Considering, the fact that classification of Wikipedia articles implemented user's opinion, it can be concluded that the system which is designed in separation or judgment articles is similar to users and site managers, style of thinking. By applying such a system, encyclopedia articles can be more quickly classified, and essays that might damage and discredit the can be promptly removed or revised validity Wikipedia and users trust quickly removed or revised. According to the results achieved by Stvilia et al. [51] in 2009 which showed that different communities have different models for measuring quality, the researchers of this paper have recommended such research to be done in other languages of Wikipedia encyclopaedia and its results utilized for raising the quality of encyclopaedia articles. The purpose of this study was to find the variables that would affect the quality of the Persian Wikipedia encyclopedia articles. It also aims at designing and implementation of a system to judge the Persian Wikipedia articles in three quality levels of high quality, cleanup needed, and deletion similar to users and administrators.

References

[1] Wikipedia. Persian Wikipedia, The Free Encyclopedia. Persian Wikipdia2013. https://en.wikipedia.org/wiki/Persian_Wikipedia (Persian Version).

[2] wikipedia. Statistics daily of persian wikipedia. Persian Wikipedia 2013. https://en.wikipedia.org/wiki/Wikipedia:Statistics (Persian Version).

[3] Liu J and Ram S. Who does what: Collaboration patterns in the wikipedia and their impact on article quality. ACM Trans Manage Inf Syst. 2011; 2: 1-23.

[4] Kittur A and Kraut RE. Harnessing the wisdom of crowds in wikipedia: quality through coordination. Proceedings of the 2008 ACM conference on Computer supported cooperative work. San Diego, CA, USA: ACM, 2008, p. 37-46.

[5] Priedhorsky R, Chen J, Lam SK, Panciera K, Terveen L and Riedl J. Creating, destroying, and restoring value in wikipedia. Proceedings of the 2007 international ACM conference on Supporting group work. Sanibel Island, Florida, USA: ACM, 2007, p. 259-68.

[6] Wilkinson DM and Huberman BA. Cooperation and quality in wikipedia. Proceedings of the 2007 international symposium on Wikis. Montreal, Quebec, Canada: ACM, 2007, p. 157-64.

[7] Ullah N. ANFIS BASED MODELS FOR ACCESSING QUALITY OF WIKIPEDIA ARTICLES. Computer Engineering. Dalarna University, 2010.

[8] Wikipedia. Wikipedia: Protection policy. 2013. https://en.wikipedia.org/wiki/Wikipedia:Protection_policy (Persian Version).

[9] Wikipedia. Policies and guidelines. 2013. https://en.wikipedia.org/wiki/Wikipedia:Policies_and_guidelin es (Persian Version).

[10] Wikipedia. Manual of Style. 2013. https://en.wikipedia.org/wiki/Wikipedia:Manual_of_Style (Persian Version).

[11] Wikipedia. Wikipedia: WikiProject Albums. 2013. http://en.wikipedia.org/wiki/Wikipedia:WikiProject_Albums.

[12] Walraven A, Brand-Gruwel S and Boshuizen HPA. How students evaluate information and sources when searching the World Wide Web for information. Comput Educ. 2009; 52: 234-46.

[13] Lim S. How and why do college students use Wikipedia? J Am Soc Inf Sci Technol. 2009; 60: 2189-202.

[14] Lucassen T and Schraagen JM. Propensity to trust and the influence of source and medium cues in credibility evaluation. J Inf Sci. 2012; 38: 566-77.

[15] Stvilia B, Twidale MB, Smith LC and Gasser L. Information quality work organization in wikipedia. J Am Soc Inf Sci Technol. 2008; 59: 983-1001.

[16] Geiger RS and Halfaker A. When the levee breaks: without bots, what happens to Wikipedia's quality control processes? Proceedings of the 9th International Symposium on Open Collaboration. Hong Kong, China: ACM, 2013, p. 1-6.

[17] Rowley J and Johnson F. Understanding trust formation in digital information sources: The case of Wikipedia. J Inf Sci. 2013; 39: 494-508.

[18] Noč M and Zumer M. The completeness of articles and citation in the Slovene Wikipedia. Program. 2014; 48: 53-75.

[19] Lih A. Wikipedia as Participatory journalism: reliable sources? metrics for evaluating collaborative media as a news resource. Proceedings of the 5th International Symposium on Online Journalism. 2004, p. 16-7.

[20] Stein K and Hess C. Does it matter who contributes: a study on featured articles in the german wikipedia. Proceedings of the eighteenth conference on Hypertext and hypermedia. Manchester, UK: ACM, 2007, p. 171-4.

[21] Wöhner T and Peters R. Assessing the quality of Wikipedia articles with lifecycle based metrics. Proceedings of the 5th International Symposium on Wikis and Open Collaboration. Orlando, Florida: ACM, 2009, p. 1-10.

[22] Saengthongpattana K and Soonthornphisaj N. Thai Wikipedia Quality Measurement using Fuzzy Logic. The 26th Annual Conference of the Japanese Society for Artificial Intelligence. Japan2012, p. ROMBUNNO. 4M1-IOS-3C-1.

[23] Anderka M, Stein B and Busse M. On the evolution of quality flaws and the effectiveness of cleanup tags in the English Wikipedia. Wikipedia Academy. 2012; 2012.

[24] Xiao K, Li B, He P and Yang X-h. Detection of Article Qualities in the Chinese Wikipedia Based on C4.5 Decision Tree. In: Wang M, (ed.). Knowledge Science, Engineering and Management. Springer Berlin Heidelberg, 2013, p. 444-52.

[25] Chai K, Hayati P, Potdar V, Chen W and Talevski A. Assessing post usage for measuring the quality of forum posts. Digital Ecosystems and Technologies (DEST), 2010 4th IEEE International Conference on. 2010, p. 233-8.

[26] Saravanan N, Cholairajan S and Ramachandran KI. Vibration-based fault diagnosis of spur bevel gear box using fuzzy technique. Expert Syst Appl. 2009; 36: 3119-35.

[27] Omid M. Design of an expert system for sorting pistachio nuts through decision tree and fuzzy logic classifier. Expert Syst Appl. 2011; 38: 4339-47.

[28] Jalali A and Mahmoudi A. Pistachio nut varieties sorting by data mining and fuzzy logic classifier. International Journal of Agriculture and Crop Sciences (IJACS). 2013; 5: 101-8.

[29] Chiu SL. Fuzzy Model Identification Based on Cluster Estimation. Journal of Intelligent and Fuzzy Systems. 1994; 2: 267-78.

[30] Yager RR and Filev DP. Approximate clustering via the mountain method. Systems, Man and Cybernetics, IEEE Transactions on. 1994; 24: 1279-84.

[31] Mathworks. Fuzzy Logic Toolbox: User's Guide (R2014a). 2014, http://www.mathworks.com/help/pdf_doc/fuzzy, pp. 2_109, 2_150, 2_156-158, 2_160-161.

[32] Gaur V, Soni A, Bedi P and Muttoo SK. Comparative Analysis Of ANFIS And ANN For Evaluating Inter-Agent Dependency Requirements. International Journal of Computer Information Systems and Industrial Management Applications. 2014; 6: 23-34.

[33] Yuan X, Khoshgoftaar TM, Allen EB and Ganesan K. An application of fuzzy clustering to software quality prediction. Application-Specific Systems and Software Engineering Technology, 2000 Proceedings 3rd IEEE Symposium on. 2000, p. 85-90.

[34] Wei L-Y, Chen T-L and Ho T-H. A hybrid model based on adaptive-network-based fuzzy inference system to forecast Taiwan stock market. Expert Systems with Applications. 2011; 38: 13625-31.

[35] Malhotra R and Sharma A. A neuro-fuzzy classifier for website quality prediction. Advances in Computing, Communications and Informatics (ICACCI), 2013 International Conference on. 2013, p. 1274-9.

[36] Afshoon R, Harounabadi A and Mir Abedini J. Assessment and Validating the Quality of Educational Web Sites using Subtractive Clustering. International Journal of Computer Applications. 2014; 98: 42-7.

[37] Wikipedia. Cleanup. 2013. https://en.wikipedia.org/wiki/Wikipedia:Cleanup (Persian Version).

[38] Saunders MN, Saunders M, Lewis P and Thornhill A. Research methods for business students, 5/e. Pearson Education India, 2011, pp. 237-240.

[39] Wikipedia. Deletion policy. 2013. https://en.wikipedia.org/wiki/Wikipedia:Deletion_policy (Persian Version).

[40] Wikipedia. Good articles. 2013. https://en.wikipedia.org/wiki/Wikipedia:Good_articles (Persian Version).

[41] Wikipedia. Good article nominations. 2013. https://en.wikipedia.org/wiki/Wikipedia:Good_article_nominations (Persian Version).

[42] Wikipedia. Wikipedia: Featured articles. Persian Wikipedia2013. https://en.wikipedia.org/wiki/Wikipedia:Featured_articles (Persian Version).

[43] Wikipedia. Wikipedia: Featured article criteria. Persian Wikipedia2013. https://en.wikipedia.org/wiki/Wikipedia:Featured_article_criteria (Persian Version).

[44] Wikipedia. Wikipedia article traffic statistics. 2013. http://stats.grok.se.

[45] Wikipedia. Wiki ViewStats. 2013. http://tools.wmflabs.org/wikiviewstats2

[46] Wikipedia. Glossary. 2013. http://en.wikipedia.org/wiki/Wikipedia:Glossary.

[47] Jang J-SR and Sun C-T. Neuro-fuzzy and soft computing: a computational approach to learning and machine intelligence. Prentice-Hall, Inc., 1997, pp. 73-74.

[48] Sivanandam SN, Sumathi S and Deepa SN. Introduction to Fuzzy Logic using MATLAB. Springer-Verlag New York, Inc., 2006, pp. 123-124.

[49] Witten, Frank and Hall. Data Mining: Practical Machine Learning Tools and Techniques, 3rd Edition. 2011, pp. 410.

[50] Bouckaert RR, Frank E, Hall M, et al. WEKA Manual for Version 3-6-2. Hamilton, New Zealand: University of Waikato, 2011, pp. 21-22.

[51] Stvilia B, Al-Faraj A and Yi YJ. Issues of cross-contextual information quality evaluation—The case of Arabic, English, and Korean Wikipedias. Library & Information Science Research. 2009; 31: 232-9.

24

Multiplex Communication with Synchronous Shift and Weight Learning in 2D Mesh Neural Network

Takuya Kamimura[1], Yasushi Yagi[2], Shinichi Tamura[1], Yen-Wei Chen[3]

[1]NBL Technovator Co., Ltd. Shindachimakino, Sennan, Japan
[2]ISIR, Osaka University, Mihogaoka, Ibaraki City, Osaka, Japan
[3]Ritsumeikan University, Nojihigashi, Kusatsu-shi, Shiga, Japan

Email address:
kamimura@nbl-technovator.jp (T. Kamimura), yagi@sanken.osaka-u.ac.jp (Y. Yagi), tamuras@nblmt.jp (S. Tamura), chen@is.ritsumei.ac.jp (Yen-Wei C.)

Abstract: We have previously proposed a multiplex communication system in a neural network. However, this system is designed to force the network to communicate in a multiplexed manner, in which "codes" or "temporal sequences" are inevitably induced. This means that the network has a main loop and coding/decoding circuits, which are somewhat artificial. In this paper, we show that it is also possible to communicate without these artificial guidance aids by multiplexing in a 2D mesh-type neural network, where learning procedures are used to find paths from an originating neuron to a destination neuron. We also provide statistics from these neural networks to show that random sequences occur more frequently than non-random sequences.

Keywords: Brain Information Processing, Neural Circuit, Pseudo-Random Sequence, M-Sequence, Multiplex Communication

1. Introduction

Investigation of the manner in which information is communicated in the brain is an important theme in brain science. We have previously developed a time-shift map for analysis of the transmission of electroencephalogram (EEG) or magnetoencephalogram (MEG) signals in the brain [1], and have shown that it is effective for diagnosis of transient global amnesia (TGA) [2]. The time-shift map is a graphical representation that shows the time difference at which the cross-correlation between a brain wave on a point and that at another point becomes a maximum. An example of a time-shift map is shown in Fig. 1. The advantageous feature of this method when compared with the magnetic resonance imaging (MRI) dipole diagram [3] is that this method can follow even small signal flows. We can see that sub-tasks are processed in each hemisphere within 5 ms, and after 10 ms, the results are exchanged between the hemispheres. The question that arises here is how the neuron cells locate the target cells to send the required signal or, alternatively, how the responsible cells can obtain the necessary signals from the due cells, even when they are at remote locations. This is a problem of finding communication links in a neuronal

network. This paper will contribute to a solution to this question, and provide a basic idea of how events are stored in the brain and how the brain can associate/recall matters related to these events.

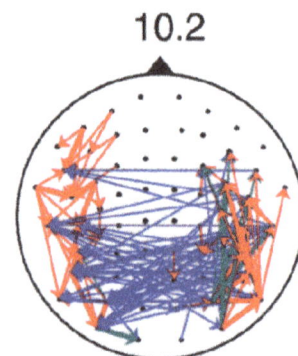

Fig. 1. Time-shift map of 10.2 Hz wave from MEG. Red < 5 ms < Green < 10 ms < Blue.

Figure 2 shows the simplest possible code multiplex communication of 2×2 communication on a 2D synchronous shift neural network, i.e., the number of input neurons is two and the number of output neurons is also two. Expanding upon this, we have previously proposed a multiplexed communication model on a 2D synchronous shift neural network that is composed of a main loop circuit and some coding/decoding neural networks [4], as shown in Fig. 3. The network is forced to communicate in a multiplexed manner, because there are only two lines between the three transmitting neurons and the three receiving neurons of the three communication channels. It was then observed that the network frequently generates and uses pseudo-random sequences, including M-sequences [5]. It is well known that in many real-world artificial communication systems, such as mobile phones, pseudo-random sequences—and particularly M-sequences—are used as almost orthogonal codes that are easy to discriminate from each other. Additionally, we have observed the M-sequence family occurring in spike trains from cultured neural networks significantly more frequently than would occur by chance [6, 7].

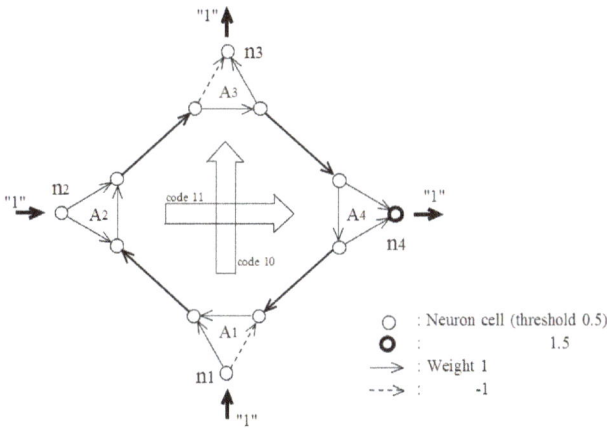

Fig. 2. Simplest example of a 2×2 multiplex communication channel by a neural network. Each neuron expressed by a small circle outputs a "1" if the weighted sum of inputs exceeds the required threshold, and outputs a "0" otherwise. If "1" is input to n_1, then it is converted into a sequence "10" by the local loop A_1 and is transferred to A_2 and A_3. Finally, "1" is output from n_3. Similarly, an input "1" to n_2 is transferred to A_4, and a "1" is output from n_4. This means that there are two independent communication channels working by temporal code multiplexing on the loop line.

However, it remains to be seen whether it is possible to communicate in a multiplexing manner in more natural neural networks without using an artificial shape to force multiplexing. Therefore, in this paper, we treat a more naturally homogeneous grid-shaped neural network. While our previous network was tested by generating numerous random networks without a learning function, here we provide a learning function for the networks to enhance the communication links.

2. Path Search Model

It is well known that a neuronal cell has more than 1000 synapses [8], and that large numbers of connection paths and loops thus exist in a neural network [9–11]. Stimulation is transmitted from one neuronal cell to another through the synapses, and the membrane electrical potential is thus increased. If the electrical potential exceeds a specific threshold, then the neuronal cell is fired, and if the potential is beneath the threshold, then the cell is not fired. The firing model [12] of a neuronal cell is as shown in Eq. (1).

$$E(t) = \begin{cases} 1 & \sum_{n=1}^{N} E_n(t-1)W_{E,E_n} > threshold \\ 0 & otherwise \end{cases} \quad (1)$$

Here, $E(t)$ represents the state of cell E at a discrete time t (where $t=0, 1, 2, \ldots$), 1 represents the firing state, and 0 represents the static state. N is the number of other cells that are connected to the cell, and W_{E,E_n} is the connection weight between E and E_n. The state of a cell is thus determined on the basis of whether the total sum of the products of the connection weight and the state of the other connected cell at the previous clock cycle time exceeds a threshold or not. In each neuron cell, the state will change with progress of t. We regard the communication as successful if an output cell is fired when external stimulation is provided to an input neuron cell and the stimulation is thus transmitted to the output cell through the neuron cells in the neural network.

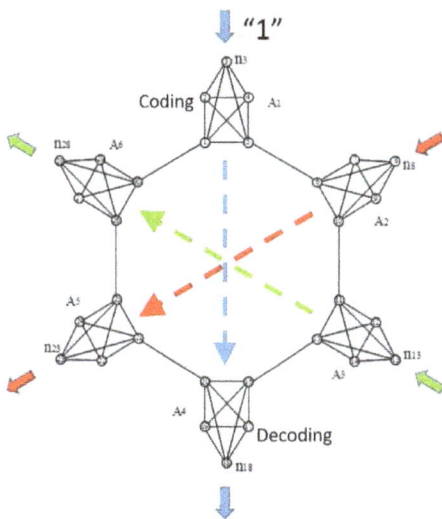

Fig. 3. A 3×3 communication model. The pulse "1" sent from neuron n_3 is supposed to be received by n_{18} only, while that sent from n_8 is supposed to be received by n_{23} only, and that sent by n_{13} is supposed to be received by n_{28} only.

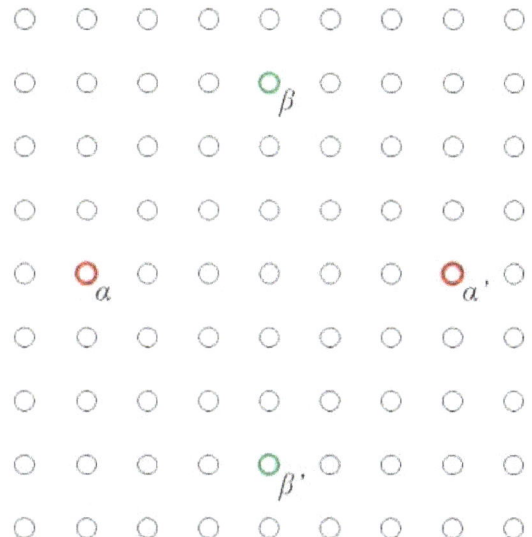

Fig. 4. Path search model with 9×9 grid.

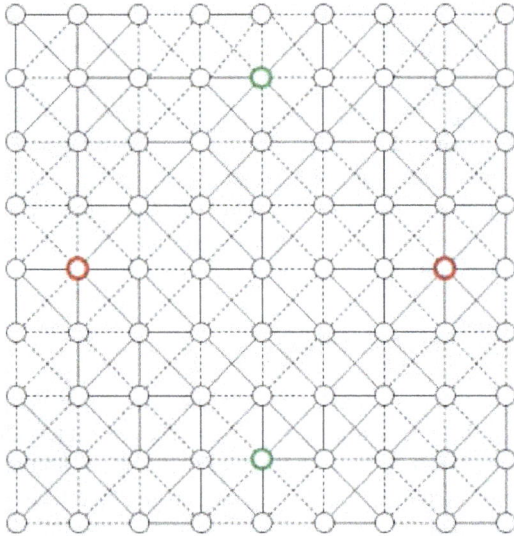

Fig. 5. *Bidirectional connection weights. Solid lines are positive weight connections, and dotted lines are negative weight connections.*

We formed a 2×2 multiplexing communication model with two input neuron cells to transmit the information and two corresponding receiving neuron cells on a 9×9 grid-shaped neural network. We call this the "path search model" [13]; the model is as shown in Fig. 5. Here, α and β are the input cells for information transmission, and α' and β' are the cells that receive information from α and β, respectively. Between a specific cell and its eight neighbouring cells, the connection weights are set randomly at time 0 within the range [−1, 1]. An example of the initial weights is shown in Fig. 5. Here, a solid line represents a positive weight connection, and a broken line represents a negative weight connection. Firing information input to input cells α and β is transmitted in the network, with the transitions of the cell states occurring according to the connection weights that are set between the cells and Eq. (1) with the progress of the time steps.

In addition, cells are sometimes fired voluntarily and randomly, irrespective of the firing transition. We call this phenomenon "random firing." With the combination of firing by the state transition and random firing, the firing cells thus move.

An example of a state transition is shown in Fig. 6. We can see that firing in Fig. 6(a) of α at time $t=0$ leads to the change into Fig. 6(b) at time $t=1$.

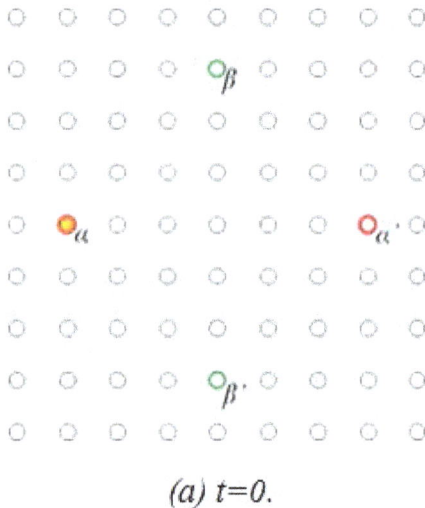

(a) t=0.

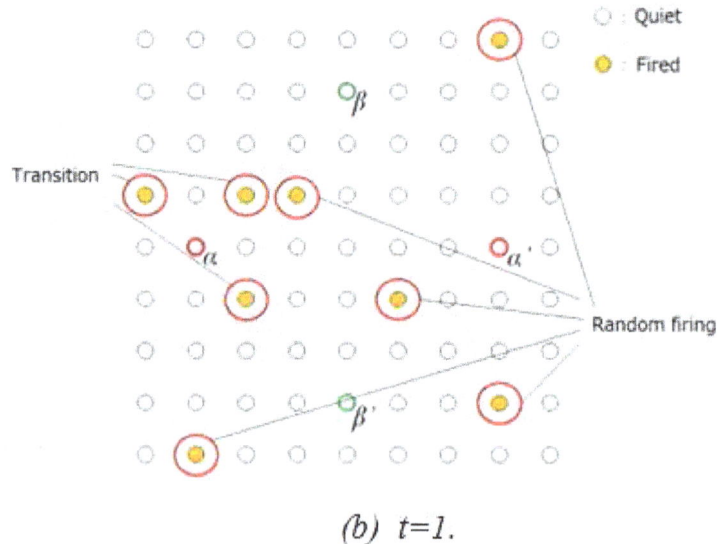

(b) t=1.

Fig. 6. *Example of a state transition.*

2.1. Updating Weights

The weight set between cells expresses the transmission efficiency of the relevant synapse. Synapse transmission efficiency in a natural neuronal network is reinforced if a post-synaptic neuron is fired by the firing of a presynaptic neuron or even by chance. This is called the Hebb rule [14]. In the path search model, the connection weights are also updated based on the Hebb rule. In this case, there are two cases of the weight updating rule of reinforcement by simultaneous firing and that based on the state one clock cycle before firing. These cases are expressed using Eq. (2) and Eq. (3), respectively.

$$\Delta w_i = \alpha s_i(t)s_{i+1}(t) + \beta_1 w_{i-1} + \beta_2 w_{i+1} \qquad (2)$$

$$\Delta w_i = \alpha s_i(t)s_{i+1}(t-1) + \beta_1 w_{i-1} + \beta_2 w_{i+1} \qquad (3)$$

Here, w_i is the attention paid weight, $s_i(t)$ is the state of cell i at time t, and i+1 represents the cell that is connected to cell i with weight w_i. Weights w_{i-1} and w_{i+1} represent those weights ahead of the extending weight along the connected direction, and have the role of an inertia term to ensure that the weight updates smoothly. In addition, α, β_1, and β_2 are coefficient parameters. Figure 7 shows a schematic diagram of these weights.

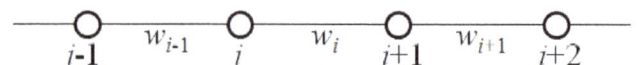

Fig. 7. *Schematic diagram of weights.*

In the case of simultaneous firing reinforcement, as shown in Eq. (2), the weights are updated by referring to the states of cell i and cell i+1 at time t, and in the case of firing reinforcement one clock cycle before firing, the weights are updated by referring to the state of cell i at time t and the state of cell i+1 at time t−1, respectively. The increment of Δw_i by a weight update is divided equally among the remaining neighbours, and the same amount is deducted from each of the weights in the neighbours. Therefore, the total sum of weights of the whole model does not change, even after the weights are updated. While we proposed two updating rules, there were no differences in the results produced by these rules, and thus we simply applied Eq. (2) in the following experiments.

2.2. Success or Failure of Communication

In a succession of state transitions, a case occurs where the firing information of the transmitting side (α, β) is transmitted to the receiving side (α', β') and the neurons are fired. In this case, we can regard the information as being transmitted and we can also regard the communication as being successful.

In case of a 2×2 multiplex communication, the information from α should be transferred to α', while that of β should be transferred to β'. Therefore, when the information is input to α at $t=0$, the number of firings in the range from $t=6$ to $t=63$ should be $N(\alpha') > N(\beta')$, and when the information is input to β at $t=0$, then the number of firings of β' should be $N(\alpha') < N(\beta')$, and only in a case such as this can we regard the communication as having succeeded fully. Here, $N(\alpha')$ represents the number of firings of α' within the corresponding period (i.e., from $t=6$ to $t=63$, or from $t=6$ to $t=31$).

2.3. Sequence That Appeared at the Cell

When we observe all 81 cells of the path search model and pay attention to each cell, we can obtain a sequence in which 0 and 1 appear with the progress of the time step t. Then, we can obtain knowledge of what contributes to the 2×2 communication process by observing and analysing these sequences.

3. Experimental and Discussion

We performed a simulation experiment for 2×2 communication using the path search model.

3.1. Experiment 1

We performed the experiment in three steps (①–③), composed of the weight learning term, an α test, and a β test, as shown in Fig. 8.
① Weight learning term: We perform up to $t=63$ path searches when setting the state of α in Fig. 8 as 1 at $t=0$. In other words, we update the connection weights based on the learning rule.
② α test: Using a model after the weight learning process, we set the state of α in Fig. 8 to be 1 at time $t=0$ and watch the state transition up to $t=63$. During this process, we do not update the weights. Here, if $N(\alpha') > N(\beta')$, then the α test is

successful, while if $N(\beta') < N(\alpha')$, we regard the α test as a failure.
③ β test: Using a model after the weight learning process, we set the state of β in Fig. 5 to be 1 at the time $t=0$ and watch the state transition to $t=63$. During this process, we again do not update the weights. Here, if $N(\beta') > N(\alpha')$, then the β test is successful, and if $N(\alpha') < N(\beta')$, then we regard the β test as a failure.

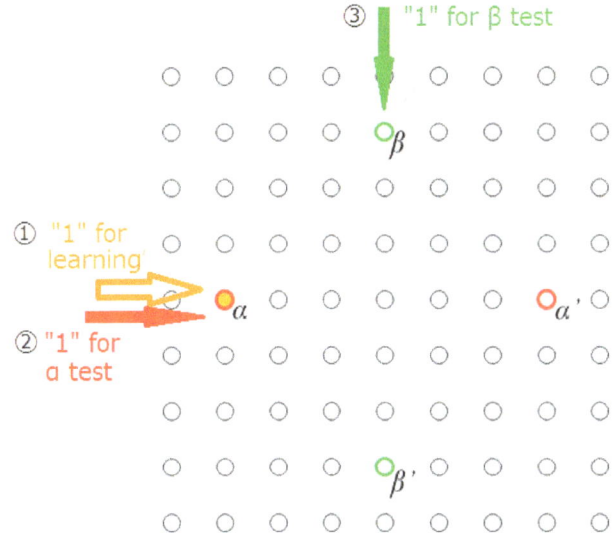

Fig. 8. Experiment 1: The firing state shown is that at $t=0$ of learning phase ①, where α is stimulated. We tried 10^6 times of a series of phase ①-③ tests.

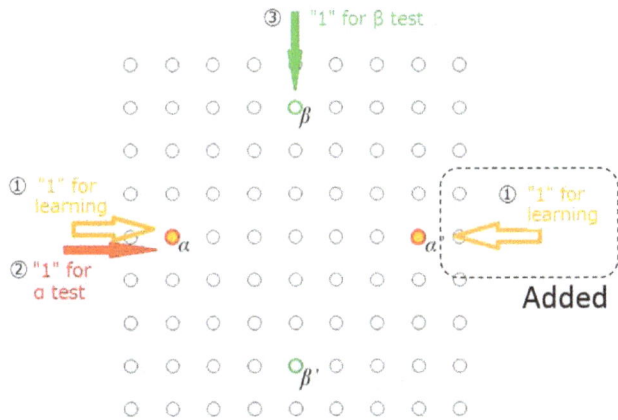

Fig. 9. Experiment 2: Bidirectional learning. The firing state shown is that at $t=0$ of learning phase ①, where α and α' are both stimulated. We tried 10^6 times of a series of phase ①-③ tests.

3.2. Experiment 2

In each weight learning term, weight updates were only performed based on the transition from α in Experiment 1, as shown in Fig. 8. Therefore, we performed weight updates based on the information from the transmission side only, and did not have the information from the cell on the reception side that was the goal for the information. In Experiment 2, as an improvement to the learning process, we formulated a bidirectional learning process, where the states of both α and α' were set as "1" at time $t=0$, and we then performed a path search up to $t=63$, as shown in Fig. 9.

3.3. Experiment 3

In the weight learning processes of Experiments 1 and 2, the learning was based solely on the information of the α side and did not have access to the information of the β side, as shown in Figs. 8 and 9. Therefore, in Experiment 3, we used both-channel learning, as shown in Fig. 10. In this case, both states α and α' are set to be "1" at time $t=0$ and the path search is then performed up to $t=31$. After that, we cleared all the states and set the states of β and β' to be "1" at time $t=0$ and again performed the path search up to $t=31$.

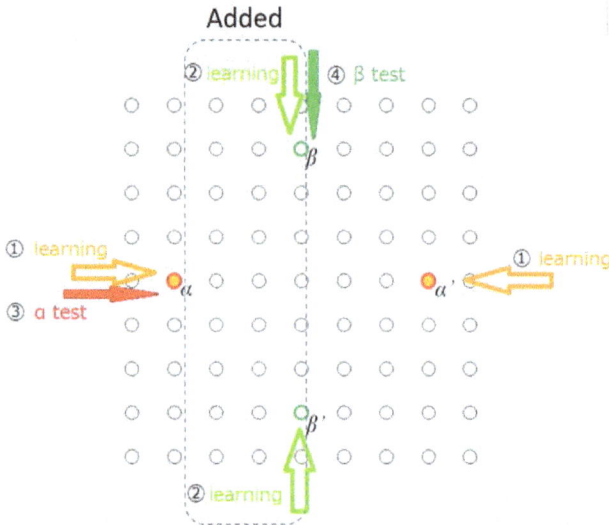

Fig. 10. *Experiment 3: Both-channel learning. The firing state shown is that at t=0 of learning phase ①, where α and α' are both stimulated. We tried 10⁶ times of a series of phase ①-③ tests.*

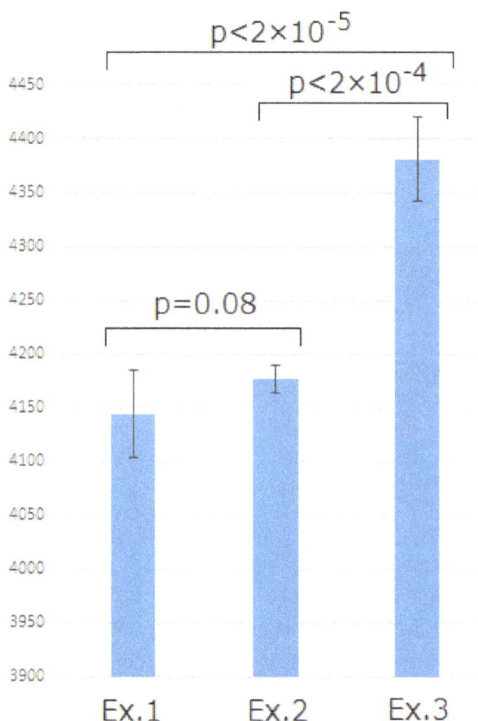

Fig. 11. *Average numbers of full successes among 10⁶ trials. Each experiment is repeated five times and the standard deviations are also shown as error bars.*

The results of Experiments 1, 2, and 3 are shown in Fig. 11. Here, full success means the case where both the α test and the β test succeeded. From each experiment, we can see that 2×2 communication can be realized in the path search model based on randomly assigned initial weights.

In addition, learning becomes easier to advance by using not only the information from the transmission side but also the information from the reception side, and we see that the number of successes increases from a comparison of Experiments 1 and 2, although it is not significant in the 5% level but is significant in the 10% level. We can also see from Experiment 3 that the number of 2×2 communication successes definitely increases by learning in both the α and β channels sequentially when compared with single channel learning in only α or β.

3.4. Experiment 4

In Experiment 3, by varying the parameters of the random firing rate or the firing threshold of the cell, and by performing reinforcement by simultaneous firing or reinforcement by a single clock period before firing, we observed the types of sequences that appeared most often when the 2×2 communication was successful or when it failed. The parameters that we varied were two kinds of random firing rates of 0.1 and 0.05, four kinds of firing thresholds of 0, 0.25, 0.5, and 0.6, and two kinds of reinforcement by simultaneous firing or by a single clock period before firing.

The target sequences to be checked are the sequences with "1" at both ends and with lengths ranging 3 to 8 digits. This gives a total of 120 kinds of target sequence, as follows:

Length 3: [1 1 1] of one kind;

Length 4: [1 0 1 1], [1 1 0 1], [1 1 1 1] of three kinds;

Length 5: [1 0 0 1 1], [1 0 1 0 1], …[1 1 1 1 1] of seven kinds;

Length 6: [1 0 0 0 1 1], [1 0 0 1 0 1], …[1 1 1 1 1 1] of 15 kinds;

Length 7: [1 0 0 0 0 1 1], [1 0 0 0 1 0 1], …[1 1 1 1 1 1 1] of 31 kinds;

Length 8: [1 0 0 0 0 0 1 1], [1 0 0 0 0 1 0 1], …[1 1 1 1 1 1 1 1] of 63 kinds.

We classify these 120 kinds of sequences into two kinds of sequences: one where the arrangement of "0s" and "1s" is a random-type arrangement and another where the arrangement is not random, as follows.

Case 1: Six output sequences composed of all "1s", i.e., [1 1 1], [1 1 1 1], …, [1 1 1 1 1 1 1 1] are regarded as non-random sequences. All others are regarded as random sequences.

Case 2: In addition to the six kinds of non-random sequences composed of all "1s", the three sequences of [1 0 1 0 1], [1 0 0 1 0 0 1], [1 0 1 0 1 0 1] are also regarded as non-random sequences. That is, totally nine sequences are regarded as non-random sequences. The three sequences of [1 0 1 0 1], [1 0 0 1 0 0 1], [1 0 1 0 1 0 1] may be called "semi-random" sequences.

For these output sequences, we compare the numbers of their appearances when the communications succeeded and failed. We divide the number of appearances when successful by the number of the appearances when the communication failed. Therefore, if these values exceed 1, then the sequence is relatively frequent when the communication is successful, and if the value is less than 1, the sequence is relatively frequent when the communication failed. We show the average rates of the detected sequences in the results for all parameters in Fig. 12 on the upper side, with an enlarged section on the lower side.

From Fig. 12, we can see that the perfectly non-random

sequences of [1 1 1], [1 1 1 1], and [1 1 1 1 1] with all "1" outputs had values of less than 1, while the other sequences tended to exceed 1. This result shows that when the 2×2 communication succeeded, random-like sequences appeared relatively frequently, and when the communication failed, the non-random sequences appeared relatively frequently. Therefore, this indicates that the random-like sequences contribute to the success of the communication.

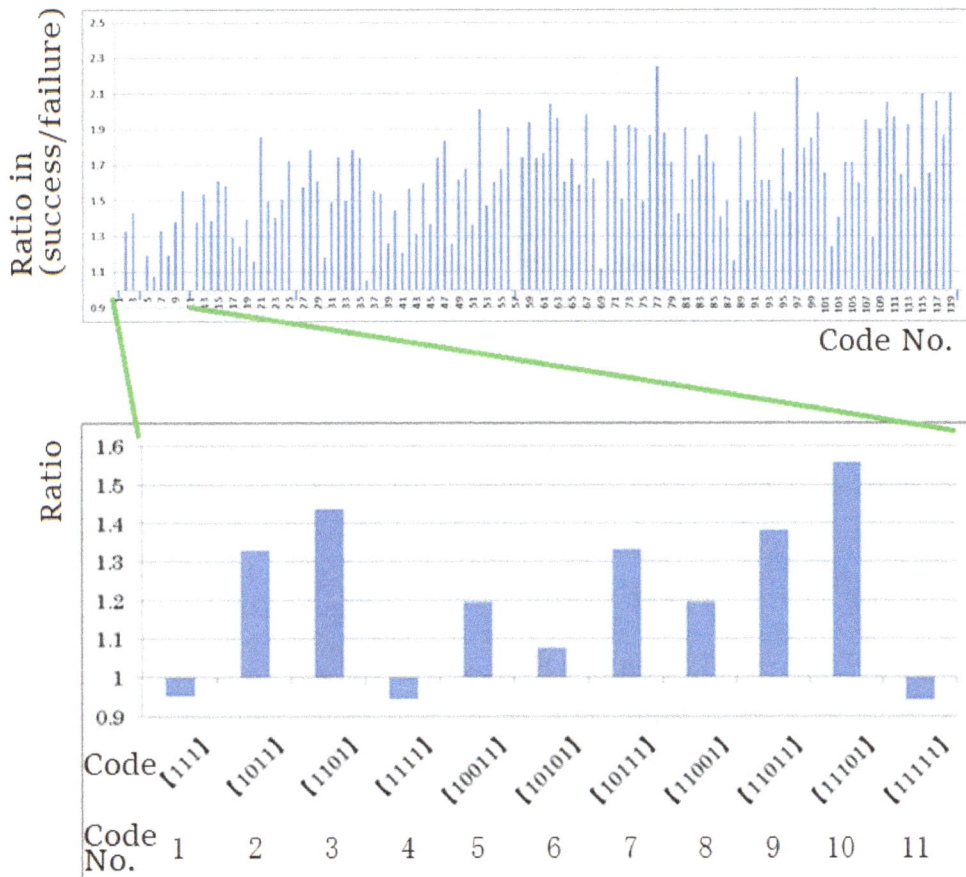

Fig. 12. *Ratio of number of sequences in the cases of success/failure that appeared in the output trains for 120 kinds of sequences.*

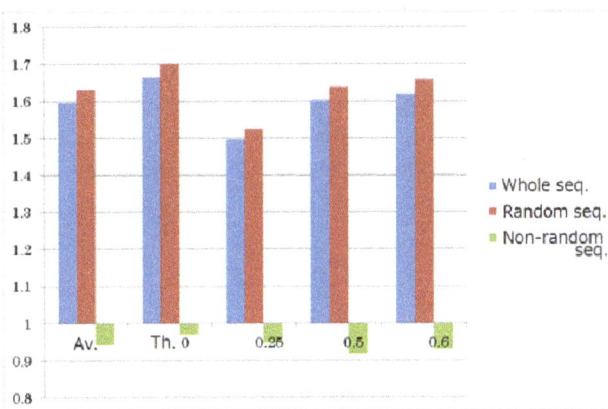

Fig. 13. *Comparison of the ratios of numbers of sequences in the cases of success/failure that appeared in the output trains when varying the threshold (Th.). Av. means average. The semi-random sequences are regarded as random sequences in this case.*

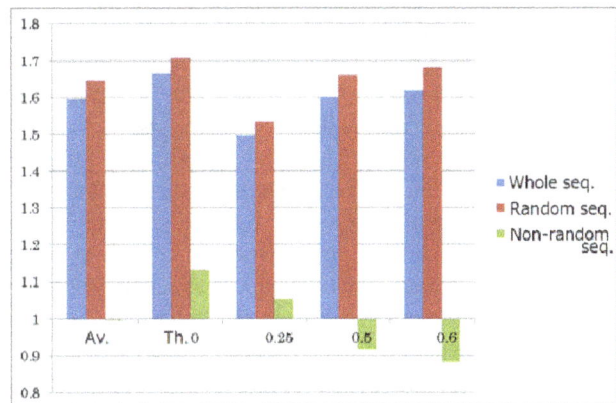

Fig. 14. *Comparison of the ratios of numbers of sequences in the cases of success/failure that appeared in the output trains when varying the threshold. The semi-random sequences are regarded as non-random sequences in this case.*

In Fig. 13, the output ratios of complete sequences, random sequences, and non-random sequences for various values of the threshold are shown. In this figure, the non-random sequences are again those composed of all "1s" (Case 1). We can see from Fig. 13 that, irrespective of the firing threshold, the random sequences appear more often when the communication succeeded than when communication failed.

In Fig. 14, the results with regard to the nine kinds of non-random sequence, including the three kinds of semi-random sequence of [1 0 1 0 1], [1 0 0 1 0 0 1], [1 0 1 0 1 0 1], are shown (Case 2).

In addition, we know from the results shown in Fig. 14 that a difference in the ratio of the number of output random sequences and non-random sequences observed in the cases of communication success and failure increases when we increase the threshold. This indicates that when the firing threshold of the cell rises, the quantity of firing decreases, and the degree of difficulty of the information transmission and communication processes increases. In such situation, the random characteristics of the sequences became more important.

As a result of the experiments, the factor that is most important to the success or failure of the communication is known to be the random characteristics of a sequence output by a cell in a path search model.

3.5. Synchronous Lateral Inhibition-Type Reception

We have confirmed in a previous loop-type multiplex communication neural network that when we know the time at which the signal is due to arrive at the receiving neuron, the lateral inhibition (LI)-type receptive field characteristic [15] was effective and its recognition rate was improved considerably [4]. This was a case where the network shape was fixed and the process was performed without learning. We also performed an experiment in the present neural network to receive a pulse train with LI-type receptive field under the assumption that the pulse train will arrive with the lowest possible delay. Because the pulse takes at least six clock cycles to arrive and never arrives at time 5 or earlier, the shape of the LI-type receptive field is set to be asymmetric, unlike the ordinary symmetric field, as shown in Fig. 15. We can see from Fig. 16 that the correct recognition rate is improved by about six times when compared with the results of Experiments 1, 2 and 3. The LI-type receptive field has characteristics in common with the living body neural network and this is also shown in this experiment, although it is assumed that the arrival time can be anticipated beforehand by the receptive neuron. This temporal anticipation ability contributes to the improved correct reception rate. Then, LI reception is classified as synchronous reception, while the previous simple counting reception methods of Experiments 1, 2, and 3 can capture the signals irrespective of the time, and are thus called asynchronous reception methods.

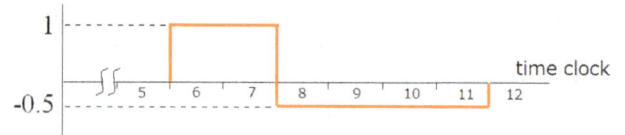

Fig. 15. *Weight of LI-type temporal receptive field.*

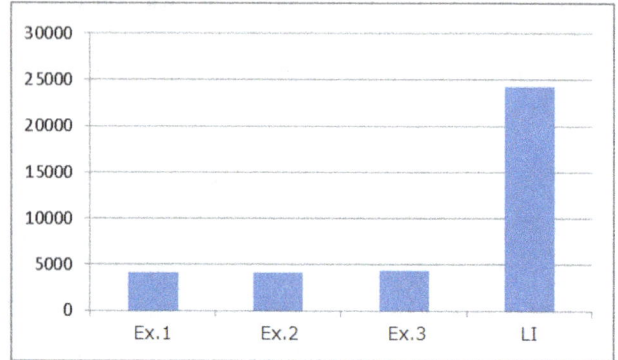

Fig. 16. *Numbers of full successes by LI-type temporal receptive field (LI: synchronous reception) among 10^6 trials shown for comparison with the results (Experiments 1, 2, 3; asynchronous reception) of Fig. 11.*

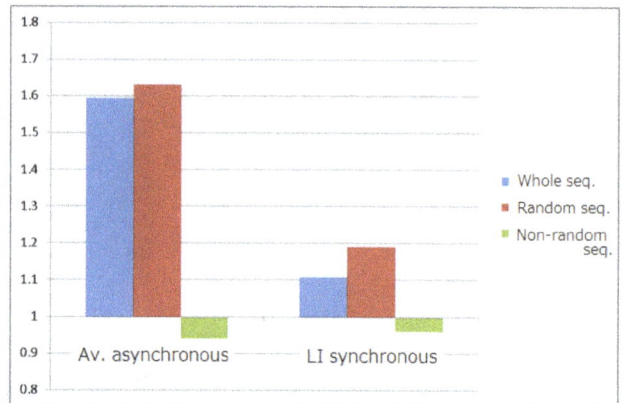

Fig. 17. *Comparison of the ratios for success and failure of the number of sequences in pulse trains on average by asynchronous reception from Fig. 11 vs. that by LI synchronous reception.*

We can see from Fig. 17 that in the LI reception case, while the ratios of the random and non-random sequences for success and failure were both close to 1, the difference between the cases of success and failure became small. However, the relative significance of the random sequences in LI reception seems to have increased. In this case, the three kinds of semi-random sequences are included in the non-random sequence (Case 2).

4. Discussion and Conclusions

In a previous paper, we proposed multiplex communication systems in neural networks. However, these systems are composed and designed such that they force the network to communicate in a multiplexed manner where "codes" or "temporal sequences" are inevitably induced. In this case, the network has a main loop and some coding/decoding circuits, and these aspects are somewhat artificial. In this paper, we have shown that without these artificial guidance aspects, it is also possible to communicate via multiplexing on a 2D mesh-type neural network, where some learning procedures are used to find the paths from an originating neuron to a destination neuron.

In this study, we built a path search model on a homogeneous mesh-type 2D neural network with a synchronous state transition; it was shown in the model that 2×2 multiplex communication was possible. In addition, given the possibility of multiplex communication, we knew that the random nature of the sequence of pulses output by a cell had an important meaning. Specifically, we found that multiplexing communication often failed when the network was observed to have many cells that produced less random firing phenomena (e.g., sequences such as "1111"), and, in contrast, we often succeeded in multiplex communication when the network was observed to have many cells that produced richly random firing phenomena (e.g., sequences such as "1101").

From these results, we knew that random sequences or strongly random sequences greatly enhanced the possibility of communication using methods such as time and space multiplexed communication, as well as artificial communication systems. We also found that when we can anticipate the signal arrival by LI-type synchronous reception, the communication success rate is greatly improved.

While we could confirm the possibility of multiplexed communication on even homogeneous neural networks as a more difficult scheme, our scheme is still limited in that there are only two communication channels and the absolute values of the correct recognition rate are less than 0.5%, despite the addition of the learning ability to the proposed scheme. These aspects should be improved as the next stage of our work. Additionally, although the state transitions of cells in the present paper are synchronous throughout the network, the work should be extended to the asynchronous actions of the cells to investigate the intelligent functions of a natural neuronal network. These aspects will be addressed hereafter in our series of papers.

Acknowledgements

The authors would like to thank Dr. Yoshi Nishitani of Osaka University for his helpful suggestions.

This work was supported in part by Grants-in-Aid for Scientific Research of Exploratory Research (21656100, 25630176) and Scientific Research (A) (22246054) from the Japan Society for the Promotion of Science.

The authors declare that there is no conflict of interest regarding the publication of this paper

References

[1] Yuko Matsumoto-Mizuno, Kozo Okazaki, Amami Kato, Toshiki Yoshimine, Yoshinobu Sato, Shinichi Tamura, and Toru Hayakawa: ``Visualization of epileptogenic phenomena using crosscorrelation analysis: Localization of epileptic foci and propagation of epileptiform discharges," *IEEE Transactions on Biomedical Engineering*, Vol. 46, No. 3, pp. 271–279, 1999.

[2] Yuko Mizuno-Matsumoto, Masatsugu Ishijima, Kazuhiro Shinosaki, Takashi Nishikawa, Satoshi Ukai, Yoshitaka Ikejiri, Yoshitsugu Nakagawa, Ryouhei Ishii, Hiromasa Tokunaga, Shinichi Tamura, Susumu Date, Tsuyoshi Inouye, Shinji Shimojo, Masatoshi Takeda: ``Transient Global Amnesia (TGA) in an MEG Study," *Brain Topography*, Vol.13, No.4, pp.269-274, 2001.

[3] Peter C Hansen, Morten L Kringelbach, Riitta Salmelin, *MEG: An Introduction to Methods*, Oxford University Press Inc. New York, 2010.

[4] Shinichi Tamura, Yoshi Nishitani, Takuya Kamimura, Yasushi Yagi, Chie Hosokawa, Tomomitsu Miyoshi, Hajime Sawai, Yuko Mizuno-Matsumoto, Yen-Wei Chen. "Multiplexed Spatiotemporal Communication Model in Artificial Neural Networks," *Automation, Control and Intelligent Systems*. Vol. 1, No. 6, 2013, pp. 121-130. doi: 10.11648/j.acis.20130106.11.

[5] S. W. Golomb, G. Gong, *Signal Design for Good Correlation: For Wireless Communication and Rader*, Cambridge University Press. 2005.

[6] Yoshi Nishitani, Chie Hosokawa, Yuko Mizuno-Matsumoto, Tomomitsu Miyoshi, Hajime Sawai and Shinichi Tamura, ``Detection of M-sequences from spike sequence in neuronal networks," *Computational Intelligence and Neuroscience - Special issue on Computational Intelligence in Biomedical Science and Engineering* Volume 2012, 2012.

[7] Shinichi Tamura, Yoshi Nishitani, Chie Hosokawa, Yuko Mizuno-Matsumoto, Takuya Kamimura, Yen-Wei Chen, Tomomitsu Miyoshi, Hajime Sawai. "M-sequence family from cultured neural circuits," *6th International Conference on New Trends in Information Science, Service Science and Data Mining (ISSDM 2012)*, Oct. 23-25, 2012.

[8] Daniel L. Schacter, Daniel T. Gilbert, Daniel M. Wegner, *Psychology (2nd ed.)*, Worth Publishers, New York, 2011.

[9] Christophe Lecerf, ”The double loop as a model of a learning neural system," *Proceedings World Multiconference on Systemics, Cybernetics and Informatics*, Vol.1, pp. 587-594, 1998.

[10] Y. Choe, "Analogical cascade: a theory on the role of the thalamo-cortical loop in brain function," *Neurocomputing* 52-54, pp.713-719, 2003.

[11] Takuya Kamimura, Yoshi Nishitani, Yen-Wei Chen, Yasushi Yagi, and Shinichi Tamura, "Copy of neural loop circuits for memory and communication," *Journal of Communications and Information Sciences*, Vol.4, No.1, pp.46-56, Jan 2014.

[12] Wulfram Gerstner and Werner Kistler, *Spiking Neuron Models. Single Neurons, Populations, Plasticity*, Cambridge University Press, 2002.

[13] Daniel Delling, Peter Sanders, Dominik Schultes, Dorothea Wagner, "Engineering Route Planning Algorithms," in Jürgen Lerner, Dorothea Wagner, Katharina A. Zweig (eds.), *Algorithmics of Large and Complex Networks, Design, Analysis, and Simulation*, 2009.

[14] Paulsen, O.; Sejnowski, T. J. (2000). "Natural patterns of activity and long-term synaptic plasticity". *Current Opinion in Neurobiology*, 10 (2): 172–179. doi:10.1016/S0959-4388(00)00076-3.

[15] Steven Yantis, *Sensation and Perception*, Worth Publishers, New York, 2014.

25

Data-Driven Models and Methodologies to Optimize Production Schedules

Prabhakar Sastri[*], Andreas Stephanides

Automation and Data Analytics Department, Isa Technologies Pvt. Ltd., Manipal, India

Email address:

prabhakar_sastri@yahoo.com (P. Sastri)

Abstract: Data driven Models based on production parameters in combination with modern optimization algorithms are shown to be useful in industry to optimize production schedules and improve profitability. Based on real data obtained from an existing facility, we have developed models for time and costs of heat treatment. Using these data statistical models have been developed and used to find an optimal solution to the Job-Shop scheduling problem using three algorithms namely Particle Filter, Particle Swarm Optimization and Genetic Algorithm. The algorithm is useful when we would like to arrive at job schedules based on a mix of both time and cost optimization. The results are compared and future work discussed with respect to the data used.

Keywords: Data-Driven Models, Optimized Production Scheduling, Job-Shop Scheduling, Time Based and/or Cost Based Production Optimization, Management Decision Tool

1. Introduction and Background

A heat treatment facility typically consists of many furnaces. The parts undergoing heat treatment can be processed in one or more furnaces. Each furnace can also process a variety of parts. A decision on price and resource allocation has to be made every time a part gets processed in a facility consisting of many furnaces. The problem becomes more complicated if only partial processing can be done in a furnace and hence a part may have to be processed in more than one furnace.

Delivery of some parts is time critical and others are cost critical. There are tactical and operational level decisions to be addressed including the cost of a single job, which may be optimized either for cost or time. Based on the type of individual jobs to be processed, it is also necessary to optimize the overall profitability of the facility and hence work out the production schedule and resource allocation.

Imagine a scenario:

A set of jobs have been booked from various clients. Some are "time critical" and the others "cost critical". Once a list of such jobs is available, it becomes necessary to work out a production schedule which accommodates all these jobs based on the current loading pattern in the furnaces of the facility.

It needs to be reiterated that
- Each furnace can process many different types of parts.
- It is possible to process multiple parts in a furnace at the same time. (Currently this is outside the scope of the present paper).
- It may be possible / necessary to process a part in either one furnace or a combination of furnaces.
- The cost of processing of the parts is different in each of the furnaces.

In our case the two main parameters to decide which resources should be used at what time are duration and cost of a job. The priorities may vary for each job. We optimize these parameters depending on the preferences for a set of jobs.

The optimal solution for a production schedule depends not only on the preferences regarding cost and duration, but also on the estimation of these parameters based on the furnace in which they are processed.

The objective of this paper is to present an algorithmic methodology of solving resource allocation and scheduling problem. The methodology takes into account that the model for cost vs time optimization in various furnaces is data driven i.e. based on the data collected from existing furnaces and the jobs processed so far. The need to update the model also depends on the number of furnaces. If the number of furnaces or the numbers of type of parts processed changes very often then it may be required to update the model more frequently. This is also true if the type of parts are different from the one based on which the model was created. This could vary from once a day to several times a month. The data driven model is generic enough to be applied to any facility and once the parameters are estimated, the model can be used only for that facility.

The paper is organized as follows:

In the second section we give a review of the literature and the work done so far to understand the importance of the current work.

In the third section we formulate empirical Models for duration and costs based on data provided by an existing facility in India.

In the fourth section we start with optimizing a single resource allocation decision and extend that to take into account multiple jobs to be scheduled.

The optimization problem becomes a variation of a Job-Shop Scheduling problem. These problems were first described in Graham [1]. Dimopoulos [2] compared different research about Job-Shop Scheduling problems and showed that recent research increasingly takes into account costs and other optimization criteria that are more relevant to real manufacturing decisions. Since we optimize a real live production facility we will consider multiple parameters for our optimization.

Finally we explore future work.

2. Review of Previous Work

The Job-Shop Scheduling problem is known to be n-p hard. Different algorithms are used to solve these problems. Mathirajan, Chandru and Sivakumar [3] discuss in detail the process of heat treatment and the operations prior to that of preparation of molten metal and its casting. They optimized the operation of two furnaces using heuristic algorithms. Their assumptions include the following:

a Due to technical reasons, it is not possible to process jobs from different families together in the same batch. We shall call these job-families incompatible. Furthermore, these jobs will have to be processed without interruption on parallel and non-identical BPs (BPs with different capacities), which are available continuously with an objective of maximizing the utilization of the BPs.

b Scheduling planning period is one week

c All batch processors are continuously available and all jobs must pass through the operation(s) to be carried out at the BPs.

They successfully used data from an existing foundry and suggested that "the way forward would be to be able to schedule the jobs at more frequent intervals than the current 24 hrs that they used as jobs arrive in the shop floor at shorter intervals"[3].

Amin Jamili & Mohammad Ali Shafia & Reza Tavakkoli-Moghaddam [4] proposed a hybrid algorithm based on particle swarm optimization and simulated annealing for a periodic job shop scheduling problem. They suggest that "evolutionary algorithms are finding more use than those for various reasons such as convergence speed". They compare the various algorithms and point out that PSO has the advantage of memory: the characteristics of the good solutions are retained by all particles even if the population has changed etc.

Recently Gomez Urrutia, Aggoune and Dauzere-Peres [5] used heuristics to solve lot-sizing and scheduling problems. The scheduling of the tasks is an operational one whereas the need to satisfy the objective of cost and time is a tactical one and the two cannot be pursued independently. We close that gap and integrate preferences for a single job in the tactical production plan as well as using the production plan to decide on facility optimization measures.

Dimopoulos [2] has extensively reviewed research using advanced optimization methods to solve scheduling problems in production. Particle Swarm Optimization (PSO) and Genetic Algorithms (GA) are widely used. Luarn [6] compared PSO to GA. Jamili, Shafia and Tavakkoli-Moghaddam [4] combined PSO with simulated annealing. Genetic algorithms are used by Wang, Yin and Wang [7] and by Pfund [8] for the scheduling problem.

For these reasons, we have considered Particle Filter, Particle Swarm Optimization and Genetic Algorithm for our work.

We introduce an approach to solve the optimization problem, using three commonly used methods. These are Particle Filter, Particle Swarm Optimization and Genetic Algorithm referred to above. We use random-key representation as described by Bean [9] of the priority of the jobs and a scheduling rule to get a robust representation that is independent of the optimization algorithm used. The main advantage of this approach is that compared to most heuristics, parameters by which the optimal decision is influenced is very clear.

3. Models for Duration and Costs

We begin by formulating the Model based on the data collected from the automation system installed in the plant.

3.1. Empirical Models for Duration

To formulate a Model for the duration of heat treatment, data from an existing facility has been used. The facility has 20 furnaces and processes about 109 different parts. The types of furnaces include Sealed Quench Furnaces, Pit Furnaces, Nitriding Furnaces, Rotary Hearth and other furnaces. The data is over a period of 22 months.

Before a detailed statistical analysis is possible, the data has to be cleaned. For the current scenario the following was done for the following reasons:

• Missing fixture weights were added: This is to ensure that fixtures which also get heated in the furnace and hence require time and energy are considered.

• Charges which are involved with rework were removed.

This is to ensure purity of data as reworking would mean quality of the end product is not acceptable.

- Errors and missing values in part weights were fixed: This is to ensure that the data is correct. The part weight is available in the plant database.
- Charges that have been processed incompletely were removed.
- Gross weight with fixed values was recalculated.
- Durations for each process step were calculated.
- Charges with more than one part were removed, and
- Only charges processed completely in automatic mode were considered.

After cleaning the data, a total of 557 batches with 109 different part numbers from 10 different furnaces have been analyzed.

To increase the data used for the model, the quality of the stored data is very important. To implement our framework one should consider optimization of the collected data. In the next phase of the work, it is planned to include data from a larger number of furnaces over a larger period of time with more number of parts. It is necessary to emphasize that quality and quantity are both critical and understanding this would further strengthen the usefulness of this work.

We give a brief review of the data here.

Figure 1 shows the process steps.

It is not necessary that all the steps be followed when processing a job. To give an idea of the complexity of the data, we present below the duration graphs of reach of the stages described above. We may mention that we are willing to provide the raw data to anyone interested as it would enable a better understanding of the Modeling possible and different viewpoints will always be welcome. The data will be in Microsoft Excel format.

The graphs depict the variety of processes taking place in the furnaces and the many complex ways of handling and heat treating various parts.

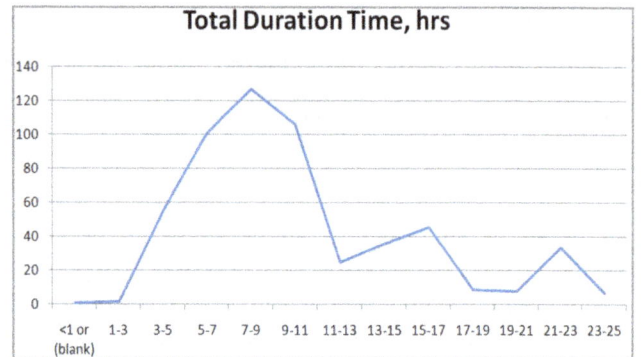

Figure 2. Distribution of Total Duration (557 data points).

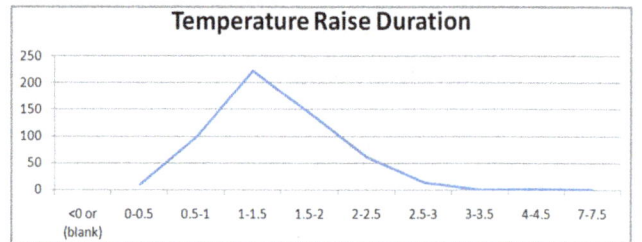

Figure 3. Distribution of Temperature Rise Duration (557 data points).

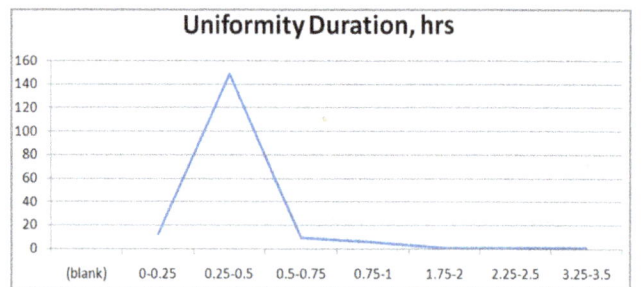

Figure 4. Distribution of Uniformity Duration (378 data points, blanks & zero omitted).

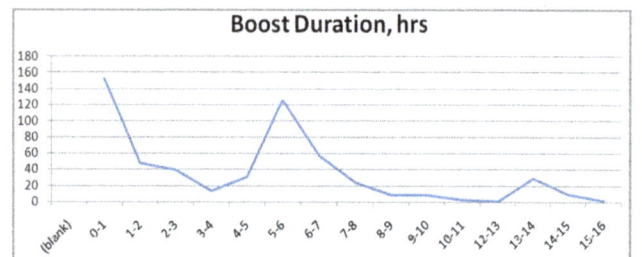

Figure 5. Distribution of Boost Duration (487 data points, less than 3 minutes omitted).

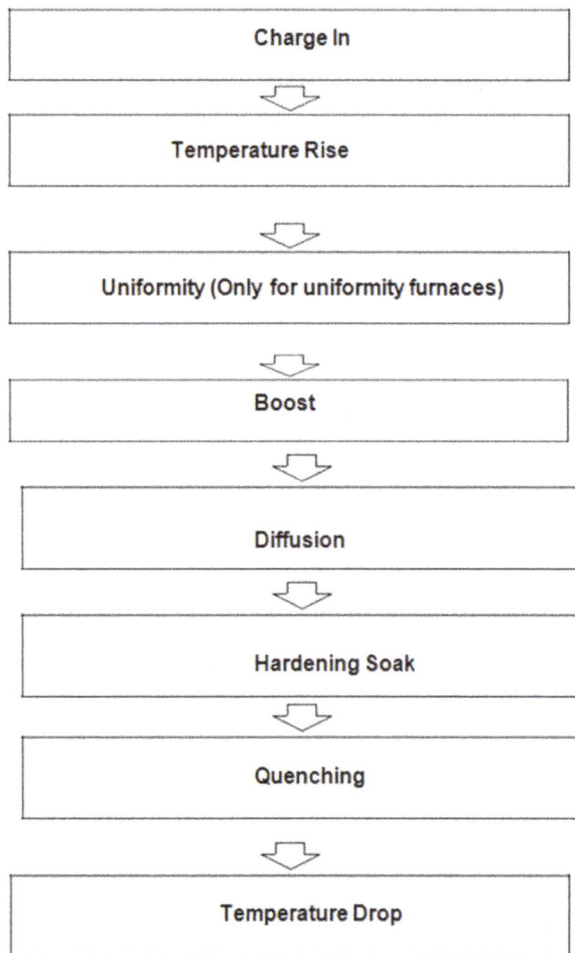

Figure 1. Process Steps.

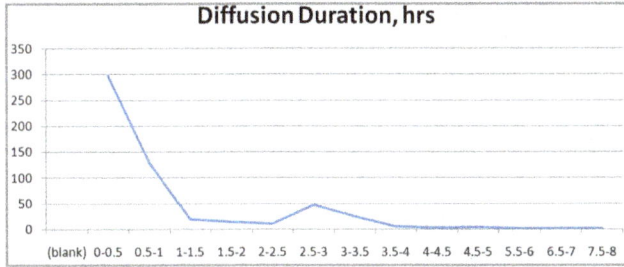

Figure 6. Distribution of Diffusion Duration (327 data points, less than 10 minutes omitted).

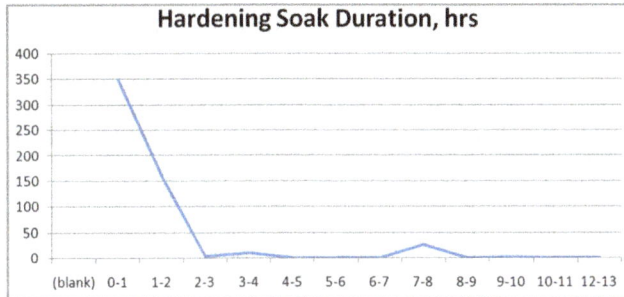

Figure 7. Distribution of Hardening Soak Duration (411 data points, less than 10 minutes omitted).

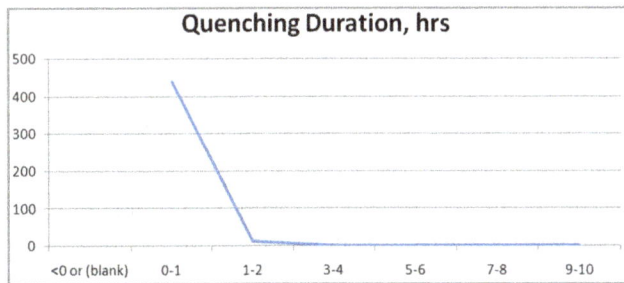

Figure 8. Distribution of Quenching Duration (454 data points, less than 5 minutes omitted).

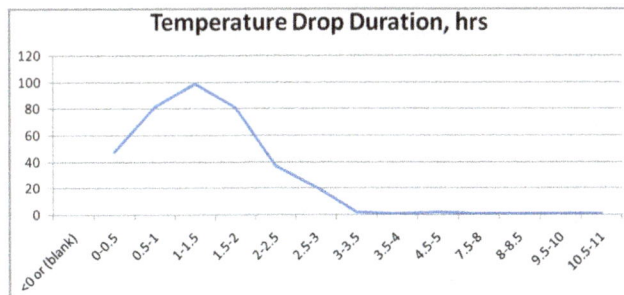

Figure 9. Distribution of Temperature Drop Duration (377 data points, blanks omitted).

We may mention here that we have been collecting data from this facility and hope to have a much larger and varied data from more number of furnaces now connected to the system processing many more different parts. The Models developed by us continue to be used and updated regularly but is limited at present to the furnaces considered above. We hope to have in place a system which will permit automatic updating of the Models presented here based on the data collected.

3.1.1. Model I

On this cleaned dataset a statistical analysis of the duration of the process was done. We proposed a linear Model with the gross weight as parameter and an offset for each furnace. The Model I for the duration T_I (Total Time for processing the job) can be written as

$$T_I = c + c_F + c_{gw} \, GW, \qquad (1)$$

with a constant general offset c, a offset for a furnace F, c_F and the parameter for the influence of the gross weight GW, c_{gw}.

Table 1. Statistical Analysis of Model I: Tests of Between-Subjects Effects.

Dependent Variable: Duration				
Source	Sum of Squares	df	Mean Square	F
Corrected Model	18842.484[a]	20	942.124	43.07
Intercept	9082.692	1	9082.692	415.3
Gross Weight	1450.404	1	1450.404	66.31
Furnace	14613.46	19	769.13	35.16
Error	74017.08	3384	21.873	
Total	495362	3405		
Corrected Total	92859.56	3404		

a. R Squared =.203 (Adjusted R Squared =.198)

Table 1 shows the statistical analysis of the general linear Model I. It can be seen, that each of the factors are statistically significant for the Model.

Since the fit of a statistical, linear model always includes uncertainties, the 95% confidence interval of the fitted parameters is considered. With those, an optimistic and a pessimistic time estimate is achieved and used as a quality criterion. Since the Model is to be used for resource allocation decisions and cost calculation, the uncertainty should be quantified. The resulting interval turns out to be typically around +/- 3 hours for the Model I.

3.1.2. Model II

The detailed model includes applying the method of Model I but for each individual process step shown in figure 1.

For each of these process steps, the model can be written as

$$T_p = c + c_{Fp} \qquad (2)$$

where c is for the entire facility and c_{Fp} is a function of the furnace and the process step. The time for "charge in" is not considered here, because in our data it is zero in most cases or close to it. This could be significant is other facilities and can then be considered for such facilities.

Uniformity time is only considered if it is a uniformity furnace. Model II for the duration T_{II} can be written as

$$T_{II} = \sum T_p \qquad (3)$$

with the duration of process step i, $T_{II;i}$. The estimates for the process steps and the parameters used in the test case are given in tables 2 and 3. The results are summarized in Table 4. This shows that for the detailed Model II the interval between the optimistic and the pessimistic estimate is larger, because the uncertainties of the parameters for each step add up and result in a more conservative estimation of the interval.

Table 2. Process step estimates, parameters as in Table 3.

Temperature Raise	01:28:03
Uniformity	00:41:54
Boost	08:45:43
Diffusion	00:32:08
Hardening Soak	00:03:05
Quenching	00:33:46
Temperature Drop	- 0:01:35
Total Duration	12:06:13 ~12:15:00

Table 2 shows the estimates on the process steps for a test case. It can be seen, that the negative estimate for the temperature drop is not an acceptable value. The value becomes negative because of the linear Model and the confidence interval and has to be interpreted as zero or close to zero estimates and the uncertainty on these estimates should be considered. For the overall estimate the values should not be removed from the calculation. Since we considered the uncertainties in Model II to be too high to use this Model in the scheduling task and we did not actually need estimates of each process step, we did not investigate this any further. However larger amount of data over a variety of conditions would reduce these uncertainties and make this Model more useful.

Table 3. Parameter for Test case.

Parameter	Value
Gross Weight	880kg
Part No	HM127415XD
Furnace	10X

3.1.3. Model III

Finally we investigate a Model including the part number (Model III), it can be written as

$$T_{III} = c + c_F + c_{partno} + c_{gw}\,GW \qquad (4)$$

where c_{partno} is the offset for each part number.

In this model the validity of the estimate is expected to be higher, because it takes into account the offset of a specific part. The main disadvantage is that the confidence interval of the parameters is higher, because the data per part is considerably smaller. To improve this model, more data will have to be collected.

3.1.4. Comparison of the 3 Models

Table 4 shows that all three Models are consistent and lead to similar estimates. The difference is the size of the intervals between optimistic and pessimistic estimates. Model I has the closest interval that can be explained by the different amount of data used in the Model to arrive at the estimates. The more detailed the Model gets, the less data can be used to estimate a certain parameter. Model III takes into account the specific part numbers and some parts have been processed very rarely. A larger set of data can lower these differences.

Table 4. Estimates for Test Case, parameters as in Table 3.

	Estimated Duration	Pessimistic	Optimistic
Model I	12:45:00	15:15:00	09:57:39
Model II	12:15:00	16:45:00	07:15:00
Model III	12:45:00	26:15:00	06:00:00

If table 5 is compared with table 4 it will be noticed that the historical data for a certain part number is consistent with the estimates of all three Models. It should however be considered, that this will not be true for a single observation because even the expected interval allows outliers. Due to the high uncertainties and the similarity of the estimates, it was decided to use the simplest Model. We therefore use Model I (equation 1) for the scheduling.

Table 5. Historical data, parameters as in table 3.

Average duration	12:27:31
No cases	23

This gives us a direction for future work. If the data is continuously collected and upgraded, it is possible to execute the Model at a regular interval and as the volume of data grows larger the parameters are expected to be estimated better.

3.2. Model for Cost

Some costs are easily quantifiable but it is hard to assign them to a single charge. Therefore the average tonnage of each furnace is used to calculate a per weight markup m for each furnace. Costs per hour can be included using the average hours per month. We used data from the same facility to calibrate our Model.

$$costs(F, GW) = m(L_F, X, C, f_{cF})\,GW \qquad (5)$$

with the parameters described in table 6.

Table 6. Parameters in Cost Model.

Parameter	Description
F	Furnace
L_F	Labour costs of furnace
C	Capital costs for furnace
X	Fixture costs
f_{cF}	Fuel costs per month

4. Optimal Production Schedule

In this section we show how an optimal production schedule can be generated, based on a Model for cost and duration. We use Model I equation (1) with slight modifications for the duration and equation (5) for costs. Although we depend on the models developed here, they can be replaced by any other suitable model.

4.1. Single Decision Step

Before we approach the overall optimization of the scheduling problem we first consider a single decision for scheduling one job.

4.1.1. Simple Optimization Problem

So far we described Models for duration and costs for a single charge. On these we can base a decision assuming that all or a given set of furnaces are available. With this proposition we can find the optimal solution based on costs

$$\min_{F} T_l (F, GW) \qquad (6)$$

or duration

$$\min_{F} costs (F, GW) \qquad (7)$$

One can also find an optimal solution regarding costs with a constraint to duration,

$$\min_{F} T_l \ costs(F, GW) \qquad (8)$$

$$costs(F, GW) < costs_{max}$$

or an optimal solution regarding duration with a constraint to cost

$$\min_{F} T_l (F; GW) \quad < Tmax \qquad (9)$$

4.1.2. Advanced Availability Constraints

The chances are that the furnaces may not be available for certain time due to preoccupation with different jobs. To consider this, one can add a schedule that takes previous decisions into account. We thus consider previous scheduled jobs to be constraints to the optimization problem. This can be achieved by calculating the waiting time for each furnace and add it to the duration.

$$T_A(F, GW) = T_l (F, GW) + T_{wait}(F) \qquad (10)$$

Table 7. Parameters for cost function.

Parameter		Value
d;j		0 .... 100^a
c;j		1/4000
pen$_d$		1000
pen$_c$	1000	(1/4000)^2

a. In almost every case 0

4.1.3. Multi-parameter Penalty Function

In this step we extend the optimization problem to include costs and duration at the same time along with the constraints. For the constraints we use the penalty method to relax the constraints and include them in the objective function.

We introduce a new objective function, which includes duration and cost for each job weighted by the preferences for the job. For one Job j the objective is to find the solution to

$$\min_{F} JF$$

with

$$JF = wd;j \ TA;j(F;GW)$$
$$+ wc;j \ costs(F;GW)$$
$$+ pen_d \ (T_{A;j}(F;GW) - T_{max;j})^2$$
$$+ pen_c \ (costs_{I;j}(F;GW) - cost_{max;j})^2 \qquad (11)$$

with pen$_d$ = 0 if the duration including waiting time is smaller than the maximum duration and pen$_c$ = 0 if costs are smaller than the maximum costs and otherwise according to table 7, wd;j is the weight of duration for job j and wc;j the weight of costs for job j, these weights should be chosen much smaller than the weight of the penalty function and relative to each other according to preferences for each job. Note that here we use the extended duration model given by (10) and the cost model (2). Since the preferences for each job are different we take into account, that for some jobs the duration is critical and for some the duration is not. The advantage of dropping the assumption that preferences are the same for every job is that it reflects the real economic situations. The resulting schedule is optimized by the priorities of each job and tries to meet as many constraints as possible.

4.2. Multiple Job Scheduling

We now schedule a list of jobs in an optimal way. In our model we can neither assume that each job will have the same parameters on each furnace nor that the decision criteria will be the same for each job. Equation (9) shows that the total penalty is the sum of the penalty for each job, by using the scheduling rule the problem can be reduced to finding an optimal order to schedule the jobs.

We propose as a scheduling rule, that if we have the jobs in an ordered list we always take the optimal single decision regarding the penalty function (8) for every job. It can be argued, that for a global optimum a non optimal decision for a single job can only be accepted if the benefits of the optimal decision for another job are higher. If that is the case the job which would benefit more should be prioritized, i.e. being scheduled first. We propose as a scheduling rule, that if we have the jobs in a ordered list we always take the optimal single decision regarding the penalty function (11) for every job.

That for optimization a prioritized list can be transformed into a schedule by schedule builders is used in many studies as shown by Dimopoulos [2] To represent the order of the list we will use random-key representation as described in Bean [9] and Uzsoy [10]

The overall objective is to find a solution to

$$\min_{F} J = \sum_{j} Jj(F) \qquad (12)$$

By these rules we get a schedule that optimizes (12). Due to the random-key representation of the ordered list, we have a very strong representation that leads for every set of random keys to a feasible solution. We are able to use any optimization algorithm to find the optimal set of random-keys, which corresponds to an optimal schedule.

In some papers related to operation scheduling, heuristics

are used to find a schedule. The argument is usually that the problem is N-P hard, e.g. in Mathirajan, Chandru, and Sivakumar [3] a heuristic for steel industry was developed. As already mentioned in section 1 on page 1, there are also a lot of references proposing genetic algorithms and particle swarm optimization to solve a Job-Shop problem.

We compare the advanced optimization techniques: particle filter, particle swarm optimization and a genetic algorithm, with the goal to find an optimal schedule. A set of jobs similar to existing data is used and certain restrictions are assumed. Taillard [11] proposed certain benchmarks for their optimization, since the main goal is to solve our specific problem the focus is on that problem.

4.2.1. Particle Filter (PF)

Each ordered list of jobs is considered a particle, with all scheduling information attached. The particle filter has with genetic algorithm in common, that both algorithms can be used to solve highly non-linear optimization problems if the penalty function can be calculated easily, but the gradient can't.

The particle filter selects the best particles and modifies them randomly. This can be achieved by adding a random number r to the key of each job in the list. The random number should be smaller than the initial interval to reduce the modification. The descendants of a particle should be similar to the particle; the optimal variation of the particle depends on how close the particle is to the optimal solution.

We propose to choose 21 particles and keep the best, particle 0, unchanged, modify particles 1-10 and remove particles 11-20. Particles 11-20 are replaced by descendants of the best particles.

The particle filter uses algorithm 1, with the random modification operator $RM(l; \sigma)$, which modifies each key of the jobs in list l by a random variable with a normal probability distribution with a variance σ.

Algorithm 1 Schedule optimization

```
Create random lists l0...... L20          . l... ordered Lists
Schedule(l0.... l20)
Sort by overall penalty function (12)
for g = 2 to 45 do                        . g... Generation
    Algorithm 2 on the next page
    Schedule(l0.... l20)
    Sort by overall penalty function (12)
end for
Select best solution l0
```

4.2.2. Genetic Algorithm (GA)

The genetic algorithm can be understood as an extension of the idea of particle filters. The algorithm 1 is modified by replacing algorithm 2 by algorithm 3. We retain the number of generations and the overall structure to make it comparable. For industry usage, the generations should be chosen dependent on the selected algorithm and the necessary convergence.

We use a statistical multi-point crossover operator (MXO) with a 0.05 probability to crossover after each gene.

Therefore the points for the crossover are random and the number of crossovers is random.

Algorithm 2 change ordered lists by particle filter

```
for i=11 to 14 do
    li ← RM(l0; 0.1)
end for
for i=15 to 16 do
    li ← RM(l1; 0.5)
end for
for i=17 to 18 do
    li ← RM(l2; 0.6)
end for
for i=19 to 20 do
    li ← RM(l3; 0.6)
end for
for i=1 to 10 do
    li RM(li; 0.5)
end for
```

Genetic algorithms are known to be useful to solve multi-objective problems and are widely used in context of realistic scheduling problems in manufacturing; Shaw et al [12] used genetic algorithms to solve multi-objective scheduling problems in batch-processing.

Algorithm 3 change ordered lists by genetic algorithm

```
for i=11 to 14 do
    j        rand(1..10)
    li = MXO(l0; lj)
    RM(li; 0:01)
end for
for i=15 to 16 do
    j        rand(1..10)
    li = MXO(l1; lj)
    RM(li; 0:1)
end for
for i=17 to 18 do
    j        rand(1..10)
    li = MXO(l2; lj)
    RM(li; 0:1)
end for
for i=19 to 20 do
    j        rand(1..10)
    li = MXO(l3; lj)
    RM(li; 0:1)
end for
for i=1 to 10 do
    RM(li; 0:4)
end for
```

4.2.3. Particle Swarm Optimization (PSO)

In Jamili, Shafia, and Tavakkoli-Moghaddam [4] Wang, Yin, and Wang [7]) and many similar studies, Particle Swarm Optimization is used to solve Job-Shop scheduling problems. Jamili, Shafia, and Tavakkoli-Moghaddam [4] combine it with simulated annealing to a hybrid optimization process. Particle swarm optimization is inspired by the behavior of animal swarms. Each particle has a speed and uses it's own best known solution and the swarms best known solution to update its speed. The keys of each job can be collected in a vector k(li), the speed of the particle v(li) is the change rate of the keys. The keys of generation $G + 1$ of a particle i, kG+1(li) are calculated by

$$k_{G+1}(l_i) = k_G(l_i) + v_{G+1}(l_i) \qquad (13)$$

The speed v_{G+1}(li) is

$$v_{G+1}(l_i) = w_a \, v_G(l_i) + w_b \, r + w_c \, (k_G(l_i) \, b_G(l_i)) \\ + w_d \, (k_G(l_i) - b_G(l_{0......}l_{20})) \qquad (14)$$

with the weights w_j, a random vector r, bG(lj.......lk), the best known key set of lists j to k.

Algorithm 4 change ordered lists by particle swarm optimization

```
for i=0 to 20 do
    r   rand()
    Update speed (11)
    Update particle key (10)
end for
```

4.3. Results

To compare the different algorithms we compiled a set of 75 jobs with different constraints, preferences, part number and weight. These jobs were to be scheduled in an optimal way using the different optimization algorithms. To make the results comparable we always used the same number of generations. The weights of the penalty functions are chosen as in table 7. Figure 10 shows that the three tested algorithms all converge to a result within the given amount of generations. The found optimum is in all cases a different one. GA found on average a better optimum than the other algorithms.

It must be noted that all the three algorithms gave similar results and one is free to choose the algorithm to use.

Thus the methodology provides an elegant process of generating the production schedule for the scenarios described in page 1.

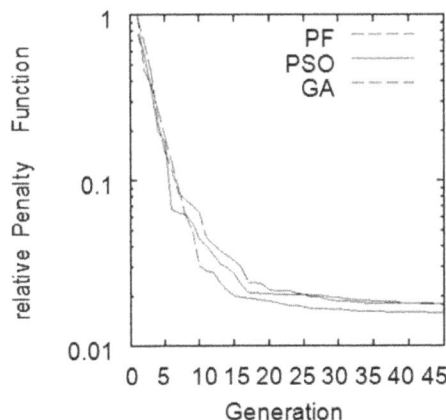

Figure 10. *Comparison of convergence of different algorithms.*

5. Conclusions and Outlook

The framework can be used in several ways. The most obvious is to get an optimal production schedule based on models driven by real production data. The models could be improved by using additional data or parameters in the Models for duration and costs. We showed in section 4.1.3 how to optimize a single scheduling decision for one job and in section 4.2 how to extend that to an optimal schedule. The Model is currently in use at the facility from which the data has been taken but is generic enough to be used at other facilities. Efforts are on to install and get similar data from another facility for the same purpose.

It must be mentioned that the method and models are generic. They can be applied to any facility. Since the data will be different, the model parameters will be decidedly different and hence the schedules will also be different. We will also mention that the models need periodic updating. The frequency depends on the changes brought about in the facility. If the changes are not many, the results are not expected to be very different. On the other hand if frequent additions/changes are made it is necessary to update the model regularly. The judgment is left to the facility manager. It is extremely easy to rework the model as the algorithm is coded in an Excel sheet to and just adding/ changing data and sending a request is adequate to update the model. The data can be directly linked to the automation system as has been done by us and this makes it easier to update the model.

We tested multiple optimization algorithms for that and found that these algorithms can be utilized to solve real life optimization problems in industry. There are other possible applications of our framework and we leave these as future work.

6. Future Work

Only a few possible applications for our framework have been shown. It could be extended to answer other questions in industry regarding resource allocation decisions. In section 3.1.4 it was decided to use Model I. The scheduling could also be investigated with Model II or Model III.

A possible improvement is to develop more detailed Models for duration and costs of the process, including extended data of furnaces, parts or the production process.

Additionally the framework could be easily tested in a real live environment with a greater set of jobs and preferences.

References

[1] Graham, Ronald L. Bounds for certain multiprocessing anomalies. Bell System Technical Journal 1966; 45 (9): 1563-1581.

[2] Dimopoulos, A. M. S., C.; Zalzala. Recent developments in evolutionary computation for manufacturing optimization: problems, solutions, and comparisons. IEEE Transactions on Evolutionary Computation 2000; 4.

[3] Mathirajan, M, V Chandru, and AI Sivakumar. Heuristic algorithms for scheduling heat-treatment furnaces of steel casting industries. Sadhana 2007; 32 (5): 479-500.

[4]   Jamili, Amin, Mohammad Ali Shaa, and Reza Tavakkoli-Moghaddam. A hybrid algorithm based on particle swarm optimization and simulated annealing for a periodic job shop scheduling problem. The International Journal of Advanced Manufacturing Technology 2011; 54 (1-4): 309-322.

[5]   Gomez Urrutia, Edwin David, Riad Aggoune, and Stephane Dauzere-Peres. Solving the integrated lot-sizing and job-shop scheduling problem. International Journal of Production Research (ahead-of-print): 2014; 1-19.

[6]   Luarn, Ching-Jong Liao; Chao-Tang Tseng; Pin. A discrete version of particle swarm optimization for flowshop scheduling problems. Computers & Operations Research 2007; 34.

[7]   Wang, Yong Ming, Hong Li Yin, and Jiang Wang. "Genetic algorithm with new encoding scheme for job shop scheduling." The International Journal of Advanced Manufacturing Technology 2009; 44 (9-10) 977-984.

[8]   Pfund, Lars Monch, Hari Balasubramanian; John W. Fowler; Michele E. Heuristic scheduling of jobs on parallel batch machines with incompatible job families and unequal ready times. Computers & Operations Research 2005; 32.

[9]   Bean, James C. Genetic algorithms and random keys for sequencing and optimization. ORSA journal on computing 1994; 6 (2): 154-160.

[10]   Uzsoy, Cheng-Shuo Wang; Reha. A genetic algorithm to minimize maximum lateness on a batch processing machine. Computers & Operations Research 2002; 29.

[11]   Taillard, E. Benchmarks for basic scheduling problems. European Journal of Operational Research 1993; 64.

[12]   Shaw, KJ, AL Nortcli e, M Thompson, J Love, PJ Fleming, and CM Fonseca. Assessing the performance of multiobjective genetic algorithms for optimization of a batch process scheduling problem. Proceedings of the 1999 Congress on Evolutionary Computation, 1999; CEC 99, Vol. 1.

26

Probiotic induce macrophage cytokine production via activation of STAT-3 pathway

Neama Y, Habil

Medical Technology Institute, Medical Technical University, Baghdad, Iraq

Email address:

neamahabil@gmail.com (N. Y, Habil)

Abstract: Macrophages are mononuclear phagocytes generated from monocyte emigrated from blood circulation. Macrophages mediated the innate and adaptive immunity through different routes, and cytokine production is one of these routes. Signal transducer and activators of transcription (STATs) are cytoplasmic transcription factors that are key mediators of cytokine and growth factor signalling pathways. STAT-3 is implicated in macrophage cytokine signalling and production. It's well reported that the microbiota is very important as it primes the immune system for the antigens encountered later in life. Probiotics defined as 'Live microorganisms which when administered in adequate amounts confer a health benefit on the host. Therefore, the aim of this project was to answer the question whether probiotics induced cytokine production via activation of STAT-3 signalling pathway. Results showed that probiotic *Lactobacillus casie strain Shirota* was successfully induced cytokine production via activation of STAT-3 by anti-inflammatory macrophages induced by TNF-α. The findings of this study will open new strategy to modulate the immune response by probiotic bacterialeading to treat the diseasesthat related with irregular cytokine production.

Keywords: Macrophages, Probiotics, Cytokines, STAT-3

1. Introduction

Historically macrophages are one of the antigen presenting cells (APCs) found in the lamina propria of the gut that can processing and presenting of antigens to T lymphocyteafter recognition of microbesleading to induce appropriate immune responses in response to microbial infection which will be determined by T cells . Macrophages involve a critical component of innate immunity; they play vital roles in eradicating pathogenic microbes and keeping tissue homeostasis. They express non-specific esterase, lysosomal hydrolases and ectoenzymes, resulting in contributing to non-specific uptake of particular materials (Hume, 2006). Macrophages also express an array of receptors for the Fc portion of immunoglobulin (Ig) and complement components.

In adaptive immunity, macrophages have the excessive competence to present antigens to T cells after phagocytise, kill, degrade microorganism materials, and process antigens for presentation to T cells on MHCII molecule; ultimately, the expression of molecules such as MHC 1 and MHC 11 and co-stimulatory molecules, B7-1 (CD80) and B7-2(CD86) by macrophages due it to involve in antigen processing and presentation leading to direct adaptive immunity (Paolillo et al., 2009). However, macrophages are able to regulate immune response via secreting various cytokines.

At the early stage of infection, the macrophages were recruited to the infected site to engulf the microbes, and when macrophages were activated after bacterial recognition by TLRs consequently they were produced highly level of proinflammatory cytokines such as IL-1β, TNF-α, IL-8, IL-6 and chemokineswhich recruits more macrophages and other immune cells e.g. neutrophils and basophils, these immune cells were collaborated together to eliminate infection and cause acute inflammation. On the other hand, at the late stage of infection, the macrophages exhibit another phenotype that contributes to resolution of inflammation and tissue repair through production of anti-inflammatory cytokines. Therefore, macrophages are exhibiting a wide range of functions which are both determined by differentiation and activation factors come upon by the cells in response to pathogenic infection.

Generally, macrophages classified as classically activated

M1 pro-inflammatory macrophage and alternatively anti-inflammatory M2 macrophages (Mantovani et al., 2007). Classically activated M1 pro-inflammatory macrophage demonstrates Th1 phenotype promoting inflammation and destructive effects of tissues, whilst alternatively M2 anti-inflammatory/regulatory macrophages demonstrate a Th2 phenotype promoting constructive effects of the tissues and resolve the inflammation (Mantovani et al., 2007, Mosser and Edwards, 2008); both of the phenotypes are important in both innate and adaptive immunity. The diversity and plasticity of macrophage depend on many factors such as the stimuli represented by tissue environment and cytokine produced by other immune cells such as T helper cells (Haller et al., 2000, Gordon, 2003). Granulocyte macrophage colony stimulating factor (GM-CSF) and macrophage colony stimulating factor (M-CSF) have been concerned in the differentiation of M1 and M2-like macrophages respectively (Verreck et al., 2004). Based on an evidenced,Foey.,(2012) reported that classically activated M1 macrophage cells differentiation are requires priming by IFNγ plus triggering with microbial LPS ,or GM-CSF, these events are followed by dramatic alteration in the secretory profile of the cells and the cells become the phenotype be like IL-12 ^{hi}, IL-23 ^{hi}, IL-10 ^{lo}, TNF-α ^{hi}, CD14+, CD86+, iNOS, STAT-1+, and professional in degradation extracellular components such as collagen, elastin, and fibrinogen by producing proteolytic enzymes including matrix metalloproteinase MMP-1, -2, -7, -9, and -12(Duffield, 2003). Among the pro-inflammatory cytokines produced by classically activated M1macrophages are TNF-α, IL- 6, IL-8, and IL-1β which have a crucial role in chemo attractant for neutrophils, immature dendritic cells, natural killer cells, and activated T cells. For alternatively M2 macrophage cells differentiation and activation does not need any priming by LPS, and IL-4, IL-13, IL-10, TGF-β, M-CSF, vitamin D_3 and immune complexes can act as sufficient stimuli to achieve this goal (Gordon, 2003). Cytokines are mainly released by macrophages whether M1 or M2., however, the production of these molecules is vital for host defence and track the adaptive immune system.The uncontrolled way of these mediators release by macrophage cells have significant roles of collateral injure on the host micro-environment that control inflammatory responses.

Recently, one member of the Stat family (STAT-3), has appeared as a negative regulator of inflammatory responses(Cheng et al., 2003). These transcriptional factors are latent in themacrophage cytoplasm until they are activated by extracellular signalling proteins (mainly cytokines and growth factors) that bind to specific cell-surface receptors. These extracellular-signalling proteins can activate various tyrosine kinases in the cell that phosphorylate STAT proteins. The activated STAT proteins accumulate in the nucleus to drive transcription. The duration and degree of gene activation are under strictly regulated by a series of negative acting proteins. There are several types of negative regulators of STAT proteins in the cell cytoplasm such as suppressors of cytokine signalling (SOCS proteins) which block further STAT activation in the cell cytoplasm

(Bromberg, 2002).STAT-3 signalling plays a critical role in the induction of antigen-specific T cell tolerance. Targeted disruption of STAT-3 signalling in APCs resulted in priming of antigen-specific CD4$^+$T cells. Cheng et al., (2003) demonstrated that manipulation of STAT-3 signalling in either direction (blockade or stimulation) influenced immune responses explaining that STAT-3 have a role in the immune activation versus immune tolerance, which critical decision with profound implications in autoimmunity, transplantation, and cancer. Constitutively active forms of STAT-3 increase transcription of anti-apoptotic and cell-cycle-progression genes such as BCLXL, Cyclin-D1, Cyclin-D2, Cyclin-D3, and Cyclin-A, Pim1, c-Myc and p19 (Donnelly et al., 1999). Takeda et al., (1999) demonstrated that the disruption of the STAT-3 gene in macrophages increased production of inflammatory cytokines such as TNF-α, IL-1, IFNγ, and IL-6 and suppressed IL-10 leading to develop chronic colitis, explaining that STAT-3 plays a critical role in the deactivation of macrophages mainly exerted by IL-10, these observation supported the concept that macrophages have a role in maintaining intestinal homeostasis through the expression of IL-10 by activation of STAT-3 (Matsukawa et al., 2005). However, it is important to note that the STAT-3 signalling pathway is shared by other cytokines such as IL-6through gp130 a common signal transducer for the IL-6 cytokine (Fukada et al., 1996), which has been shown to be vital for intestinal epithelial homeostasis (Yang et al., 2007). However, overproduction of IL-6 mediated epithelial cell cancer. Grivennikov et al., (2009) reported that STAT-3 have been mediated initiating tumour of macrophage cells through the proliferative and survival effects of IL-6.

Probiotic bacteria are friendly bacteria that have important roles in modulating the immune system and confer health benefits prophylactically to many of diseases such as allergies, inflammatory pathologies and cancer(Winkler et al., 2007, Dotan and Rachmilewitz, 2005). However, these immunomodulation roles of probiotic are partly attributable to immune cell phenotype being studied, its environment and the strain of probiotic being used. Current understanding of probiotic modulations of such important gut mucosal cells like macrophages mediated immune responses relevant to mucosal homeostatic and inflammatory pathological environments is relatively poorly understood. Therefore, the current study was performed to evaluate the potential effects of probiotic on modulation of cytokine production via such a specific mechanism of cell signalling involve modulation of transcriptional factor activation of STAT-3 pathway.

2. Materials and Methods

2.1. Isolation and Culture of Peripheral Blood Monocytes and Macrophage Induction

Peripheral blood was obtained from healthy volunteers by venipuncture, using heparin to prevent coagulation in accordance with approved ethical guidelines. Blood was diluted 1:1 in un-supplemented media (α-MEM). Based on methods ofBuechler et al. (2000), mononuclear cells were isolated by centrifuging 15 ml of αMEMblood suspension over 25 ml of Histopaque-1077 (Sigma-Aldrich, UK), at 700 ×g for30 min at 4°C. The buffy layer containing monocytes was removed and washed in 10ml of non-supplemented α-MEM then centrifuged at 400 ×g for 10 min at 4°C. The cell pellet was re-suspended in culture medium containing 10% FCS and redcells lysed using a 10% acetic acid solution. For the induction of macrophage formation, 1×10^4 PBMCs were cultured in 96 well plates containing α-MEM supplemented with 50 ng/ml M-CSF for three days at 37°C in a humidified atmosphere of 5% CO_2 to differentiate monocyte cells to the anti-inflammatory M2 macrophage cells.

2.2. Protein Blotting

After resolving the proteins by SDS-PAGE, gels were transferred to polyvinylidene fluoride (PVDF) membranes using an electro blotter system (Criterion blotter, BIO-RAD, UK). For detection phosphorylated STAT-3 (pSTAT-3) protein (Tyr 705), western blotting (WB) technique was used, and developing colour for protein detection was performed by using ECL-Plus to detect peroxidase activity from HRP-conjugated antibody.

2.3. Bacterial Culture and Preparation of Heat Killed Bacteria

Lactobacillus casei strain Shirota (LcS) probiotic bacteria were obtained from commercially available Yakult drink (Yakult, UK), and *Lactobacillus fermentum* strain MS15 (LF) was isolated from the crop of a chicken (Savvidou, 2009) and obtained from internal microbiology stocks at the University of Plymouth (UK). Probiotic bacterial cell culture and preparation of heat killed bacteria wereperformed according to Habil et al. (2012).

2.4. Activation of Macrophage Cytokine Production and STAT-3 Activation

The anti-inflammatory M2 macrophages were stimulated with10 ng/ml of TNF-α or 5 ng/ml IL-1β for 6 hours. The supernatants were harvested and stored at -20 °C until required for assay by sandwich ELISA, whereas the cell pelletswere harvested, lysed then total proteins were resolved by SDS-PAGE followed by plotting proteins using Western blotting technique.

2.5. Regulatory Effect of Probiotics

To investigate probiotic regulation of macrophage cytokineproduction and STAT-3 activation, heat killed LcS (HK-LcS) or (HK-LF)were added in culture to final concentrations of 3×10^8 bacterial cells/ml, as a pretreatment for 6 hours prior to cytokine stimulation (either TNF-α or IL-1β) for a further 6hours in a humidified environmentat 37°C, 5% CO_2 as described by Habil et al. (2011).

2.6. Cytokine Measurement

Macrophage cell production of the inflammatory cytokines,TNFα and IL-6, and anti-inflammatory cytokine IL-10 were analysed by sandwich ELISA using commercially available capture and detection antibodies from BD-Pharmingen (Oxford, UK). Protocols were performed based on methods ofHabil et al. (2014),colorimetric development was measured spectrophotometrically by an OPTIMax tuneable microplate reader at 450 nm and analysed by Softmax Pro version 2.4.1 software (Molecular Devices Corp., Sunnyvale, CA, USA).

2.7. Statistical Analysis

Statistical significance was analysed using a balanced analysis of variance (General Linear Model, Minitab version 16) followed by a multiple comparison test (LSD, least significant difference). Significance was set at $p < 0.05$ (*$p < 0.05$, **$p < 0.01$ and ***$p < 0.001$).

3. Results

3.1. Probiotic Bacteria Selectively Modulate Anti-Inflammatory Macrophage Cell Cytokine Production

Stimulation anti-inflammatory macrophages by TNF-α or IL-1β was successfully inducing significant level of TNF-α, IL-6 and IL-10 (Figure 1). Treatments of stimulated anti-inflammatory macrophages with HK-LcS or HK-LF showed that probiotic treatments selectively modulated anti-inflammatory macrophage cytokine production of TNF-α, IL-6 and IL-10. Results showed that HK-LcS suppressed TNF-αsignals by 26%, whereas HK-LF augmented TNF-αsignalling by 49%, resulting in modulation of TNF-α cytokine production of anti-inflammatory macrophages. Using IL-1β to induce TNF-α, HK-LcS augmented IL-1β signalling by44%, and HK-LF suppressed it by 18%(Figure 1A). Multi-faced IL-6 macrophage cytokine production was significantly induced by TNF-α or IL-1β signals. HK-LcS or HK-LF expresses different signals in modulation of TNF-α or IL-1β resulting in modulation of IL-6 macrophage cytokine production. HK-LcS augmented TNF-α signal by 72%, while HK-LF suppressed TNF-α inducing IL-6 by 111%. Using IL-1β to induce IL-6 macrophage cytokine production, HK-LcS suppressed IL6 signalling by 59%, whereas, HK-LF augmented it by 66% (Figure 1B).The anti-inflammatory cytokine IL-10 expression by anti-inflammatory macrophage was successfully induced by the signals of TNF-α or IL-1β, and probiotic bacteria were selectively modulated these signals resulting in modulation of macrophage cytokine production. HK-LcSupregulated TNF-α signal by 50%, whereas suppressed IL-1β signal by 14%. In contrast, HK-LcS suppressed IL-1β signal by 25%, and HK-LF increased it by 10% resulting in modulation of macrophage IL-10 cytokine production (Figure 1C).

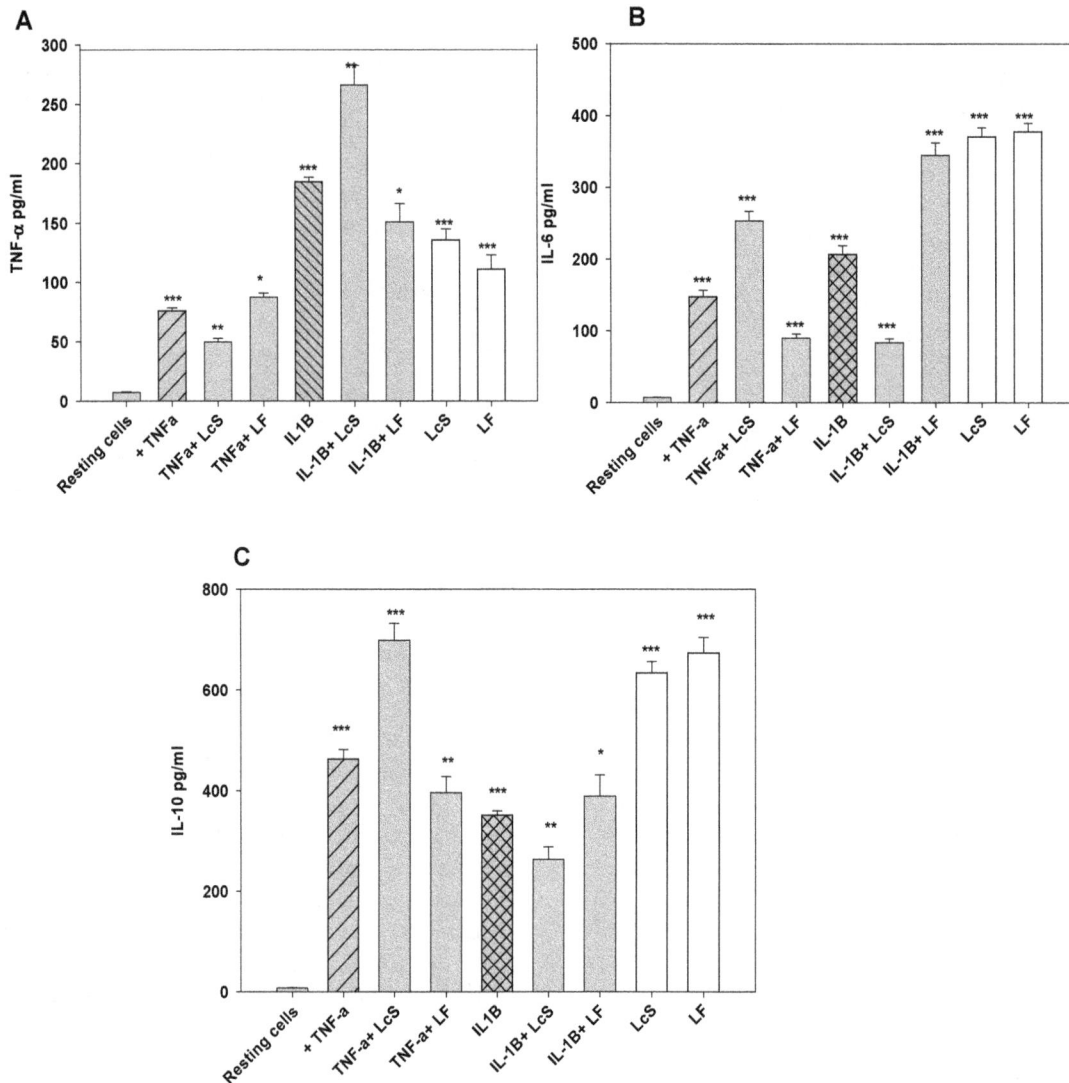

Figure (1). *TNF-α, IL-6 and IL-10 cytokine expression by anti-inflammatory macrophage cells differentially modulated by probiotic bacteria induced by TNF-α or IL-1β.*

Anti-inflammatory macrophage cell M2s were pre-treated with heat killed (HK) of *L.casie strain Shirota* (LcS),or *L.fermentum* (LF) at cell concentration of 3×10^8 CFU/ml followed by stimulation with 5ng/ml IL-1β or 10ng/ml TNF-α. Data representative of three independent experiments with triplicate sample. Significant value display as P<0.5 *, P<0.01 **, P<0.005 ***.

3.2. STAT-3 Activation Selectively Regulated by Probiotic Bacteria

STAT-3 pathway is essential for IL-6 and IL-10 macrophage cytokine production. Stimulation macrophage cells with TNF-α or IL-1β were successfully activated STAT-3 which correlated with cytokine production of IL-6 and IL-10 (Figure 1, 2).The selectively of probiotic activation of STAT-3 was clear, when LcSupregulated TNF-α signal

leading to activate STAT-3 pathway (Figure 2), whereas LcS failed to activate IL-1β signal. HK-LF failed to activate signal of TNF-α or IL-1β related with STAT-3 activation (Figure 2). The modulation style of STAT-3 activation of anti-inflammatory macrophages by probiotic bacteria was depend on type of signal and type of probiotic bacteria. These results demonstrated that phosphorylation of STAT3 pathway was essential to IL-6 and IL-10 cytokine production of macrophages after stimulation by TNF-α or IL-1β.

Full cell differentiated of anti-inflammatory macrophages were pre-treated with probiotic bacteria followed by stimulation with 5ng/mlIL-1β or 10ng/ml TNF-α. Cells lysed and cell lysates were analysed by SDS/PAGE and Western blotting using phospho- specific antibodies for STAT3 Tyr705 and an antibody that recognized the total STAT3.

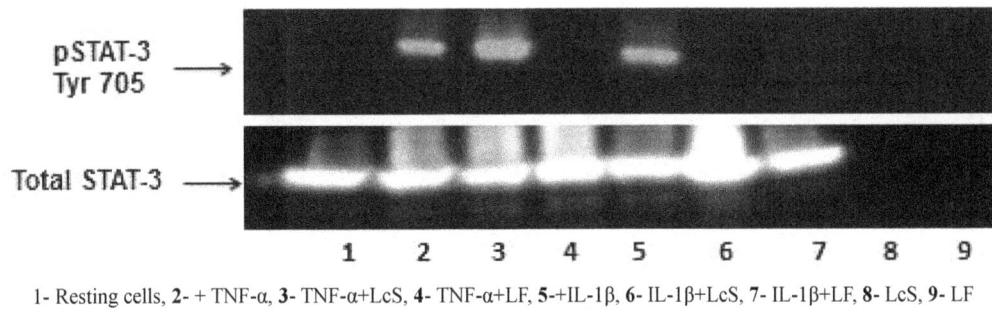

1- Resting cells, **2**- + TNF-α, **3**- TNF-α+LcS, **4**- TNF-α+LF, **5**-+IL-1β, **6**- IL-1β+LcS, **7**- IL-1β+LF, **8**- LcS, **9**- LF

Figure (2). Probiotic stimulates phosphorylation of STAT-3 on Tyrosine 705 by macrophage cells.

4. Discussion

There is a general concept that STAT-3 is required to mediate the anti-inflammatory activity of IL-10 (Williams et al., 2004). STAT-3 is activated by phosphorylation at Tyrosine705, which induces dimerization, also further phosphorylation at site Serine 727, followed by nuclear translocationand DNA binding(Darnell et al., 1994). Biethahn et al. (1999) reported that STAT-3have two isoform STAT-3a (86 kDa) and STAT-3b (79 kDa).O'Rourke and Shepherd (2002) demonstrated that transcriptional activation of STAT-3 was regulated by phosphorylation at Tyrosine 705 and Serine727 in macrophages via the Mitogen-activated protein kinases(MAPKs). It is well documented that STAT-3 expression level depends on celltype, ligand exposure or maturation stage of the cells. Several reports highlighted that probiotics mediated anti-inflammatory activity through enhancement of IL-10 by macrophages (Steidler et al., 2000, Madsen et al., 2001, Galdeano and Perdigón, 2006, Shida et al., 2011, Habil et al., 2013). The present study demonstrates that IL-6 and IL-10 derived from anti-inflammatory macrophages are increased after treating induced macrophages with TNF-α by probiotic HK-LcS, which correlated with phosphorylation of STAT-3 transcriptional factor. It is well reportedthat STAT-3 mediates IL-10R signalling exerts anti-inflammatory activity by inhibiting proinflammatory cytokines such as TNF-α and IL-1β (Bromberg, 2002 , Williams et al., 2007, Williams et al., 2004). The STAT-3 signalling pathway is shared by IL-6 through gp130 a common signal transducer for the IL-6 cytokine (Fukada et al., 1996, Ahmed and Ivashkiv, 2000). IL-6/STAT-3 signals show a key role in the pathogenesis of IBD (Matsumoto et al., 2010, Matsumoto et al., 2005). IL-6 cytokine is one of the important cytokine mediated gut pathology especially in the setting of inflammatory bowel diseases (IBD) such as Crohn's disease (CD) and ulcerative colitis (UC). Many theories behind IBD pathology, and one of them is initiating the inflammation as a results of the interaction between antigen-presenting cells (APCs) and the local bacterial flora which contributes to an uncontrolled activation of mucosal CD4[+]T lymphocytes called Th17. This type of T cells (Th17) described as high production of proinflammatory cytokines such as TNF-α, IL-6, IL-12, IL-

23, IL-27(Harrington et al., 2005). It is well documented that both of CD and UC diseases, CD4[+]T cells at the location of inflammation are vitally dependent on anti-apoptotic IL-6 signalling. By this means, IL-6 induces the transcription factor STAT-3 via trans-signalling (activation of a cell lacking membrane-bound IL-6 receptor via soluble IL-6 receptor)(Matsumoto et al., 2010). STAT-3 itself induces the anti-apoptotic factors bcl-2 and bcl-xL, thus resulting in T-cell resistance against apoptosis and become flooded with accumulated T cells (Fukada et al., 1996, Kovalovich et al., 2001). Therefore, the build-up of the nasty circle of Tcells mediated by apoptosis resistance was leading to chronic inflammation, which can be inhibited by down-regulation of IL-6 receptor. Anti-inflammatoryIL-10 cytokine production was upregulated by HK-LcS correlated with phosphorylation of STAT-3 when macrophages stimulated with TNF-α only. These novel results showed that HK-LcS was drive phosphorylation of STAT-3 for TNF-α not for IL-1β signal. HK-LF treatment augmented TNF-α macrophage cytokine production induced by TNF-α. It seems to be that lipoteichoic acids (LTA) from HK-LF which consider as one of the important member of the *Lactobacillus* family is the main cause for inducing pro-inflammatory cytokines by macrophage cells via NF-KB activation(Matsuguchi et al., 2003). Data suggest that probiotic bacteria drive proinflammatory process via activation of NF-KB resulting in inducing proinflammatory cytokines such as TNF-α and IL-6(Matsuguchi et al., 2003), on the other hand, probiotic drive anti-inflammatory process via activation the phosphorylation of STAT-3 pathway resulting in inducing anti-inflammatory cytokines such as IL-10 and suppressing the proinflammatory cytokine such as TNF-α (Matsumoto et al., 2005, Kim et al., 2006). It seems to be that the complex of polysaccharide–peptidoglycan (PSPG) in the bacterial cell wallof HK-LcS was mediated the anti-inflammatory process (Matsumoto et al., 2009).The results of this study provide good opportunity to inhibit IL-6 produced by macrophages activated by TNF-α or IL-1β by inhibiting the phosphorylation of STAT-3 leading to inhibit the gut inflammation. However, Matsukawa et al. (2005) reported that STAT-3 function as a repressor protein in resident macrophages. In conclusion, there is a specific structure on the probiotic bacteria cell wall plays a crucial role in controlling either proinflammatory or anti-inflammatory process via activation of NF-KB or STAT-3, therefore choosing specific probiotic strain in probiotic treatment is very important to control inflammation and treating the diseases related with cytokine production and T cell apoptosis.

References

[1] AHMED, S. T. & IVASHKIV, L. B. 2000. Inhibition of IL-6 and IL-10 Signaling and Stat Activation by Inflammatory and Stress Pathways. *The Journal of Immunology,* 165, 5227-5237.

[2] AKHTAR, M., WATSON, J. L., NAZLI, A. & MCKAY, D. M. 2003. Bacterial DNA evokes epithelial IL-8 production by a MAPK-dependent, NF-[kappa]B-independent pathway. *FASEB J.,* 17, 1319-1321.

[3] BIETHAHN, S., ALVES, F., WILDE, S., HIDDEMANN, W. & SPIEKERMANN, K. 1999. Expression of granulocyte colony-stimulating factor– and granulocyte-macrophage colony-stimulating factor–associated signal transduction proteins of the JAK/STAT pathway in normal granulopoiesis and in blast cells of acute myelogenous leukemia. *Experimental hematology,* 27, 885-894.

[4] BROMBERG, J. 2002 Stat proteins and oncogenesis. *J Clin Invest. :,* 09(9), 1139-1142.

[5] BUECHLER, C., RITTER, M., ORSÓ, E., LANGMANN, T., KLUCKEN, J. & SCHMITZ, G. 2000. Regulation of scavenger receptor CD163 expression in human monocytes and macrophages by pro- and antiinflammatory stimuli. *Journal of Leukocyte Biology,* 67, 97-103.

[6] CHENG, F., WANG, H.-W., CUENCA, A., HUANG, M., GHANSAH, T., BRAYER, J., KERR, W. G., TAKEDA, K., AKIRA, S., SCHOENBERGER, S. P., YU, H., JOVE, R. & SOTOMAYOR, E. M. 2003. A Critical Role for Stat3 Signaling in Immune Tolerance. *Immunity,* 19, 425-436.

[7] DARNELL, J., KERR, I. & STARK, G. 1994. Jak-STAT pathways and transcriptional activation in response to IFNs and other extracellular signaling proteins. *Science,* 264, 1415-1421.

[8] DONNELLY, R., DICKENSHEETS, H. & FINBLOOM, D. 1999. The interleukin-10 signal transduction pathway and regulation of gene expression in mononuclear phagocytes. *J Interferon Cytokine Res,* 19(6), 563-73.

[9] DOTAN, I. & RACHMILEWITZ, D. 2005. Probiotics in inflammatory bowel disease: possible mechanisms of action. *Current Opinion in Gastroenterology,* 21, 426-430.

[10] DUFFIELD, J. S. 2003. The inflammatory macrophage: a story of Jekyll and Hyde. *Clin. Sci.,* 104, 27-38.

[11] FOEY, A. 2012. Mucosal macrophages: phenotype and functionality in homeostasis and pathology. In. Handbook of macrophages: life cycle, functions and diseases. *Eds. Takahashi, R. and Kai, H. Nova Science Publishers Inc., NY, USA. Invited review chapter.*

[12] FUKADA, T., HIBI, M., YAMANAKA, Y., TAKAHASHI-TEZUKA, M., FUJITANI, Y., YAMAGUCHI, T., NAKAJIMA, K. & HIRANO, T. 1996. Two Signals Are Necessary for Cell Proliferation Induced by a Cytokine Receptor gp130: Involvement of STAT3 in Anti-Apoptosis.

Immunity, 5, 449-460.

[13] GALDEANO, C. M. & PERDIGÓN, G. 2006. The Probiotic Bacterium Lactobacillus casei Induces Activation of the Gut Mucosal Immune System through Innate Immunity. *Clinical and Vaccine Immunology,* 13, 219-226.

[14] GORDON, S. 2003. Alternative activation of macrophages. *Nat Rev Immunol,* 3, 23-35.

[15] GRIVENNIKOV, S., KARIN, E., TERZIC, J., MUCIDA, D., YU, G., VALLABHAPURAPU, S., SCHELLER, J., ROSE-JOHN, S., CHEROUTRE, H., ECKMANN, L. & KARIN, M. 2009. IL-6 and STAT3 are required for survival of intestinal epithelial cells and development of colitis associated cancer. *Cancer Cell,* 3; 15(2), 103-113.

[16] HABIL, N., ABATE, W., BEAL, J. & FOEY, A. D. 2014. Heat-killed probiotic bacteria differentially regulate colonic epithelial cell production of human β-defensin-2: dependence on inflammatory cytokines. *Beneficial Microbes,* 5, 483-495.

[17] HABIL, N., BEAL, J. & FOEY, A. 2011. Probiotic bacterial strains differentially modulate macrophage cytokine in a strain-depent and cell subset-specific manner. *Beneficial Microbe,* 2(4), 283-293.

[18] HABIL, N., BEAL, J. & FOEY, A. 2012. Lactobacillus casei strain Shirota selectively modulates macrophage subset cytokine production. *Int. J. Probiotics & Prebiotics* 7(1), 1-12.

[19] HABIL, N., FOEY, A. & BEAL, J. 2013. Probiotic Modulation of Mucosal Immune Responses in an In Vitro Co-Culture Model.

[20] HALLER, D., BODE, C., HAMMES, W. P., PFEIFER, A. M. A., SCHIFFRIN, E. J. & BLUM, S. 2000. Non-pathogenic bacteria elicit a differential cytokine response by intestinal epithelial cell/leucocyte co-cultures. *Gut,* 47, 79-87.

[21] HARRINGTON, L. E., HATTON, R. D., MANGAN, P. R., TURNER, H., MURPHY, T. L., MURPHY, K. M. & WEAVER, C. T. 2005. Interleukin 17-producing CD4+ effector T cells develop via a lineage distinct from the T helper type 1 and 2 lineages. *Nat Immunol,* 6, 1123-1132.

[22] HUME, D. A. 2006. The mononuclear phagocyte system. *Current Opinion in Immunology,* 18, 49-53.

[23] KIM, S. O., SHEIKH, H. I., HA, S.-D., MARTINS, A. & REID, G. 2006. G-CSF-mediated inhibition of JNK is a key mechanism for Lactobacillus rhamnosus-induced suppression of TNF production in macrophages. *Cellular Microbiology,* 8, 1958-1971.

[24] KOVALOVICH, K., LI, W., DEANGELIS, R., GREENBAUM, L. E., CILIBERTO, G. & TAUB, R. 2001. Interleukin-6 Protects against Fas-mediated Death by Establishing a Critical Level of Anti-apoptotic Hepatic Proteins FLIP, Bcl-2, and Bcl-xL. *Journal of Biological Chemistry,* 276, 26605-26613.

[25] MADSEN, K., CORNISH, A., SOPER, P., MCKAIGNEY, C., JIJON, H., YACHIMEC, C., DOYLE, J., JEWELL, L. & DE SIMONE, C. 2001. Probiotic Bacteria Enhance Murine and Human Intestinal Epithelial Barrier Function. *Gastroenterology,* 121, 580-591.

[26] MANTOVANI, A., SICA, A. & LOCATI, M. 2007. New vistas on macrophage differentiation and activation. *European Journal of Immunology,* 37, 14-16.

[27] MATSUGUCHI, T., TAKAGI, A., MATSUZAKI, T., NAGAOKA, M., ISHIKAWA, K., YOKOKURA, T. & YOSHIKAI, Y. 2003. Lipoteichoic Acids from Lactobacillus Strains Elicit Strong Tumor Necrosis Factor Alpha-Inducing Activities in Macrophages through Toll-Like Receptor 2. *Clinical and Diagnostic Laboratory Immunology*, 10, 259-266.

[28] MATSUKAWA, A., KUDO, S., MAEDA, T., NUMATA, K., WATANABE, H., TAKEDA, K., AKIRA, S. & ITO, T. 2005. Stat3 in Resident Macrophages as a Repressor Protein of Inflammatory Response. *The Journal of Immunology*, 175, 3354-3359.

[29] MATSUMOTO, S., HARA, T., HORI, T., MITSUYAMA, K., NAGAOKA, M., TOMIYASU, N., SUZUKI, A. & SATA, M. 2005. Probiotic Lactobacillus-induced improvement in murine chronic inflammatory bowel disease is associated with the down-regulation of pro-inflammatory cytokines in lamina propria mononuclear cells. *Clinical & Experimental Immunology*, 140, 417-426.

[30] MATSUMOTO, S., HARA, T., MITSUYAMA, K., YAMAMOTO, M., TSURUTA, O., SATA, M., SCHELLER, J., ROSE-JOHN, S., KADO, S.-I. & TAKADA, T. 2010. Essential Roles of IL-6 Trans-Signaling in Colonic Epithelial Cells, Induced by the IL-6/Soluble IL-6 Receptor Derived from Lamina Propria Macrophages, on the Development of Colitis-Associated Premalignant Cancer in a Murine Model. *The Journal of Immunology*, 184, 1543-1551.

[31] MATSUMOTO, S., HARA, T., NAGAOKA, M., MIKE, A., MITSUYAMA, K., SAKO, T., YAMAMOTO, M., KADO, S. & TAKADA, T. 2009. A component of polysaccharide peptidoglycan complex on Lactobacillus induced an improvement of murine model of inflammatory bowel disease and colitis-associated cancer. *Immunology*, 128, e170-e180.

[32] MOSSER, D. M. & EDWARDS, J. P. 2008. Exploring the full spectrum of macrophage activation. *Nat Rev Immunol*, 8, 958-969.

[33] O'ROURKE, L. & SHEPHERD, P. R. 2002. Biphasic regulation of extracellular-signal-regulated protein kinase by leptin in macrophages: role in regulating STAT3 Ser727 phosphorylation and DNA binding. *Biochem. J.*, 364, 875-879.

[34] PAOLILLO, R., ROMANO CARRATELLI, C., SORRENTINO, S., MAZZOLA, N. & RIZZO, A. 2009. Immunomodulatory effects of Lactobacillus plantarum on human colon cancer cells. *International Immunopharmacology*, 9, 1265-1271.

[35] SAVVIDOU, S. 2009. Selection of a chicken Lactobacillus strain with probiotic properties and its application in poultry production. . *PhD thesis, University of Plymouth, Plymouth, UK*.

[36] SHIDA, K., NANNO, M. & NAGATA, S. 2011. Flexible cytokine production by macrophages and T cells in response to probiotic bacteria: A possible mechanism by which probiotics exert multifunctional immune regulatory activities. *Gut Microbes*, 2, 109-114.

[37] STEIDLER, L., HANS, W., SCHOTTE, L., NEIRYNCK, S., OBERMEIER, F., FALK, W., FIERS, W. & REMAUT, E. 2000. Treatment of Murine Colitis by Lactococcus lactis Secreting Interleukin-10. *Science*, 289, 1352-1355.

[38] TAKEDA, K., CLAUSEN, B. E., KAISHO, T., TSUJIMURA, T., TERADA, N., FÖRSTER, I. & AKIRA, S. 1999. Enhanced Th1 Activity and Development of Chronic Enterocolitis in Mice Devoid of Stat3 in Macrophages and Neutrophils. *Immunity*, 10, 39-49.

[39] VERRECK, F. A. W., DE BOER, T., LANGENBERG, D. M. L., HOEVE, M. A., KRAMER, M., VAISBERG, E., KASTELEIN, R., KOLK, A., DE WAAL-MALEFYT, R. & OTTENHOFF, T. H. M. 2004. Human IL-23-producing type 1 macrophages promote but IL-10-producing type 2 macrophages subvert immunity to (myco)bacteria. *Proceedings of the National Academy of Sciences of the United States of America*, 101, 4560-4565.

[40] WILLIAMS, L., BRADLEY, L., SMITH, A. & FOXWELL, B. 2004. Signal Transducer and Activator of Transcription 3 Is the Dominant Mediator of the Anti-Inflammatory Effects of IL-10 in Human Macrophages. *The Journal of Immunology*, 172, 567-576.

[41] WILLIAMS, L. M., SARMA, U., WILLETS, K., SMALLIE, T., BRENNAN, F. & FOXWELL, B. M. J. 2007. Expression of Constitutively Active STAT3 Can Replicate the Cytokine-suppressive Activity of Interleukin-10 in Human Primary Macrophages. *Journal of Biological Chemistry*, 282, 6965-6975.

[42] WINKLER, P., GHADIMI, D., SCHREZENMEIR, J. & KRAEHENBUHL, J.-P. 2007. Molecular and Cellular Basis of Microflora-Host Interactions. *J. Nutr.*, 137, 756S-772.

[43] YANG, J., LIAO, X., AGARWAL, M. K., BARNES, L., AURON, P. E. & STARK, G. R. 2007. Unphosphorylated STAT3 accumulates in response to IL-6 and activates transcription by binding to NFkappaB. *Genes & development*, 21, 1396-1408.

27

Determination of the Aerodynamic Characteristics of Flying Vehicles Using Method Large Eddy Simulation with Software ANSYS

Nguyen Quang Vinh

Department of Radio-Electronic, Institute of Science and Technology, Cau Giay District, Hanoi, Vietnam

Email address:

Vinhquang2808@yahoo.com

Abstract: Nowadays, together with the huge development in information technology, the use of numerical method is no longer an obstacle. It provides a lot of advantages such as providing high precise solutions, reducing time and cost for designing and manufacturing flying vehicle models and dealing with extremely complex problems by using physical models compared to the usual analytical methods. The operation of airplanes represents a critical aerodynamic practical problem due to the wing. Interference this article presents a method to identify the aerodynamics coefficients of flying vehicles for isolate wing using software ANSYS. CFX and compares obtained results to the with those received by analytical method.

Keywords: Aerodynamic Characteristics, Wing, Airfoil Section, Angle of Attack

1. Introduction

One of the most practical problems in flight structure design and operation, is the wing ground interference. The aerodynamics characteristics of wings are changed in collision phenomena. In the present paper, the isolate wing is investigated analyzability and numerically. The investigation takes into account the solution of flow around a isolate wing which has profile represented in Fig.1; and a simulation is carried out at 20 angle of attack (AOA). Experimental results are presented for a Mach number ranged from 0.6 to 2.0 at averaged Reynolds number of 0.89.

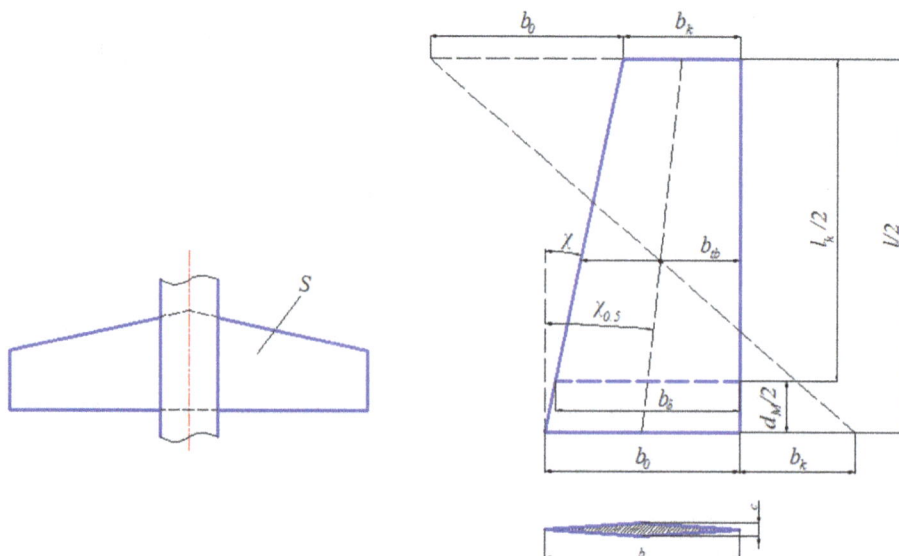

Fig. 1. The model profile of isolate wing for simulating.

Where b0 = 1 m; c = 0.064 mm; l = 1.5m; bk = 0.6 m, the flow around an isolate wing is modeled as incompressible at a Re number. Numerical simulations are performed with the commercial software package ANSYS. CFX and GAMBIT. Simulated results using numerical simulations compared with those using analytical method which is given in [3].

2. Experimental Procedure

2.1. Numerical Simulation Method

In all efforts aerodynamic studies, no matter whether they are theoretical, experimental or computational are normally aimed at a common objective: to determine the aerodynamic forces and moments acting on a body moving through air. The main purpose of employing Computer Fluid Dynamics (CFD) is to predict the aerodynamic forces: lift and drag, and the pitching moments, acting on the wing, in a way that allows to reduce time and cost when comparing with the real experiment tests.

An algorithm for modeling isolate wing with the profile shown in Fig. 1 is described on schemata Fig. 2 below.

Within the scope of this problem, we will isolate with the following assumptions:

- The surface is smooth channel (bypass roughness, surface profile);
- Wings endothermic (heat exchanger bypass between wing panels and the environment);
- The flow of air to flow under small angle of attack with different velocities corresponding Mach number M = 0.5, 0.6, 0.7, 0.8, 0.9, 1.0, 1.1, 1.2, 1.3, 1.4, 1.5, 2.0, 2.5, 3.
- The nature of the airflow to be taken according to GOST 4401-81: temperature $T_\infty = 298.15K$; Pressure $p_\infty = 101330\,Pa$; dynamic viscosity $v_\infty = 1.46 \cdot 10^{-5}\, m^2/s$; density $\rho_\infty = 1.225\, kg/m^3$.

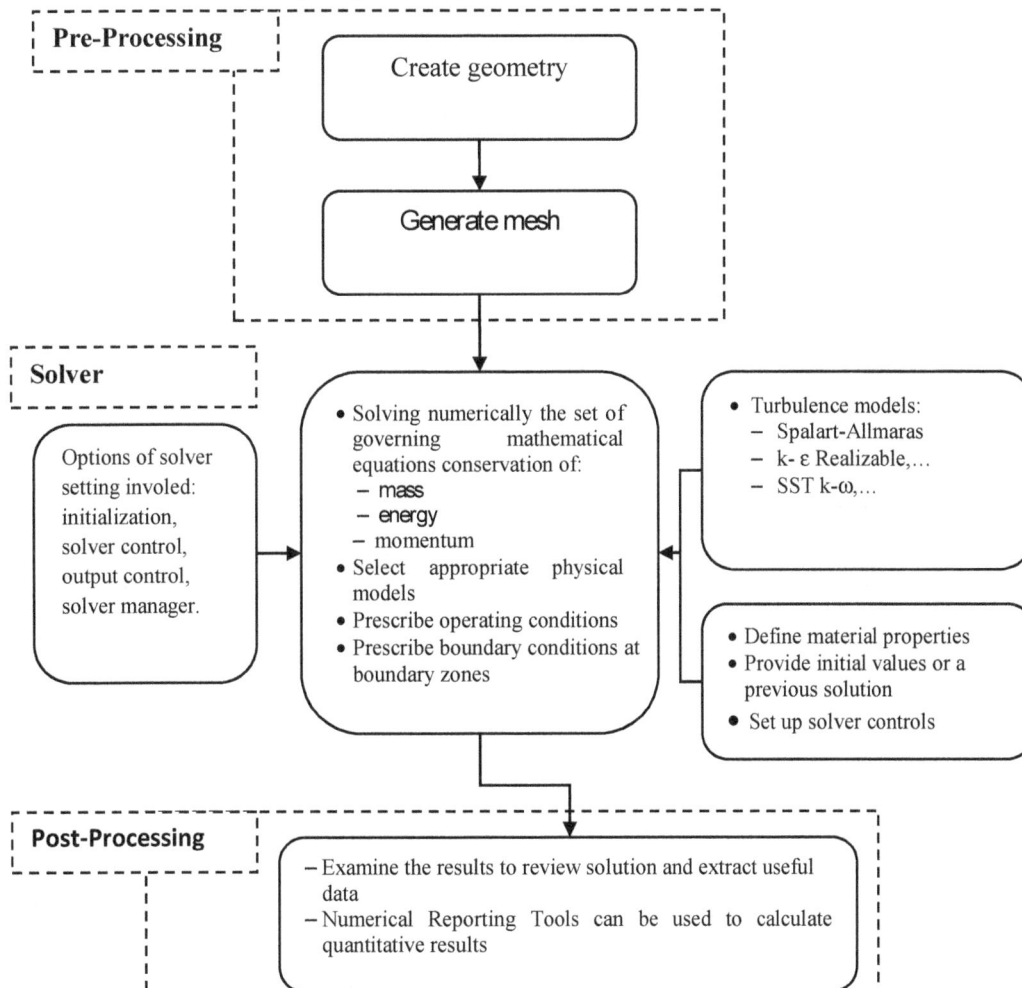

Fig. 2. Scheme of process to simulate considered model.

2.2. Boundary and Initial Conditions

In this investigation, considered isolate wing is supposed to have:

- Smooth surface (ignoring the surface roughness).

- Isothermal wing (excepted for heat transfer between plane and environment)
- The flow around wing carried out at 20 AOA and the free stream witch Mach number ranged from 0.6 to 2.0

at an averaged Reynolds number of 5.0e5–5.2e5.
- The wind tunnel operates at approximately $L \times H \times W = (10b_0 \div 15b_0) \times (6b_0 \div 8b_0) \times (6b_0 \div 10b_0)$ for subsonic and $L \times H \times W = (20b_0 \div 70b_0) \times (10b_0 \div 30b_0) \times (10b_0 \div 30b_0)$ for supersonic

2.3. Turbulent Model

Within the scope of this problem, we use the equation of state based on the ideal gas model. Then:

For incompressible gas flow:

$$\rho = \rho_{spec} = const; \quad dh = c_p dT + \frac{1}{\rho} dp;$$

$$c_p = c_p(T); \quad c_p - c_v = R$$

or compressed air lines are:

$$\rho = \frac{p_{abs}}{RT}; \quad dh = c_p dT; \quad c_p = c_p(T); \quad c_p - c_v = R$$

When conducting the calculation, we can choose the expression depends on the temperature of the heat capacity at constant pressure, c_p according to the methods already available to support the following:
- Perfect Calorically $c_p = const$
- Multi-standard mode or Zero Pressure NASA Format [1].

For a three dimensional, steady, incompressible, and turbulent viscous flow, the mass and momentum equations could be written in the Reynolds-Averaged Navier-Stokes:

Continuity equation: $\dfrac{\partial \rho}{\partial t} + \dfrac{\partial}{\partial x_j}(\rho U_i) = 0$

Momentum conservation equation:

$$\frac{\partial(\rho u_i)}{\partial t} + \frac{\partial}{\partial x_j}(\rho u_i u_j) =$$

$$= -\frac{\partial p'}{\partial x_i} + \frac{\partial}{\partial x_j} \cdot \left\{ \left(\tau_{ij} - \left\langle \overline{\rho u_i u_j} \right\rangle \right) \right\} + S_M$$

Conservation equation of total energy

$$\frac{\partial(\rho h_{tot})}{\partial t} - \frac{\partial p}{\partial t} + \frac{\partial}{\partial x_j}(\rho u_j h_{tot}) =$$

$$= \frac{\partial}{\partial x_j}\left(\lambda \frac{\partial T}{\partial x_j} + \frac{\mu_t}{Pr_t}\frac{\partial h}{\partial x_j} \right) + \frac{\partial}{\partial x_j} \cdot \left(u_j \cdot \left(\tau_{ij} - \left\langle \overline{\rho u_i u_j} \right\rangle \right) \right) +$$

$$+ u_j \cdot S_M + S_E$$

Here: p' – Pressure has been adjusted, taking into account elements of pressure caused by the turbulent kinetic energy and the effective viscosity:

$$p' = p + \frac{2}{3}\rho k + \frac{2}{3}\mu_{eff}\nabla \cdot U$$

τ_{ij} – stress tensor: $\tau_{ij} = \mu_{eff}\left(2S_{ij} - \frac{2}{3}\frac{\partial u_k}{\partial x_k}\delta_{ij} \right)$

S_{ij}, δ_{ij} – respectively, and strain rate tensor function Kronecker function:

$$S_{ij} = \frac{1}{2}\left(\frac{\partial u_i}{\partial x_j} + \frac{\partial u_j}{\partial x_i} \right); \quad \delta_{ij} = \begin{cases} 1 & i = j \\ 0 & i \neq j \end{cases}$$

$\left\langle \overline{\rho u_i u_j} \right\rangle$ – Reynold stress codes. In theory, the eddy viscosity Reynolds stresses depends on the average velocity gradient and diffusion gradient swung by the following expression:

$$-\left\langle \overline{\rho u_i u_j} \right\rangle = 2\mu_t S_{ij} - \frac{2}{3}\left(\mu_t \frac{\partial u_k}{\partial x_k} + \rho k \right)\delta_{ij}$$

μ_{eff} – the actual viscosity of the flow. μ_{eff} quantity is determined by the formula: $\mu_{eff} = \mu + \mu_t$

μ – viscosity is determined according to the formula on the basis of the model is supported by ANSYS. CFX such as Ideal Mixture, Non-Newtonian Model, Sutherland's Formula …

μ_t – turbulent viscosity (or viscosity vortex) generated by the turbulence and is determined according to the formula depending on the model of turbulence that we choose.

In the solution of the conservation equations, the turbulence term should be replaced by suitable equivalent model; otherwise, the equations are not solvable. The model used to solve the simulation is SST $k - \omega$. The SST $k - \omega$ turbulence model is a two-equation eddy-viscosity model that is used for many aerodynamic applications. It is a hybrid model combining the Wilcox k-omega and the k-epsilon models.

GAMBIT 2-4.6 and ANSYS. CFX 12.1 are used for modeling and simulation respectively.

3. Semi-empirical Method

3.1. Determine the Lift Coefficient

When small angle of attack and angle helical wing $\alpha \leq 100; \delta = 0$ twisted wing is the lift coefficient depends on angle of attack is almost linear. Therefore, the lift coefficient can be expressed as follows [5]:

$$c_{y.wing} = c_{y.wing}^{\alpha}\alpha = \left(c_{y1.wing}^{\alpha} - \frac{c_{x0.wing}}{57.3} \right)\alpha$$

$c_{y1.wing}^{\alpha}$ – partial lifting force coefficient $c_{y1.wing}$ according to the angle of attack in the coordinate system $Ox_1y_1z_1$. $c_{y1.wing}^{\alpha}$ depending on the number M_{∞} and shape on plane wing: elongation λ_{wing}, spasm of wings η_{wing} and swing-wing corner χ. $\bar{c}$ – relatively thick wing profile, as measured in section parallel to the plane wing symmetry of flying machines. The dimensionless quantity is determined according to the following formula:

$$\bar{c} = \frac{c}{b}; \qquad \eta_{wing} = \frac{b_0}{b_k}; \qquad \lambda_{wing} = \frac{l^2}{S}.$$

$c_{y1.wing}^{\alpha}$ - defined by the expression [4, 5]:

$$\frac{c_{y1.wing}^{\alpha}}{\lambda_{wing}} = f\left(\lambda_{wing}\sqrt{|M^2-1|}; \lambda_{wing} tg\chi_{0,5}; \lambda_{wing}\sqrt[3]{\bar{c}}\right)$$

$c_{x0.wing}$ − coefficient of drag, lift the front as zero

3.2. Identify the Front Drag Coefficient

Resistance coefficient frontally $c_{x.wing}$ is defined by the expression [4, 5]:

$$c_{x.wing} = c_{x0.wing} + c_{xi.wing}.$$

$c_{xi.wing}$ − induced drag coefficient, which is generated by the lift, which is determined by the formula:

$$c_{xi.wing} = c_{y.wing}^{\alpha}\left(K_{\alpha\alpha} - 57,3\xi\bar{c}_F c_{y.wing}^{\alpha} k_{\alpha\alpha}^2\right)\frac{\alpha^2}{57,3}$$

$K_{\alpha\alpha}, k_{\alpha\alpha}$ − interference coefficient between the wings and fuselage. In isolated cases wing $K_{\alpha\alpha} = k_{\alpha\alpha} = 1$.

$\bar{c}_F, \xi$ − its corresponding absorption coefficient and its efficiency coefficient. For wings with sharp edges before then $\bar{c}_F = \xi = 0$.

$c_{x0.wing}$, code for the next case without jail wing after wing, which is determined by the formula [4.5]:
$c_{x0.wing} == c_{xp.wing} + c_{xb.wing}$

$c_{xp.wing}$ − coefficient of drag caused by the viscosity of the gas stream to flow. This coefficient is determined by the following formula: $c_{xp.wing} = (2f)_{M=0}\,\eta_M\eta_c$.

$\eta_M, \eta_c, (2f)_{M=0}$ is determined according to experimental results in [4,5].

$c_{xb.wing}$ − A coefficient of drag caused by the compressibility of the flow. The value of this coefficient, depending on the Mach number, the plane wing shape, thickness contour wing and wing shape, which is determined from the expression dependent [4, 5]:

$$\frac{c_{xb.wing}}{\lambda_{wing}\bar{c}^2} = f\left(\begin{array}{l}\lambda_{wing}\sqrt{|M^2-1|}; \lambda_{wing} tg\chi; \\ \eta_{wing}; \ wing's\ profine\end{array}\right)$$

Results determined coefficient method and empirical theory is presented in the form of graphs in the document [5].

4. Results and Discussion

The simulated results are obtained using numerical simulation with software package ANSYS. CFX and GAMBIT. The Fig. 3 to Fig. 6 are the plots of the pressure contour effort to wing surface with different Mach numbers.

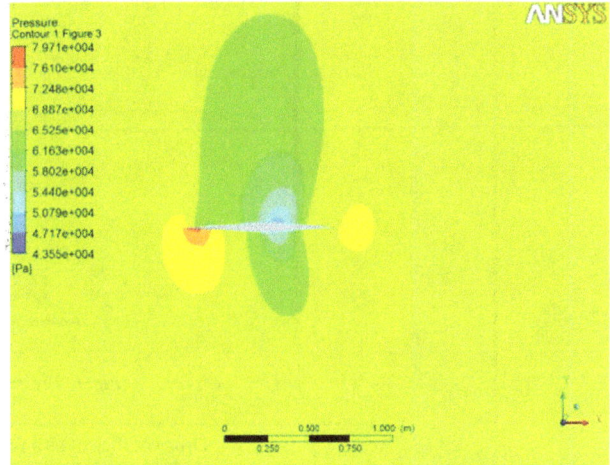

Fig. 3. *Pressure contour at M = 0.8.*

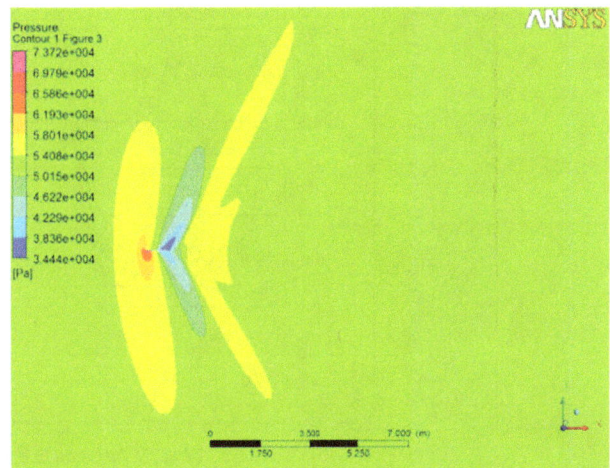

Fig. 4. *Pressure contour at M = 1.0.*

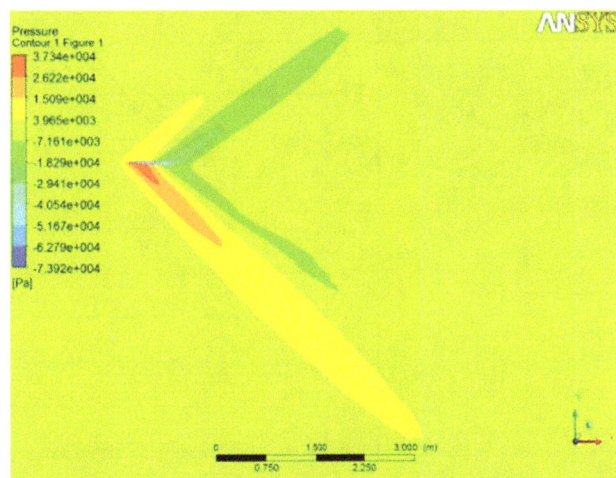

Fig. 5. *Pressure contour at M = 1.5.*

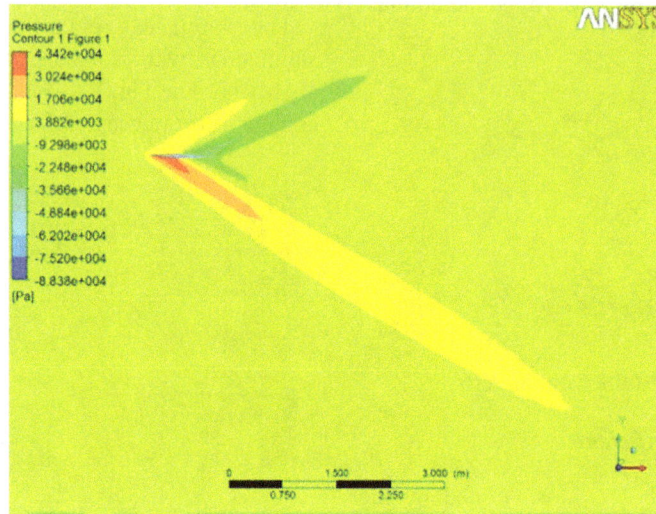

Fig. 6. Pressure contour at M = 2.0.

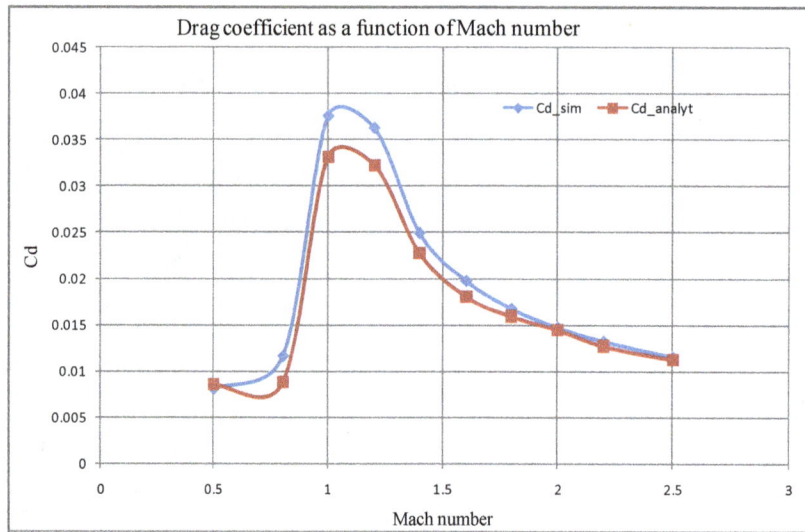

Fig. 7. Variation of drag coefficient as function of M.

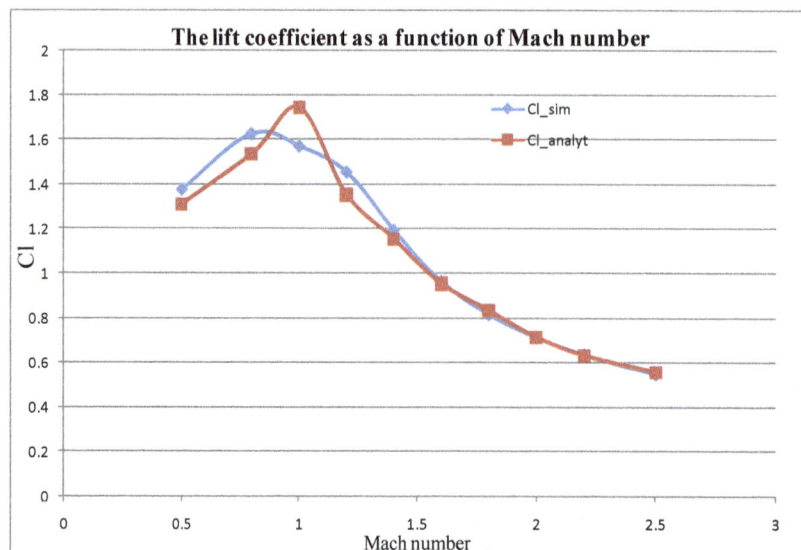

Fig. 8. Variation of drag coefficient as function of M.

In Fig. 3 to Fig. 6 can be observed that the total pressure increases when Mach number increases from 0.5 to about 1.2, and decreases when the Mach number is over 1.2

Figure 7 are showing the results of the drag, lift coefficients at 20 AOA and the Mach number varies in the range from 0.6 to 2.5, respectively as obtained from the analytical and numerical simulation. Quite good agreement between the analytical and numerical results could be realized. The variation of the lift coefficient between the analytical and numerical simulation cause the assumption we have represented above (include smooth surface, isothermal wing ...).

5. Conclusions

The collision phenomena with influence on the aerodynamic characteristics of isolate wing operation were investigated analytically and numerically. On the numerical part, the conservations of mass and momentum equations are solved by software package ANSYS. CFX using finite volume technique. To close the domain, the turbulence part of the flow is approximated by using SST $k - \omega$ model. The aerodynamic parameters, lift, drag coefficients were measured analytically and predicted numerically at 20 and various Mach number in the range from 0.6 to 2.5. The results are compared with the analytical method and good agreements is achieved.

Acknowledgements

The authors would like to thank colleagues in Vietnam Academy of Military Science and Technology for their assistance during the use of ANSYS's laboratory to produce the wing model.

References

[1]   Ansys. CFX Release 12.0: ANSYS CFX-Solver Theory Guide.

[2]   Ansys. Gambit Helps.

[3]   Zakharchenko VF, et al., Determination of total aerodynamic characteristics of various configurations of aircraft. M: Publishing house of the MSTU. Bauman, 1993.

[4]   AA Lebedev, Chernobrovkin LS Flight Dynamics: unmanned aerial vehicles. M: Mechanical Engineering, 1973.

[5]   Process and research on the method of guidance a maneuvering target with scholastic prediction about its motion and with nonlinearities in the control circle" (A report on the scientific researching work)/MSTU named after Bauman. The supervisor: Pupkov K. A. The executors: Nguyen Quang Vinh, Bobkov A. V., Ustiuzanin A. D. / Moscow, 2006, 75 pages.

28

Notes on "Some Properties of *L*-fuzzy Approximation Spaces on Bounded Integral Residuated Lattices"

Yuan Wang[1,*], Keming Tang[1], Zhudeng Wang[2]

[1]College of Information Engineering, Yancheng Teachers University, Yancheng, People's Republic of China

[2]School of Mathematics and Statistics, Yancheng Teachers University, Jiangsu, People's Republic of China

Email address:

yctuwangyuan@163.com (Yuan Wang), tkmchina@126.com (Keming Tang), zhudengwang2004@163.com (Zhudeng Wang)

[*]Corresponding author

Abstract: In this note, we continue the works in the paper [Some properties of *L*-fuzzy approximation spaces on bounded integral residuated lattices", Information Sciences, 278, 110-126, 2014]. For a complete involutive residuated lattice, we show that the *L*-fuzzy topologies generated by a reflexive and transitive *L*-relation satisfy $(TC)_L$ or $(TC)_R$ axioms and the *L*-relations induced by two *L*-fuzzy topologies, which are generated by a reflexive and transitive *L*-relation, are all the original *L*-relation; and give out some conditions such that the *L*-fuzzy topologies generated by two *L*-relations, which are induced by an *L*-fuzzy topology, are all the original *L*-fuzzy topology.

Keywords: Involutive Residuated Lattice, *L*-relation, *L*-fuzzy Topology, *L*-fuzzy Approximation Space

1. Introduction

A residuated lattice (see [1, 10]) is an algebra $L = (L, \wedge, \vee, \cdot, \rightarrow, \leftarrow, 0, 1)$ of type (2, 2, 2, 2, 2, 0, 0) satisfying the following conditions:

(L1) $(L, \wedge, \vee)$ is a lattice,

(L2) $(L, \cdot, 1)$ is a monoid, i.e., is associative and $x \cdot 1 = 1 \cdot x = x$ for any $x \in L$,

(L3) $x \cdot y \leq z$ if and only if $x \leq y \rightarrow z$ if and only if $y \leq x \leftarrow z$ for any $x, y, z \in L$.

A residuated lattice with a constant 0 is called an *FL*-algebra. If $x \leq 1$ for all $x \in L$, then L is called integral residuated lattice. An *FL*-algebra L, which satisfies the condition $0 \leq x \leq 1$ for all $x \in L$, is called an FL_w-algebra or a bounded integral residuated lattice (see [1]).

We adopt the usual convention of representing the monoid operation by juxtaposition, writing ab for $a \cdot b$.

Let L be a bounded integral residuated lattice. Define two negations on L, $\neg^L$ and $\neg^R$:

$$\neg^L x = x \rightarrow 0, \quad \neg^R x = x \leftarrow 0 \quad \forall x \in L.$$

A bounded residuated lattice L is called an involutive residuated lattice (see [3]) if

$$\neg^L \neg^R x = \neg^R \neg^L x \quad \forall x \in L.$$

In the sequel, unless otherwise stated, L always represents any given complete involutive residuated lattice with maximal element 1 and minimal element 0.

Definition 1.1 (see Liu and Luo [5]). Let $\tau \subseteq L^X$ and J be an index set. If τ satisfies the following three conditions:

(LFT1) $0_X, 1_X \in \tau$,

(LFT2) $\mu, \nu \in \tau \Rightarrow \mu \wedge \nu \in \tau$,

(LFT3) $\mu_j \in \tau \, (j \in J) \Rightarrow \vee_{j \in J} \mu_j \in \tau$,

then τ is called an L-fuzzy topology on X and (L^X, τ) L-fuzzy topological space. Every element in τ is called an open subset in L^X.

When $L = [0, 1]$, an L-fuzzy topological space (L^X, τ) is *also called an F-topological space.*

Let $\tau'_L = \{\neg^L \mu | \mu \in \tau\}$ and $\tau'_R = \{\neg^R \mu | \mu \in \tau\}$. The elements of τ'_L and τ'_R are, respectively, called left closed subsets and right closed subsets in L^X (see Wang et al. [12]).

Definition 1.2 (Wang and Liu [11], Wang et al. [12]). Let τ be an L-fuzzy topology on X and μ L-fuzzy subset of X. The interior, left closure and right closure of μ w.r.t τ are, respectively, defined by

$$\text{int}(\mu) = \vee\{\eta | \eta \leq \mu, \eta \in \tau\},$$

$$cl_L(\mu) = \wedge\{\xi | \mu \leq \xi, \xi \in \tau'_L\},$$

$$cl_R(\mu) = \wedge\{\zeta | \mu \leq \zeta, \zeta \in \tau'_R\}.$$

int , cl_L and cl_R *are, respectively, called the interior, left closure and right closure operators.*

For the sake of convenience, we denote $\text{int}(\mu)$, $cl_L(\mu)$ and $cl_R(\mu)$ by μ^o, μ_L^- and μ_R^-, respectively.

Zhang et al. [14, 15] studied some properties of rough sets and rough approximation operators, Ouyang et al. [6, 7] investigated some generalized models of fuzzy rough sets, Liu and Lin [4] considered the intuitionistic fuzzy rough set model, Wu et al. [13] discussed the axiomatic characterizations of fuzzy rough approximation operators, Radzikowska and Kerre [9] studied L-fuzzy rough sets and lower (upper) L-fuzzy approximation based on commutative residuated lattices. Recently, Wang et al. [12] discussed the notion of left (right) lower and left (right) upper L-fuzzy rough approximation based on complete bounded integral residuated lattices.

Definition 1.3 (Wang et al. [12]). Let R be an L-relation on X. A pair (X, R) is called an L-fuzzy approximation space. Define the following four mappings $R \downarrow_L$, $R \uparrow_L$, $R \downarrow_R$, $R \uparrow_R : L^X \to L^X$, called a left lower, left upper, right lower, and right upper L-fuzzy rough approximation operators, respectively, as follows: for every $\mu \in L^X$ and $x \in X$,

$$R \downarrow_L (\mu)(x) = \wedge_{y \in X} (R(x, y) \to \mu(y)),$$

$$R \uparrow_L (\mu)(x) = \vee_{y \in X} \mu(y)R(y, x),$$

$$R \downarrow_R (\mu)(x) = \wedge_{y \in X} (R(y, x) \leftarrow \mu(y)),$$

$$R \uparrow_R (\mu)(x) = \vee_{y \in X} R(x, y)\mu(y).$$

$R \downarrow_L (\mu)$, $R \uparrow_L (\mu)$, $R \downarrow_R (\mu)$ and $R \uparrow_R (\mu)$ are called left lower, left upper, right lower, and right upper L-fuzzy rough approximations of μ, respectively.

A pair $(\lambda, \xi) \in L^X \times L^X$ such that $\lambda = R \downarrow_L (\mu)$ ($\lambda = R \downarrow_R (\mu)$) and $\xi = R \uparrow_L (\mu)$ ($\xi = R \uparrow_R (\mu)$) for some

$\mu \in L^X$, is called a left (right) L-fuzzy rough set in (X, R) .

When $L = [0, 1]$, L-fuzzy rough approximation operators, L-fuzzy approximation space and left (right) L-fuzzy rough sets are, respectively, called fuzzy rough approximation operators, fuzzy approximation space and left (right) fuzzy rough sets.

2. The *L*-fuzzy Topologies Generated by a Reflexive and Transitive *L*-relation

In this section, we supplement some properties of the L-fuzzy topologies generated by a reflexive and transitive L-relation.

If R is a reflexive and transitive L-relation on X, then it follows from Theorem 6.1 in [12] that

$$\tau_1 = \{\xi | R \downarrow_L (\xi) = \xi, \xi \in L^X\},$$

$$\tau_2 = \{\xi | R \downarrow_R (\xi) = \xi, \xi \in L^X\}$$

are all L-fuzzy topologies on X and $R \downarrow_L$ and $R \downarrow_R$ are just the interior operators w.r.t τ_1 and τ_2, respectively. Here, τ_1 and τ_2 are called the L-fuzzy topologies generated by the L-relation R or by left lower L-fuzzy rough approximation operator $R \downarrow_L$ and right lower L-fuzzy rough approximation operator $R \downarrow_R$, respectively.

Theorem 2.1. If R is a reflexive and transitive L-relation on X, then

$$\tau_1 = \{\neg^L R \uparrow_R (\xi) | \xi \in L^X\},$$

$$\tau_2 = \{\neg^R R \uparrow_L (\xi) | \xi \in L^X\},$$

$R \uparrow_R$ *and* $R \uparrow_L$ *are, respectively, the right closure operator w.r.t. τ_1 and the left closure operator w.r.t. τ_2.*

Proof. When L is an involutive residuated lattice, $\neg^R \neg^L \mu = \neg^L \neg^R \mu = \mu$ for any $\mu \in L^X$.

If R is a reflexive and transitive L-relation on X, then it follows from Theorem 4.1(5) and Remark 5.2 in [12] that

$$R \downarrow_L (\neg^L R \uparrow_R (\xi)) = \neg^L (R \uparrow_R (R \uparrow_R (\xi)))$$
$$= \neg^L (R \uparrow_R (\xi)) \quad \forall \xi \in L^X,$$

i.e., $\neg^L R \uparrow_R (\xi) \in \tau_1$ for any $\xi \in L^X$. If $\xi \in \tau_1$ and $\mu \in L^X$, then it follows from Theorems 3.1(3) and 4.1(5) in [12] that

$$\xi = R \downarrow_L (\xi) = R \downarrow_L (\neg^L \neg^R \xi)$$
$$= \neg^L (R \uparrow_R (\neg^R \xi)) = \neg^L R \uparrow_R (\eta),$$

$$\mu_R^- = \neg^R (\neg^L \mu)^o$$
$$= \neg^R (R \downarrow_L (\neg^L \mu)) = \neg^R (\neg^L R \uparrow_R (\mu))$$
$$= \neg^R \neg^L (R \uparrow_R (\mu)) = R \uparrow_R (\mu),$$

where $\eta = \neg^R \xi$. So, $\tau_1 = \{\neg^L R \uparrow_R (\eta) | \eta \in L^X\}$ and $R \uparrow_R$ is the right closure operator w.r.t. τ_1.

Similarly, we can show that $\tau_2 = \{\neg^R R \uparrow_L (\eta) | \eta \in L^X\}$ and $R \uparrow_L$ is the left closure operator w.r.t. τ_2.

The theorem is proved.

Recently, Qin et al. [2, 8] studied the topogical properties of fuzzy rough sets. The following left and right (TC) axioms are generalizations of (TC) axiom in [8].

$(TC)_L$ axiom: for any $x, y \in X$ and $\mu \in \tau$ there exists $\mu^* \in \tau$ such that $\mu^*(x) = 0$ and

$$\mu^*(y) \to \mu^*(x) \le \mu(y) \to \mu(x).$$

$(TC)_R$ axiom: for any $x, y \in X$ and $\nu \in \tau$ there exists $\nu^* \in \tau$ such that $\nu^*(y) = 0$ and

$$\nu^*(x) \leftarrow \nu^*(y) \le \nu(x) \leftarrow \nu(y).$$

Theorem 2.2. If R is a reflexive and transitive L-relation on X, then the L-fuzzy topologies τ_1 and τ_2, generated by R, satisfy $(TC)_R$ and $(TC)_L$ axioms, respectively.

Proof. For any $x, y \in X$ and $\mu \in \tau_1$, let

$$\mu^* = \neg^L (R_{\tau_1}^R \uparrow_R (1_{\{y\}})),$$

then

$$\mu^*(y) = \neg^L (R_{\tau_1}^R \uparrow_R (1_{\{y\}}))(y)$$
$$= \neg^L R_{\tau_1}^R (y, y) = \neg^L 1 = 0,$$

$$\mu^*(x) \leftarrow \mu^*(y) = \neg^L (R_{\tau_1}^R \uparrow_R (1_{\{y\}}))(x) \leftarrow 0$$
$$= \neg^R \neg^L R_{\tau_1}^R (x, y) = R_{\tau_1}^R (x, y)$$
$$= \wedge_{\xi \in \tau_1} (\xi(x) \leftarrow \xi(y)) \le \mu(x) \leftarrow \mu(y),$$

i.e., τ_1 satisfies $(TC)_R$ axiom; for any $\nu \in \tau_2$, let

$$\nu^* = \neg^R (R_{\tau_2}^L \uparrow_L (1_{\{x\}})),$$

Then

$$\nu^*(x) = \neg^R (R_{\tau_2}^L \uparrow_L (1_{\{x\}}))(x)$$
$$= \neg^R R_{\tau_2}^L (x, x) = \neg^R 1 = 0,$$

$$\nu^*(y) \to \nu^*(x) = \neg^R (R_{\tau_2}^L \uparrow_L (1_{\{x\}}))(y) \to 0$$
$$= \neg^L \neg^R R_{\tau_2}^L (x, y) = R_{\tau_2}^L (x, y)$$
$$= \wedge_{\xi \in \tau_2} (\xi(y) \to \xi(x)) \le \nu(y) \to \nu(x),$$

i.e., τ_2 satisfies $(TC)_L$ axiom.

The theorem is proved.

3. The L-relations Induced by an L-fuzzy Topology

In this section, we supplement some properties of the L-relations induced by an L-fuzzy topology.

Let τ be an L-fuzzy topology on X. For any $x, y \in X$, let

$$R_\tau^L (x. y) = \wedge_{\mu \in \tau} (\mu(y) \to \mu(x)),$$
$$R_\tau^R (x. y) = \wedge_{\mu \in \tau} (\mu(x) \leftarrow \mu(y)).$$

Clearly, R_τ^L and R_τ^R are reflexive L-relations on X. Moreover, it follows from Theorem 2.1(5) in [12] that

$$R_\tau^L (x, y) R_\tau^L (y, z)$$
$$= \{\wedge_{\mu \in \tau} (\mu(y) \to \mu(x))\} \{\wedge_{\nu \in \tau} (\nu(z) \to \nu(y))\}$$
$$\le \wedge_{\mu \in \tau} \{\mu(y) \to \mu(x)\} \{(\mu(z) \to \mu(y)\}$$
$$\le \wedge_{\mu \in \tau} (\mu(z) \to \mu(x))$$
$$= R_\tau^L (x, z) \quad \forall x, y, z \in X,$$

$$R_\tau^R (x, y) R_\tau^R (y, z)$$
$$= \{\wedge_{\mu \in \tau} (\mu(x) \leftarrow \mu(y))\} \{\wedge_{\nu \in \tau} (\nu(y) \leftarrow \nu(z))\}$$
$$\le \wedge_{\mu \in \tau} \{\mu(x) \leftarrow \mu(y)\} \{(\mu(y) \leftarrow \mu(z)\}$$
$$\le \wedge_{\mu \in \tau} (\mu(x) \leftarrow \mu(z))$$
$$= R_\tau^R (x, z) \quad \forall x, y, z \in X.$$

Thus, R_τ^L and R_τ^R are all transitive L-relations on X. Let

$$R_\tau(x, y) = R_\tau^R (x, y) \wedge R_\tau^L (x, y)$$
$$= \wedge_{\mu \in \tau} \{\mu(x) \leftarrow \mu(y)\} \wedge \{\mu(y) \to \mu(x)\} \forall x, y \in X.$$

It is easy to see that $R_\tau = R_\tau^R \wedge R_\tau^L$ is also a reflexive and transitive L-relations on X.

Theorem 3.1. If R is a reflexive and transitive L-relation on X, then

$$R = R_{\tau_1}^R = R_{\tau_2}^L.$$

Proof. For any $x, y \in X$, by virtue of Definitions 1.2 and 1.3 and Theorem 2.1, we see that

$$R(x, y) = R \uparrow_R (1_{\{y\}})(x) = (1_{\{y\}})_R^- (x)$$
$$= \wedge \{\neg^R \xi(x) | 1_{\{y\}} \le \neg^R \xi, \xi \in \tau_1\}$$
$$= \wedge \{\neg^R \xi(x) | \neg^R \xi(y) = 1, \xi \in \tau_1\}$$
$$= \wedge \{\xi(x) \leftarrow 0 | \xi(y) = 0, \xi \in \tau_1\}$$
$$= \wedge \{\xi(x) \leftarrow \xi(y) | \xi(y) = 0, \xi \in \tau_1\}$$
$$\ge \wedge \{\xi(x) \leftarrow \xi(y) | \xi \in \tau_1\} = R_{\tau_1}^R (x, y);$$

$$R(x, y) = R \uparrow_L (1_{\{x\}})(y) = (1_{\{x\}})_L^- (y)$$
$$= \wedge \{\neg^L \xi(y) | 1_{\{x\}} \le \neg^L \xi, \xi \in \tau_2\}$$
$$= \wedge \{\neg^L \xi(y) | \neg^L \xi(x) = 1, \xi \in \tau_2\}$$
$$= \wedge \{\xi(y) \to 0 | \xi(x) = 0, \xi \in \tau_2\}$$
$$= \wedge \{\xi(y) \to \xi(x) | \xi(x) = 0, \xi \in \tau_2\}$$
$$\ge \wedge \{\xi(y) \to \xi(x) | \xi \in \tau_2\} = R_{\tau_2}^L (x, y).$$

Thus, $R \geq R_{\tau_1}^R$ and $R \geq R_{\tau_2}^L$.

On the other hand, $R \downarrow_L$ and $R \uparrow_R$ are, respectively, the interior and right closure operators w.r.t. τ_1 and $R \downarrow_R$ and $R \uparrow_L$ are, respectively, the interior and left closure operators w.r.t. τ_2. Thus, by virtue Theorem 3.1(3) and Remark 5.2 in [12], we can see that

$$R \uparrow_R \{\neg^R(R \downarrow_L (\mu))\} = \{\neg^R(R \downarrow_L (\mu))\}_R^-$$
$$= \neg^R\{\neg^L\neg^R(R \downarrow_L (\mu))\}^o = \neg^R\{R \downarrow_L (\mu)\}^o$$
$$= \neg^R\{R \downarrow_L (R \downarrow_L (\mu))\}$$
$$= \neg^R(R \downarrow_L (\mu)) \ \forall R \downarrow_L (\mu) \in \tau_1,$$

$$R \uparrow_L \{\neg^L(R \downarrow_R (\mu))\} = \{\neg^L(R \downarrow_R (\mu))\}_L^-$$
$$= \neg^L\{\neg^R\neg^L(R \downarrow_R (\mu))\}^o = \neg^L\{R \downarrow_R (\mu)\}^o$$
$$= \neg^L\{R \downarrow_R (R \downarrow_R (\mu))\}$$
$$= \neg^L(R \downarrow_R (\mu)) \ \forall R \downarrow_R (\mu) \in \tau_2.$$

So, it follows from the proof of Theorem 7.2 in [12] that $R \leq R_{\tau_1}^R$ and $R \leq R_{\tau_2}^L$.

Therefore, $R = R_{\tau_1}^R = R_{\tau_2}^L$.

The theorem is proved.

This result shows that the reflexive and transitive L-relations $R_{\tau_1}^R$ and $R_{\tau_2}^L$ induced by, respectively, the L-fuzzy topologies τ_1 and τ_2 are all the original reflexive and transitive L-relation.

For any $\mu \in L^X$ and $R \in L^{X \times X}$,

$$\mu = \vee_{x \in X} (\mu(x)_X \wedge 1_{\{x\}}).$$

Thus, by Definition 1.3 and Theorem 4.1(3) in [12], we see that

$$R \uparrow_L (\mu) = \vee_{x \in X} R \uparrow_L (\mu(x)_X \wedge 1_{\{x\}})$$
$$= \vee_{x \in X} \mu(x)_X \cdot R \uparrow_L (1_{\{x\}}),$$

$$R \uparrow_R (\mu) = \vee_{x \in X} R \uparrow_R (\mu(x)_X \wedge 1_{\{x\}})$$
$$= \vee_{x \in X} R \uparrow_L (1_{\{x\}}) \cdot \mu(x)_X.$$

Theorem 3.2. Let τ be an L-fuzzy topology on X and J index set. Then the following properties hold.

(1) If τ satisfies $(TC)_L$ *axiom and the left closure operator w.r.t. τ satisfies the following two conditions:*

(CL1) $(\vee_{j \in J} \mu_j)_L^- = \vee_{j \in J} (\mu_j)_L^- \ \forall \mu_j \in L^X$,

(CL2) $(a \wedge 1_{\{x\}})_L^- = a \cdot (1_{\{x\}})_L^- \ \forall a \in L, x \in X$,

then $R_\tau^L \uparrow_L$ and $R_\tau^L \downarrow_R$ are, respectively, just the left closure operator and the interior operator w.r.t. τ and

$$\tau = \{\xi \mid R_\tau^L \downarrow_R (\xi) = \xi, \xi \in L^X\}.$$

(2) If τ satisfies $(TC)_R$ *axiom and the right closure operator w.r.t. τ satisfies the following two conditions:*

(CR1) $(\vee_{j \in J} \mu_j)_R^- = \vee_{j \in J} (\mu_j)_R^- \ \forall \mu_j \in L^X$,

(CR2) $(a \wedge 1_{\{x\}})_R^- = (1_{\{x\}})_R^- \cdot a \ \forall a \in L, x \in X$,

then $R_\tau^R \uparrow_R$ and $R_\tau^R \downarrow_L$ are, respectively, just the right closure operator and the interior operator w.r.t. τ and

$$\tau = \{\xi \mid R_\tau^R \downarrow_L (\xi) = \xi, \xi \in L^X\}.$$

Proof. We only prove (1).

If τ satisfies $(TC)_L$ axiom and the left closure operator w.r.t. τ satisfies the conditions (CL1) and (CL2), then it follows from Definition 1.3 and the proof of Theorem3.1 that

$$R_\tau^L \uparrow_L (1_{\{x\}})(y) = R_\tau^L(x, y) = \wedge_{\mu \in \tau} (\mu(y) \to \mu(x))$$
$$= \wedge\{\mu(y) \to \mu(x) \mid \mu(x) = 0, \mu \in \tau\}$$
$$= (1_{\{x\}})_L^-(y) \ \forall x, y \in X,$$

i.e., $(1_{\{x\}})_L^- = R_\tau^L \uparrow_L (1_{\{x\}})$ for any $x \in X$. Thus, for any $\mu \in L^X$, we have that

$$\mu_L^- = \{\vee_{x \in X} (\mu(x)_X \wedge 1_{\{x\}})\}_L^-$$
$$= \vee_{x \in X} (\mu(x)_X \wedge 1_{\{x\}})_L^- = \vee_{x \in X} \mu(x)_X \cdot (1_{\{x\}})_L^-$$
$$= \vee_{x \in X} \mu(x)_X \cdot R_\tau^L \uparrow_L (1_{\{x\}})$$
$$= \vee_{x \in X} R_\tau^L \uparrow_L (\mu(x)_X \wedge 1_{\{x\}})$$
$$= R_\tau^L \uparrow_L \{\vee_{x \in X} (\mu(x)_X \wedge 1_{\{x\}})\}$$
$$= R_\tau^L \uparrow_L (\mu),$$

i.e., $R_\tau^L \uparrow_L$ is just the left closure operator w.r.t. τ. By Theorems 3.1(2) and 4.1(5) in [12], we see that

$$\mu^o = \neg^R(\neg^L \mu)_L^- = \neg^R R_\tau^L \uparrow_L (\neg^L \mu)$$
$$= R_\tau^L \downarrow_R (\neg^R \neg^L \mu) = R_\tau^L \downarrow_R (\mu) \ \forall \mu \in L^X,$$

i.e., $R_\tau^L \downarrow_R$ is just the interior closure operator w.r.t. τ. Therefore,

$$\tau = \{\xi \mid R_\tau^L \downarrow_R (\xi) = \xi, \xi \in L^X\}.$$

The theorem is proved.

This result shows that the L-topologies generated by two reflexive and transitive L-relations R_τ^L and R_τ^R, which are induced by an L-topology τ, on X are all the original L-topology τ when τ satisfies some conditions.

Moreover, if τ satisfies (CL1) or (CR1), then it follows from Remark 2.1 and Theorem 3.1(2) in [12] that

$$(\wedge_{j \in J} \mu_j)^o = \neg^R(\neg^L(\wedge_{j \in J} \mu_j))_L^-$$
$$= \neg^R(\vee_{j \in J} \neg^L \mu_j)_L^- = \neg^R\{\vee_{j \in J} (\neg^L \mu_j)_L^-\}$$
$$= \wedge_{j \in J} \neg^R(\neg^L \mu_j)_L^- = \wedge_{j \in J} (\mu_j)^o \ \forall \mu_j \in L^X,$$

i.e., the interior operator int of τ distributes over arbitrary intersection of L-fuzzy sets. Thus, the intersection of arbitrarily many open subsets is still an open subset.

4. Conclusions and Future Work

In this note, we continue the works in [12]. For a complete involutive residuated lattice, we have supplemented some properties of the L-fuzzy topologies generated by a reflexive and transitive L-relation; showed that the L-fuzzy topologies generated by a reflexive and transitive L-relation satisfy $(TC)_L$ or $(TC)_R$ axioms; and given out some conditions such that the L-fuzzy topologies generated by two L-relations, which are induced by an L-fuzzy topology, are all the original L-fuzzy topology.

In a forthcoming paper, we will discuss the relationships between the L-fuzzy topological spaces and the L-fuzzy rough approximation spaces on a complete involutive residuated lattice.

Acknowledgements

The authors wish to thank the anonymous referees for their valuable comments and suggestions.

This work is supported by the National Natural Science Foundation of China (61379064).

References

[1] K. Blount and C. Tsinakis, "The structure of residuated lattices", International Journal of Algebra and Computation, 13, 437-461, 2003.

[2] Z. W. Li and R. C. Cui, "Similarity of fuzzy relations based on fuzzy topologies induced by fuzzy rough approximation operators", Information Sciences, 305, 219-233, 2015.

[3] L. Z. Liu and K. T. Li, "Boolean filters and positive implicative filters of residuated lattices", Information Sciences, 177, 5725-5738, 2007.

[4] Y. Liu and Y. Lin, "Intuitionistic fuzzy rough set model based on conflict distance and applications", Applied Soft Computing, 31, 266-273, 2015.

[5] Y. M. Liu and M. K. Luo, "Fuzzy Topology", World Scientific Publishing, Singapore, 1997.

[6] Y. Ouyang, Z. D. Wang and H. P. Zhang, "On fuzzy rough sets based on tolerance relations", Information Sciences, 180, 532-542, 2010.

[7] D. W. Pei, "A generalised model of fuzzy rough sets", International Journal of General Systems, 34, 603-613, 2005.

[8] K. Y. Qin and Z. Pei, "On the topogical properties of fuzzy roughsets", Fuzzy Sets and Systems, 151, 601-613, 2005.

[9] A. M. Radzikowska and E. E. Kerre, "Fuzzy rough sets based onresiduated lattices", in: J. F. Peter et al. (Eds.), Transactions on Rough Sets II, LNCS 3135, 278-296, 2004.

[10] Z. D. Wang and J. X. Fang, "On v-filters and normal v-filters of a residuated lattice with a weak vt-operator", Information Sciences, 178, 3465-3473, 2008.

[11] Z. D. Wang and X. J. Liu, "L-topological spaces based on residuated lattices", Advances in Pure Mathematics, 2, 41-44, 2012.

[12] Z. D. Wang, Y. Wang and K. M. Tang, "Some properties of L-fuzzy approximation spaces on bounded integral residuatedlattices", Information Sciences, 278, 110-126, 2014.

[13] W. Z. Wu, Y. H. Xu, M. W. Shao and G. Y. Wang, "Axiomatic characterizations of (S, T)-fuzzy rough approximation operators", Information Sciences, 334–335, 17-43, 2016.

[14] X. H. Zhang, J. H. Dai and Y. C. Yu, "On the union and intersection operations of rough sets based on various approximation spaces", Information Sciences, 292, 214-229, 2015.

[15] N. L. Zhou and B. Q. Hu, "Axiomatic approaches to rough approximation operators on complete completely distributive lattices", Information Sciences, 348, 227-242, 2016.

29

Silicon dioxide ultraviolet to visible phosphors material

Abdulkadhum Jaafar Alyasiri [*], **Abdulrahman Saleh Ibrahim**, **Amel Salih Merzah**

President of Foundation of Technical Education (F.T.E), Baghdad, Iraq.
Technical College, Baghdad, Iraq.

Email address:
Dr.a.alyasiri@gmail.com (A. J. Alyasiri)

Abstract: The present work include preparation of phosphors materials from rare earth (Eu) ion doping Silicon Oxide (SiO_2), and study the characteristics of phosphors for ultraviolet to visible conversion. The phosphor materials have been synthesized by two steps, the first was preparing the powder by solid state method using SiO_2, and Eu_2O_3 with doping concentration of 5% and different calcination temperature (1000, 1200 and 1400 $^{\circ}$C), the second step is the preparing of the colloid by dispersing the produced powder in a polyvinyl alcohol solution (4%). Powder preparation achieved by mixing the powder according to weight percentage, milling then calcinning the mixture at the above temperatures. The produced powder was characterized by X-Ray diffraction .The X-Ray resultsshow a mismatch with the standard peaks. The difference may be due to the variation in purity of the materials and the phases before calcination. Colloid preparation achieved by dissolving the PVA in water (4%) then dispersing the powder into the solution by using the hot plate magnetic stirrer and ultrasonic bath. The produced powder was characterized by using Fourier transform infrared (FTIR) and Photoluminescence spectra (PL). The results of photoluminescence spectra show that samples doped with Europium ions emit red color with wave length 612nm.The intensity of emission was increased with increasing calcination temperature.

Keywords: Nano Smart Material, Chemical Doping Processor, Fourier Transform Infrared, Photoluminescence Spectra

1. Introduction

Luminescence is an optical phenomenon, whereby a material is excited with high energy radiation (typically ultraviolet light, but other forms of energy such as beta rays can also be used). The phenomenon of luminescence has been known to mankind for over a thousand years. Descriptions have been found of ancient Chinese paintings that remained visible during the night, by mixing the colors with a special kind of pearl shell [1].

Until the end of the 20th century, very little research was done on the phenomenon of luminescence. For many decades, zinc sulfide (ZnS) doped with copper (and later codoped with cobalt) was the most famous and widely used persistent phosphor. It was used in many commercial products including watch dials, luminous paints and glow-in-the-dark toys[2,3]. In recent years, considerable research has been done on the synthesis and characterization of large band gap oxide materials such as Y_2O_3 and SiO_2 doped with rare-earth elements using different chemical methods such as sol-gel precipitation and combustion[4].

There are many advantages of luminescent materials, Precise material deposition on substrate at well-defined positions, Small material consumption , Less material losses, Less material losses and It has excellent chemical stability.

There are many applications of luminescent materials which can be used in many fields, Security ink[5], Widely used in Cathode ray tube displays[5], Plasma display panel[6], High radiation energy detection (such as to make films for X-ray detection) applications[7] and white light emitting diode (LED)[6] The aim of this work is to prepare luminescent ink from rare earth materials by solid state method.

2. Experimental

The experimental part is formed from two steps the first is the preparation of the powder and the second step is preparation of the colloid.

Raw materials used in the present work were: Nano

SiO_2 and Eu_2O_3 Their properties and sources were listed in table (1) below:

Table (1). Raw Materials and the properties

Item	Purity	Source	characterization
SiO_2	99.5%	Sigma Aldrich	Nano powder <50nm
PVA	99%	Sigma Aldrich	n=89000-98000
Eu_2O_3	99.5%	Sigma Aldrich	Nano powder <150nm

2.1. Methods of Preparation

2.1.1. Powder Preparation

The powder is prepared by using solid state reaction. The starting materials and weights were listed in table (2) below:

Table (2). The percentages of starting materials.

Batch No.	Item	Weight Percentage %	Weight (gm.)
1	SiO_2	95	3.819
	Eu_2O_3	5	1.179

Raw materials were weighed according to the above table in an electronic balance (five digits) inside a Nano-filtration system .The powders feeds in a poly vinyl chloride bottles with yttrium stabilized zirconia (YSZ) balls as milling media, then the mixture was milled by using a ball mill for 24 hrs with 64.5 rpm, this process achieved with the addition of propanol to achieve homogenization.

After 24hr take off the mixture and poured inside stainless steel pan then dry it in dryer for 4hr in 80 °C to evaporate propanol. The dried sample calcined in an electrical furnace using an alumina crucible as a boat at a temperature 1000, 1200 and 1400 °C for 9 hrs.

The powder after calcination was secondly milled in the same previous way for 24 hrs with 64.5 rpm in order to prevent agglomeration.

X- Ray diffraction was used to characterize the powder after preparation by solid state method.

2.1.2. Colloid Preparation

To prepare the security ink, the powder was produced as colloid, this can be achieved by dispersing the powder in a poly vinyl alcohol $[CH_2CHOH]_n$, (n=89000-98000) inside a Nano-filtration system. First a solution of polyvinyl alcohol in water (4%) was prepared with the aid of magnetic stirrer, homogenizer and heater.

The powder added to the PVA solution 0.2 gm. for 20 ml of the solution. The suspend also mixed and heated in a magnetic stirrer and homogenizer with heating up to 50°C for 1 hr .Then mixing without heating for 2 hrs with the addition of 10 ml ethanol, ultrasonic bath was used for third stage mixing and homogenization, then the fourth stage of mixing and homogenization was achieved by magnetic stirrer and homogenizer for 2 hrs, the final product is a white solution which then tested by photoluminescence spectra and Fourier transform infrared.

2.2. Characterization

The characterization of powder achieved by X-Ray diffraction.

2.3. X-Ray Diffraction

The X-ray powder diffraction patterns presented in this work were measured using a theta to thetadiffractometer (Philips), equipped with a Cu Kα source (generator: 40 kV and 40mA) with wave length 1.5418A, a scintillation detector with pulse-height analysis, and a variable knife-edge collimatorfor high resolution X-ray diffractometry. The best achievable instrumental resolution was 0.001° in 2θ.

2.4. Colloid Characterization

The prepared colloid characterization achieved by measuring its photoluminescence spectra, in order to predict its capability to emit color (luminescence).

2.5. Photoluminescence Spectra

Photoluminescence spectra is measured by using Fluoro Max-4 spectrofluorometer. The excitation wave length for all samples were 254 nm (at the ultraviolet region) produced by Xenon flash lamp then the emitted wave detected by signal detector (photomultiplier tube and housing).

2.6. Fourier Transform Infrared

Fourier transform infrared is measured by NEXUS FTIR.
The variables of the instrument was
Number of scan: 64
Resolution: 4
Data spacing: $1.928cm^{-1}$

3. Result and Discussion

3.1. X-Ray Diffraction

The analysis of X-ray diffractions for all synthesized samples were shown below according to their calcination temperature:

SiO_2: Eu 5% Calcined at 1000°C:

Figures (1), (2)and (3) shows the XRD patterns for the system SiO_2: Eu 5% Calcined at 1000, 1200 and 1400∘C respectively. There was mismatch with the standard peaks. The difference may be due to the variation in purity of the materials and the phases before calcination.

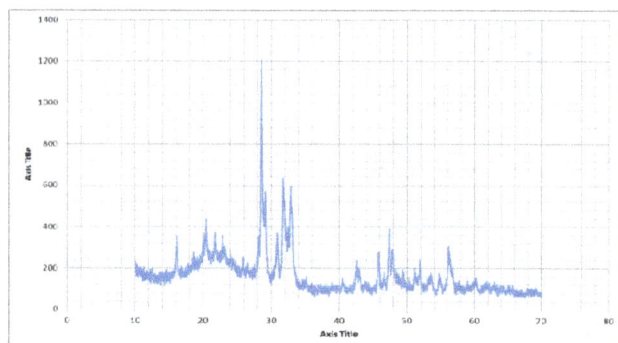

Figure (1). XRD patterns of SiO_2 : Eu 5% Calcined at 1000°C

SiO₂: Eu 5% Calcined at1200°C:

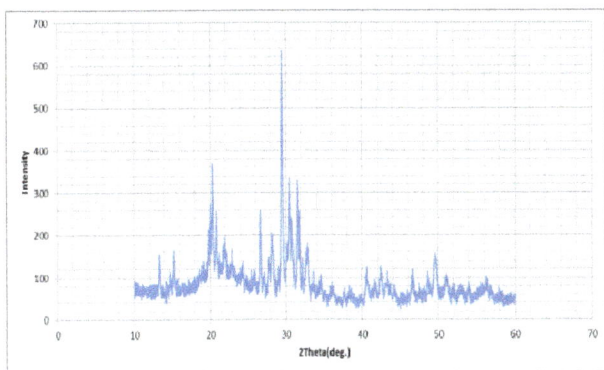

Figure (2). *XRD patterns of SiO₂: Eu 5% Calcined at1200°C*

SiO₂: Eu 5% Calcined at 1400°C:

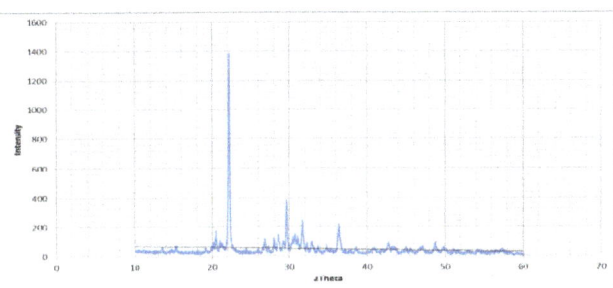

Figure (3). *XRD patterns of SiO₂: Eu 5%Calcined at 1400°C*

3.2. Fourier Transform Infrared

The analysis of Fourier transform infraredfor all synthesized colloid was shown in figure (4) can be discussed by comparing it with the data of infrared absorption of pure materials according to the wave number. The broad peak at around 3340 cm^{-1} is originated from O-H stretching in hydroxyl groups. Both PVA and alcohol contain hydroxyl group. C-H stretching was observed at around 2980 cm^{-1}.The peak at around 2300-2400cm^{-1} it is coming from C≡C group.

The peak at around 1600-1800 cm^{-1} was coming from C=O group. Absorption peak at around 1213-1420 cm^{-1} might originated from C-H bend and C-C stretching. C-O stretching was observed at around 1020-1100 cm^{-1}.

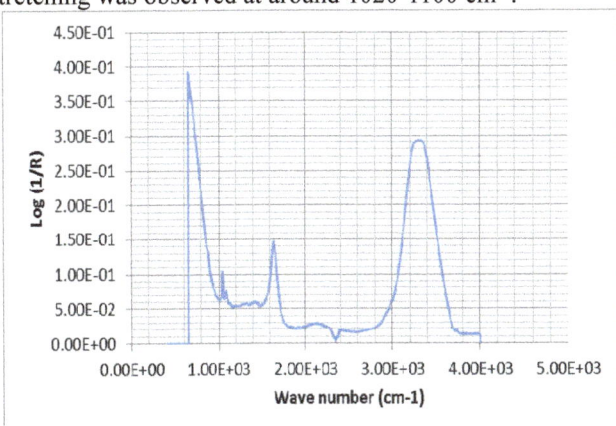

Figure (4-a). *The FTIR of SiO₂: Eu 5% calcined at 1000°C*

The absorption peak at around 890-1000cm^{-1} was originated from interaction between PVA chain and the surface of Y₂O₃:Eu₂O₃ .Figures for all samples are shown below according to the calcination temperature :
SiO₂: Eu 5% calcined at 1000°C
SiO₂: Eu 5% 1200°C:

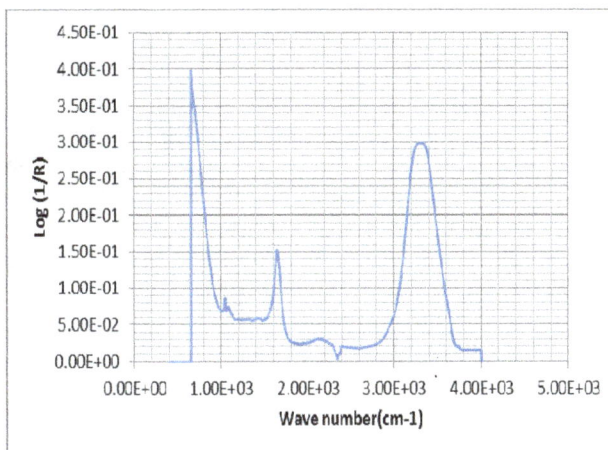

Figure (4-b). *The FTIR of SiO₂: Eu 5% 1200°C*

SiO₂: Eu 5% calcined at1400°C

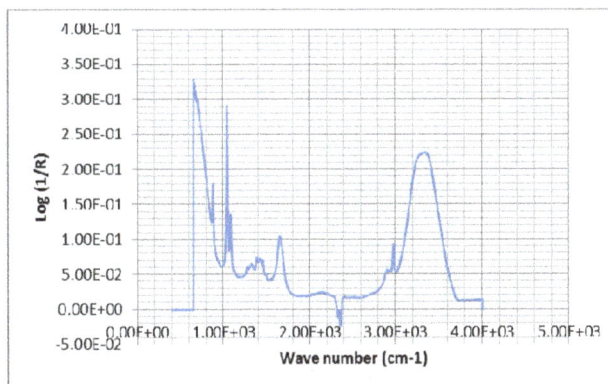

Figure (4-c). *The FTIR of SiO₂: Eu 5% calcined at1400°C*

3.3. Photoluminescence Spectroscopy

The analysis ofPhotoluminescence spectroscopy for all synthesized colloid are shown below according to the calcination temperature as shown in figure (5):

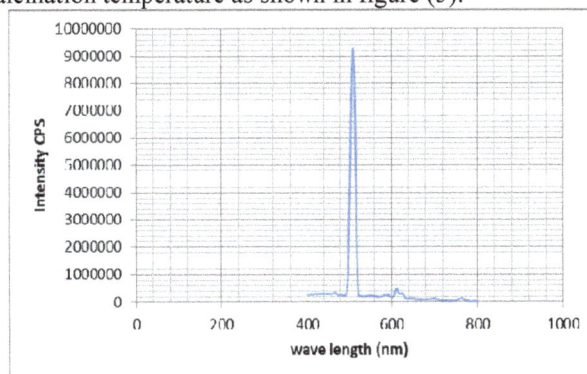

Figure (5-a). *Photoluminescence spectra of SiO₂:Eu calcined at 1000°C*

SiO₂ : Eu 5% calcined at1000˚C

From figure (5-a) below there are three peaks , the first one at wave length 508 nm it comes from 2λ where the applied λ was 254 nm. The second tiny peak at wave length 612 nm comes from electron transition in Eu^{+3} ion from $^5D_0 \rightarrow {}^7F_2$. The third peak at wave length 760 nm it comes from 3λ.

SiO₂: Eu 5% calcined at 1200˚C

From figure (5-b) below there are four peaks , the first one at wave length 585 nmcomes from electron transition in Eu^{+3} ion from $^5D_0 \rightarrow {}^7F_1$. The second peak at wave length 612 nm comes from electron transition in Eu^{+3} ion from $^5D_0 \rightarrow {}^7F_2$. It is clear that photoluminescence increased with increasing calcination temperature .The third peak at wave length 700 nm comes from electron transition in Eu^{+3} ion from $^5D_0 \rightarrow {}^7F_3$ .The fourth peak at wave length 762nm comes from 3λ where the applied wave length was 254nm.

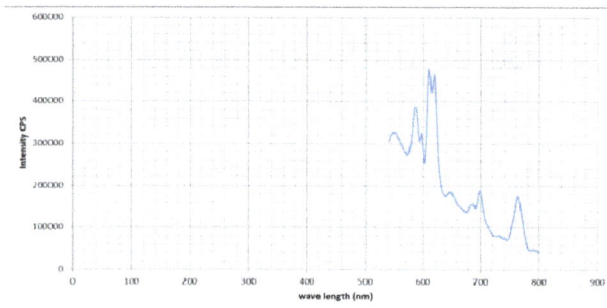

Figure (5-b). *Photoluminescence spectra of SiO2:Eu calcined at 1200ºC*

SiO₂: Eu 5% calcined at 1400˚C

From figure(5-c) below there are four peaks , the first one at wave length 585 nm comes from electron transition in Eu^{+3} ion from $^5D_0 \rightarrow {}^7F_1$. The second peak at wave length 612 nm comes from electron transition in Eu^{+3} ion from $^5D_0 \rightarrow {}^7F_2$ .It is clear that photoluminescence increased with increasing calcination temperature .The third peak at wave length 700 nm comes from electron transition in Eu^{+3} ion from $^5D_0 \rightarrow {}^7F_3$ .The fourth peak at wave length 762nm comes from 3λ where the applied wave length was 254nm.

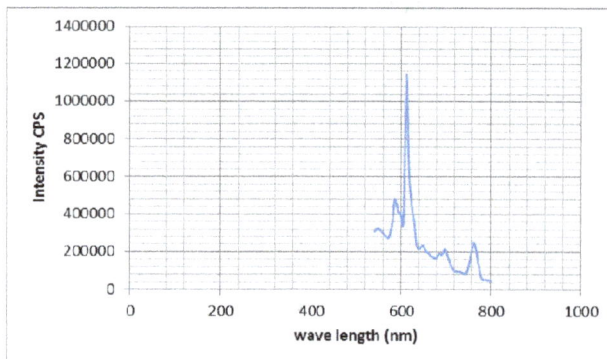

Figure (5-c). *Photoluminescence spectra of SiO₂:Eu calcined at 1000ºC*

4. Conclusions

Stable luminescent colloid was produced from powder of SiO₂:Eu powder dispersed in poly vinyl alcohol solution. SiO₂:Eu played a major role as luminescent centers in the colloid which emitted red luminescence. This colloid can meet several application s in optoelectronics such as for production of light emitting devices , light sensor, and luminescent displays in a very definitive sizes and shapes. Increasing calcination temperature lead to increase luminescent intensity.

References

[1]   Harvey, E.N. *A History of Luminescence from the Earliest Times until 1900*; American Philosophical Society: Philadelphia, PA, USA, 1957.

[2]   Hoogenstraaten, W.; Klasens, H.A. Some properties of zinc sulfide activated with copper and cobalt. *J. Electrochem. Soc.* 1953, *100*, 366–375.

[3]   Yen, W.M.; Shionoya, S.; Yamamoto, H. *Phosphor Handbook*, 2nd ed.; CRC Press/Taylor and Francis: Boca Raton, FL, USA, 2007.

[4]   D.H Aguilar, L.C Torres-Gonzalez and L.M Torres-Martinez, *Journal of Solid State Chemistry* 158(2000) 349-357.

[5]   Astuti, Mikrajuddin Abdullah, and Khairurrijal," Synthesis of Luminescent Ink from Europium-Doped Y₂O₃Dispersed in Polyvinyl Alcohol Solution", Hindawi Publishing Corporation, Article ID 918351(2009).

[6]   Yu-Chun Li, Yen-Hwei Chang, Yu-Feng Lin, Yee-Shin Changb and Yi-Jing Lin, Synthesis and luminescent properties of Ln³⁺ (Eu³⁺, Sm³⁺, Dy³⁺)-doped lanthanum aluminum germanate LaAlGe₂O₇ phosphors, Elsevier, Journal of Alloys and Compounds 439 (2007) 367–375.

[7]   M. Abdullah, K. Okuyama, I. W. Lenggoro, and S. Taya, "A polymer solution process for synthesis of (Y, Gd)₃Al₅O₁₂:Ce phosphor particles," *Journal of Non-Crystalline Solids*, vol. 351, no. 8-9, pp. 697–704, 2005.

30

Filter Structures Based on Basic, Dynamic Nonstationary Elements

Roman Kaszyński, Adrian Sztandera, Katarzyna Wiechetek

Department of Systems, Signals and Electronic Engineering Faculty of Electrical West Pomeranian University of Technology, Szczecin, Poland

Email address:
roman.kaszynski@zut.edu.pl (R. Kaszyński), adrian.sztandera@gmail.com (A. Sztandera), wiechetekkatarzyna@gmail.com (K. Wiechetek)

Abstract: The results of basic dynamic elements research of filters with time-varying parameters which are low-pass filters, high-pass filters, and band-stop filters are presented. With the use of those elements more complex nonstationary structures were created. This provided the capability of forming spectral characteristics. Examples of filtration presented in this paper showed the reduction of transient state. Presented study assumes, that it is possible to create complex structures with required spectral properties.

Keywords: Transient State, Time-Varying Parameters, Spectral Properties, Basic Nonstationary Elements, Complex Structures

1. Introduction

Presented study assumes that the dynamic, nonstationary high-pass and low-pass first order elements are basic structures. Inertial element was modified with gain function $k(t)$ and time constant function $T(t)$ [6, 11]. To the high-pass element, which is an derivative real term, functions varying the gain and time in such way as it is changed in inertial element, at the same time leaving the derivative time constant were implemented. The values of time-varying parameters are settling with time approaching to infinity. After terminating the transient state with demanded accuracy it provides required spectral characteristic [1, 7, 10, 11].

The fact, that the time-varying parameters settle their values in time is important in term of stationarity of filters output signals. Then, it is possible to determine their frequency characteristics and establishing whether all assumptions are fulfilled. In the dynamic systems described by the differential equations, nonstationarity is a feature of transients states. It can be significantly shorter in nonstationary filters with time-varying parameters than in the structures with time-invariant parameters.

During previous studies on series-parallel structures the analysis of the stability of the time-varying parameters was conducted. It can be stated, that the process of settling parameters values over time is a strong factor stabilizing individual elements of the filter. Combining serial and parallel stable elements in one structure without any feedback allows to keep stability of the newly created structure. Stability tests conducted using Lyapunov's second method showed, that only fast, periodic changes of the filter parameters may result in instability. In analyzed filters such phenomenon does not occur.

For that reason, those basic dynamic nonstationary elements are worth to be examined and analyzed.

2. Low Pass Elements with Time-Varying Parameters

The simplest dynamic low-pass element with time-varying parameters is a structure described by the following differential equation:

$$T_\infty \cdot f_T^{-1}(t) \cdot \frac{dy(t)}{dt} + y(t) = k_\infty \cdot f_k(t) \cdot x(t), \qquad (1)$$

for which we obtain easier implementation of the entire system.

Functions $f_T^{-1}(t)$ and $f_k(t)$ are described by dependencies

$$f_k(t) = f_k(t) \cdot [d_k - (d_k - 1) \cdot h_k(t)]) \qquad (2)$$

$$f_T^{-1}(t) = f_T^{-1}(t) \cdot [d_T - (d_T - 1) \cdot h_T(t)], \qquad (3)$$

where: $h_k(t)$ and $h_T(t)$ are step responses of second order elements with constant parameters.

Functions $f_T^{-1}(t)$ and $f_k(t)$ by varying the filter parameters allow to create gain function $k(t) = k_\infty \cdot f_k(t)$ and time constant function $T^{-1}(t) = T_\infty^{-1} \cdot f_T^{-1}(t)$, where $d_T = \frac{T^{-1}(0)}{T^{-1}(\infty)}$ and $d_k = \frac{k(0)}{k(\infty)}$ are the multiplicity of their value change. Gain function and time constant function must achieve the values specified in the spectrum assumptions in time not longer than the duration of the transient state. It can be assured if element generating varying function has damping factor value between $0{,}7l < \beta < 1$.

Because of the fact that each time at the beginning of filtration it is necessary to generate functions varying parameters, it was decided to use the step characteristics of second order oscillatory model with parameters k, β, ω_0 for this purpose. In spite of relatively simple implementation, it has a great capability of shaping desired functions.

Series connection of first order elements creates a filter, which order depends on number of elements. At this point the question arises: Could it be possible to unify all of the functions to one form and generate only one function without any significant properties deterioration in relation to optimal filters? The first results related to this topic were published in paper [6, 8]. Adopting these assumptions will reduce complexity of the structure without deteriorating properties in the time domain. That kind of filter, which is a connection of elements with the same time-varying functions, is described by following set of equations:

$$T_\infty \cdot f^{-1}(t) \cdot \frac{dy_1(t)}{dt} + y_1(t) = k_\infty \cdot f(t) \cdot x(t)$$

$$T_\infty \cdot f^{-1}(t) \cdot \frac{dy_2(t)}{dt} + y_2(t) = k_\infty \cdot f(t) \cdot y_1(t) \qquad (4)$$

$$\vdots$$

$$T_\infty \cdot f^{-1}(t) \cdot \frac{dy_i(t)}{dt} + y_i(t) = k_\infty \cdot f(t) \cdot y_{i-1}(t).$$

The conducted study confirmed a very good properties of filters with the same variation of parameters. Examples of settling times with the input signal for step response and sinusoidal signal with admissible value $\alpha = 0{,}05$ limiting the amplitudes of the frequency response are presented in Fig.1.

One can easily notice, that shortening the transient state of step response extends time of sinusoid attenuation and conversely. Furthermore, a solution can be found, where time of sinusoid attenuation and settling time for step input are equal. Fig.1 shows exact points of percolating planes presenting both settling times.

Fig.2 shows an example of random signal subjected to filtration. Fig.3 presents an example filtration of the random signal with the use of first order filter with constant and variable parameters. Fig.4 depicts filtration with the use of two first order filters connected in series (forming a second order filter).

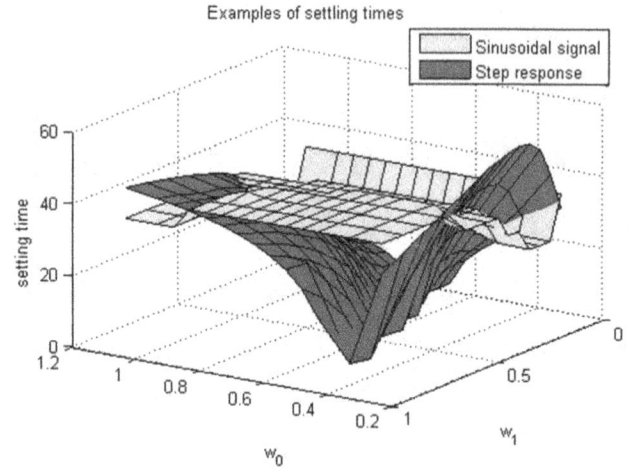

Figure 1. *Examples of settling times for step response and sinusoidal signal for the first order filter ($\alpha = 5\%$, $d=3$, $T=20$).*

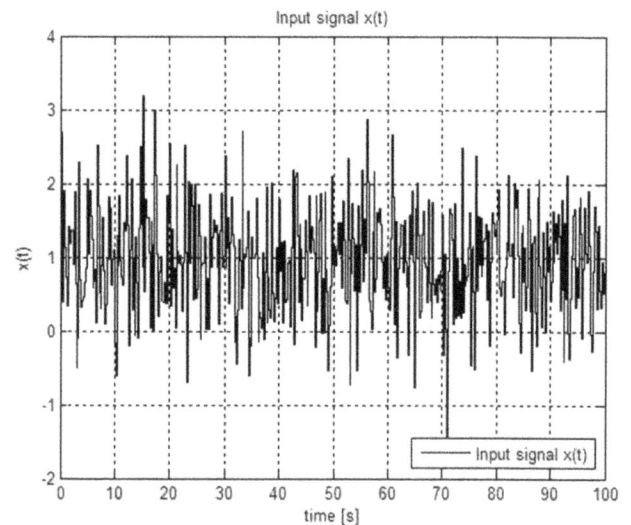

Figure 2. *Random input signal.*

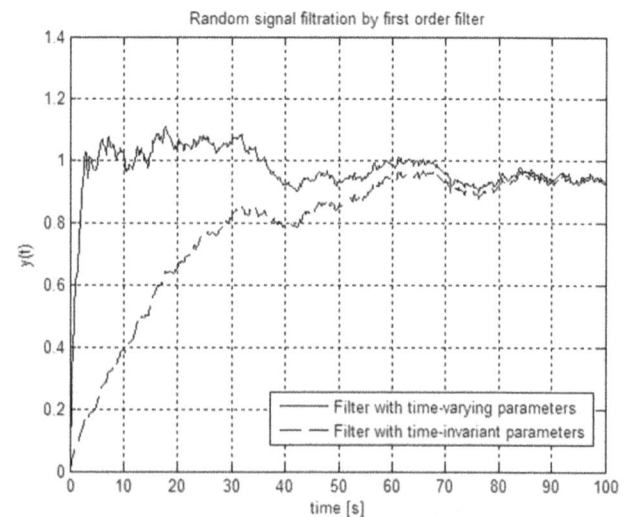

Figure 3. *Random signal filtration by first order filter.*

One can easily notice that with such selected parameters a single filter quite badly suppresses components with the lowest frequencies included in the filtered signal. Adding second equal element (Fig.4) provides an opportunity of better averaging the input signal. Simultaneously it can be seen, that filters with variable parameters have much shorter transient states compared to filters with constant parameters [3, 4]. Varying all parameters in all elements with one function allows to avoid making any mistakes related to elements connection order varied by different functions (in systems with time-varying parameters chain connection order has great importance in the duration of the transient state).

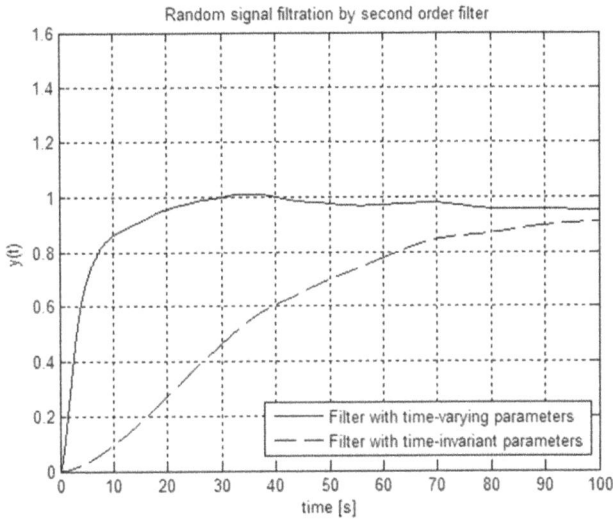

Figure 4. *Random signal filtration by second order filter.*

3. High Pass Element with Time-Varying Parameters

High-pass single element with time-varying parameters is described by the following equations:

$$T_{g_n} \cdot f_T^{-1}(t) \cdot \frac{dy(t)}{dt} + y(t) = k_\infty \cdot f_k(t) \cdot x(t) \qquad (5)$$

$$z(t) = T_g \cdot \frac{dy(t)}{dt}, \qquad (6)$$

where: $z(t)$ – high-pass element output signal, T_g – differentiation time.

Settling time of high-pass elements with time-varying parameters are not longer than the settling time of low-pass elements, from which they are built. The conclusion drawn from set of equations (3) is that the signal $z(t)$ at the output of high-pass element is proportional to the derivative of the signal $y(t)$ at the low-pass output of the same element with varying parameters. It can be assumed that the duration of the transient state depends of the lower limit of the high-pass band and can be phrased by the multiplicity of cutoff frequency period of the passband. Single high-pass elements with varying parameters can be a part of more complex structures. That would allow to create non-standard spectral characteristics of filters.

4. Band Stop Element with Time-Varying Parameters

Parallel connection of low-pass and high-pass elements depending on values of parameters can give diverse frequency characteristics. It can become phase shift corrector, band amplifying element or band-stop filter. This includes elements with time-invariant and time-varying parameters. Band-stop filter is obtained when low-pass element has lower limit frequency of the passband than the high-pass element. Elements with variable parameters provide a number of possibilities to accelerate the filtration.

Parallel connection of the low-pass and high-pass elements allows to obtain a signal, which is the sum of signals after low-pass and high-pass filtering

$$u(t) = y_d(t) + z(t). \qquad (7)$$

Fig.5 presents signal filtration, which is a sum of constant component and sinusoid with stopband range frequency.

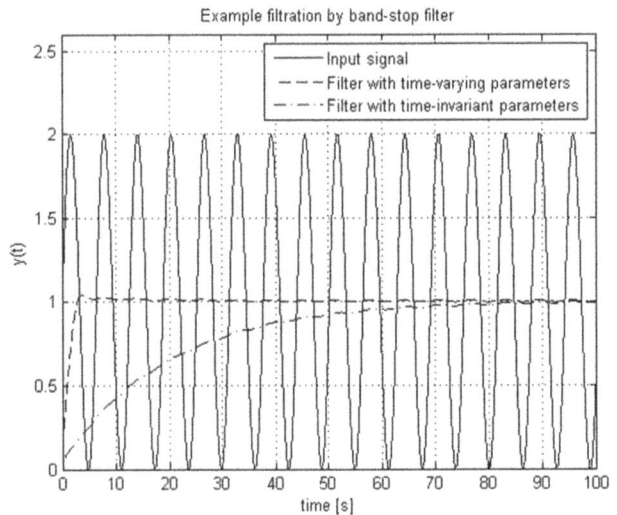

Figure 5. *Signal filtration by band-stop filter.*

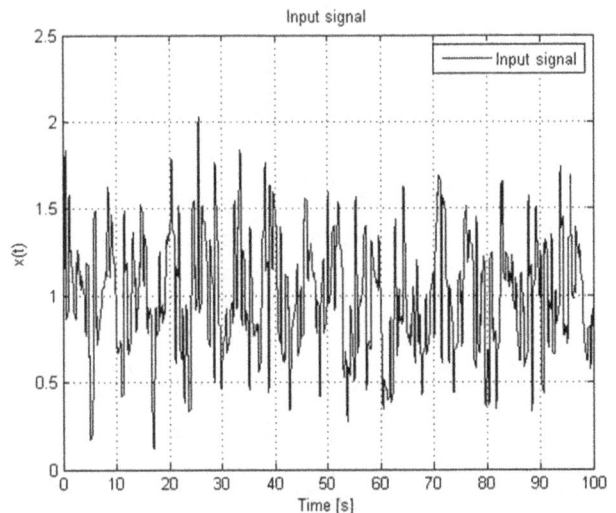

Figure 6. *Random input signal.*

Example filtration by time-invariant filter

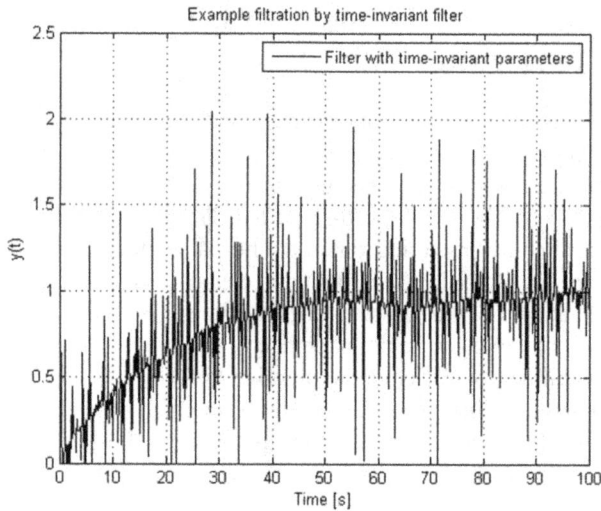

Figure 7. Filtration by filter with time-invariant parameters.

Example filtration by time-varying filter

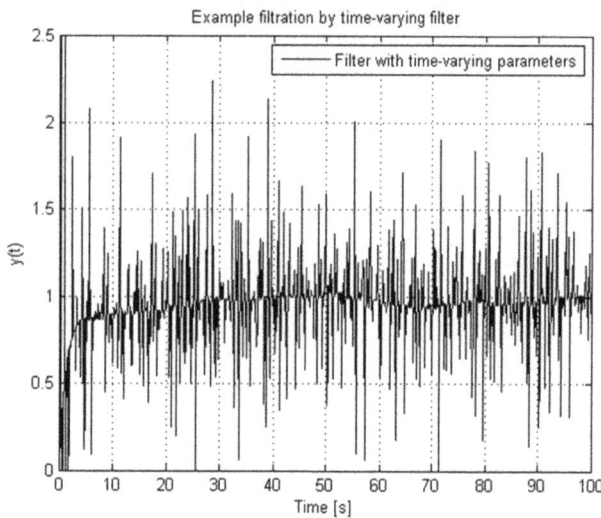

Figure 8. Filtration by filter with time-varying parameters.

As one can notice, the filter with time-varying parameters is characterized by much shorter transient state than the time-invariant filter. Based on the sample filtration of signal, which contains constant component, dominant sinusoid (e.g. derived from resonance phenomenon) and random noise, one can observe how the filtration with constant and varying parameters proceeds.

Fig.6 shows a signal subjected to filtration. Fig.7 presents filtration of that signal by using band-stop filter with fixed parameters. Fig.8 depicts filtration by filter with time-varying parameters. Presented structure of filter is different from the notch type of structure, which is initially based on second order element, in which a zero amplitude for a single frequency can be reached. However, it is covered with a great oscillation - the greater, the bigger quality factor is. In the following examples it is clear, that filter manages to suppress the sinusoid component and the transmission of other components.

Connecting in series similar pairs of elements creates a parallel-serial structure. It allows to obtain the frequency characteristics with more narrow slopes as shown in Fig.9.

Bode Diagram

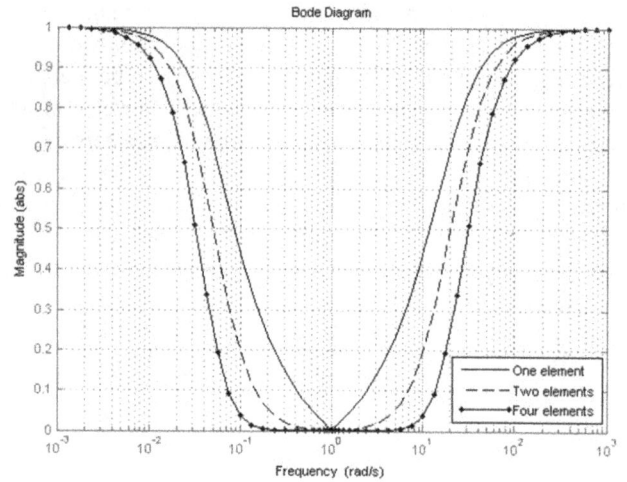

Figure 9. The amplitude characteristics of series connected filters.

Presented basic elements with time-varying parameters can be arranged in more complex structures. Each of low-pass element contains in its structure a high-pass output. One of the main disadvantage of those simplest elements with constant parameters are very long transient states. By introducing the time-varying parameters to those structures, the transient states can be reduced. They are characterized by very good properties in the time domain. Spectral characteristics can be improved and arbitrarily formed through the expansion of the structure. For example, the combination of two or more band-stop components with different values of stopband range frequency can be used in filtering signals with many resonant frequencies generated in circuits. Parallel-series structures in order to achieve non-standard spectral characteristics while maintaining good dynamic properties of these structures can be created. By using single nonstationary elements one can create structures providing even more demanding frequency characteristics. At the same time they do not suffer from many drawbacks, which filters with constant parameters are characterized with. Created structures shown in Fig.10 (or similar), that have a layered or any other parallel-series structure can be subjected to learn required properties.

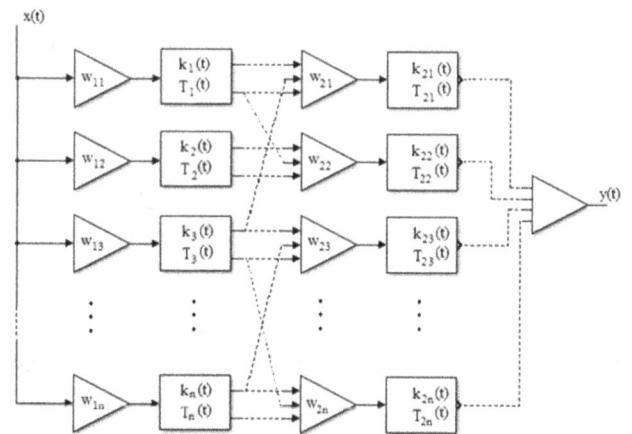

Figure 10. Example of parallel-series structure.

5. Conclusions

Conducted study has shown, that the same functions varying different parameters of the time-varying structures may significantly simplify the whole design and can provide reduced transient state time. Such simplification of the individual elements allows to combine many of them, creating expanded filter structures with unique spectral characteristics. Presented examples of filtration provide a comparison of efficiency between time varying and time-invariant filters with the same spectral properties. Creating a parallel-serial structure gives a possibility to form various frequency characteristics. Obtained structures are characterized by very high stability resulting from the simplest basic elements and the lack of feedback inside the whole system structure.

References

[1]   Anjum S.R., Bhattacharya S., Srirastava G., Effect of Phase Compensation on the performance of Classic Butterworth filter, *Proc.ICAESM*, pp.217-221, 2012.

[2]   Carusone A., Johns D.A., Analogue adaptive filters: past and present. *IEE Proc.-Circuits Devices Systems*, vol. 147, No. 1, pp. 82-90, Feb. 2000.

[3]   Chang-Siu E., Tomizuka M., and Kong K., Time-Varying Complementary Filtering for Attitude Estimation, *Proc. IEEE/RSJ*, pp. 2474-2480, 2011.

[4]   Chen W.K., *The Circuits and Filters Handbook*, CRC Press, IEEE Press, Boca Raton, 2003.

[5]   Gutierrez de Anda M.A., Sarmiento Reyes L.A., The analytic determination of the PPV for second-order oscillators via time-varying eigenvalues, *IEEE Transactions on Cirtuits and Systems II: Express Briefs*, vol. 53, (2006), no. 11, pp.1225-1229.

[6]   Gutierrez de Anda M.A., Sarmiento Reyes L.A., Martinez L.H., Piskorowski J. and Kaszyński R., The reduction of the duration of the transient response in a class of continuous-time LTV filters, *IEEE Transsactions on Circuits and Systems-II: Express Briefs*, vol. 56, (2009), no. 2, pp. 102-106.

[7]   Jafaripanah M., Al.-Hashimi B.M., White N.M., Application of analog adaptive filters for dynamic sensor compensation, *IEEE Transactions on Instrumentation and Measurement*, vol. 54, No. 1, pp. 245-251, Feb. 2005.

[8]   Lee H., Bien Z., Linear time-varying filter with variable bandwidth, in *Proc. IEEE International Symphosium on Circuits and Systems ISCAS* (2006), pp. 2493-2496.

[9]   Piskorowski J., Kaszynski R. Analytical synthesis of parameter-varying filter of constant component with application to swiching systems, *Metrology and Measurement Systems*, vol. XVIII (2011) no. 3, pp. 471-480.

[10] Piskorowski J., Phase-Compensated Time-Varying Butterworth Filter, *Analog Integrated Circuits and Signal Processing*, Springer US, 2013.

[11] Toledo K., Torres J. A Variational Approach for Designing Digital Filters with Time-Varying Parameters, *IEEE Transsactionson Circuits and Systems II:* Express Brieffs, vol. X No. X, Jan. 2015.

[12] Walczak J., Piwowar A., Cascade connection of parametric sections and its properties, *Electrical Review*, Vol. 86, No. 1, pp. 56-58, Jan. 2010.

31

Feedback Control of the Lottery System in Theme Park

Noriaki Sakamoto

Faculty of Economics, Hosei University, Tokyo, Japan

Email address:

nsaka@hosei.ac.jp

Abstract: In Tokyo Disney Park, the guests must reserve the seats by a lottery to view some shows. This paper proposes the robust feedback control system to solve the problem of the lottery system. The controlled variable is winning / losing to the guests who draw the lottery, and the control logic is ON/OFF-Type Discrete Variable Structure Controller, to compensate the uncertainty of a simulation to reproduce the lottery. The simulation that input data are made using many real data shows the effectiveness of the proposed method. Next, Neural Network Model predicts the controlled result. If the bad result is predicted, the staff of the lottery system is able to take an effective measure.

Keywords: Discrete Variable Structure Control, Lottery System, Theme Park, Uncertainty, Robust Control

1. Introduction

In Tokyo Disney Park (Disneyland and DisneySea) [1], the seats of some shows ("Once Upon a Time", "One Man's Dream II", "Big Band Beat") are reserved by a lottery. For example, to view "Once Upon a Time" in the viewing area, the guests are required the free reserved seat tickets. The tickets are distributed by the lottery from Park Opening Time until 30 minutes before the show start time. Each party can attempt the lottery only once for each show a day.

The difficulties of this lottery system are;

Requirement (1):

When the parties draw the lottery, at that moment, they know they are Winner or Loser.

Requirement (2):

The vacant seats must become zero as possible at 30 minutes before the show starts. Because, if all the seats are reserved early, all parties who come later become losers.

Requirement (3):

The guests in a party must be assigned the seats which are in the same line without a crossing the aisle.

The general lottery methods or research are classified roughly into two [2]. One is that it accepts the application of a lottery in a fixed time (or period) and then the lottery is conducted after the deadline [3,4]. The other is a lottery machine (for example, it is performed in a shopping mall). We cannot apply the former method to this lottery system, because the guests have to wait for the result. In addition, as the result affects the plan of the day, it gives restrictions to

the guest's schedule. We cannot apply the latter lottery machine as well. If each party thinks that the vacant seats become zero early, they try to draw the lottery as fast as possible and it causes congestion. Also, it is unfair for the guests who enter in the evening.

The technique used in Tokyo Disney Park decides winning or losing based on the winning probability that are written in a learning database [5]. We think that the present system has two problems;

Problem (1):

SNS (Twitter, etc.) has come to be used widely, so many guests know that the lottery system has several minutes of alternate winning-time and losing-time (See Figure 1). As a result, the parties who draw the lottery stay in the place where the lottery machines are lining up until the winning-time begins. Their behaviors disturb the flow of the lottery queue and the work of the cast members who deal with the lottery machine trouble, etc.

Problem (2):

The weather in Japan often change suddenly. We think that the database does not have enough the similar situation data. In fact, the lottery is conducted even if there is a rainstorm or typhoon coming. Very few parties attempt the lottery in such a weather, but not all the parties are winners, so there are many vacant seats left when the show starts.

Therefore, we propose that the new lottery system does not use a database to meet *Requirement* and to solve *Problem*. We apply a robust feedback control to compensate for the uncertainties of a simulation data. As one of the robust controllers, we use ON/OFF-Type Discrete Variable Structure Controller [6, 7, 8].

However, this system doesn't solve *Problem* (2), in addition, we propose that our new lottery system has NN (Neural Network Model) learning all simulation results. NN estimates the number of final vacant seats by using a morning data. The output can warn the lottery staff of a possibility that many seats are remained to be vacant at the show time.

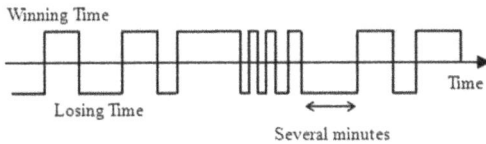

Figure 1. Several minutes of alternate winning-time and losing-time.

2. Simulation System

We simulate the lottery system for "Once Upon a Time" by the control block diagram which is shown in Figure 2. Table 1 is a list of symbols.

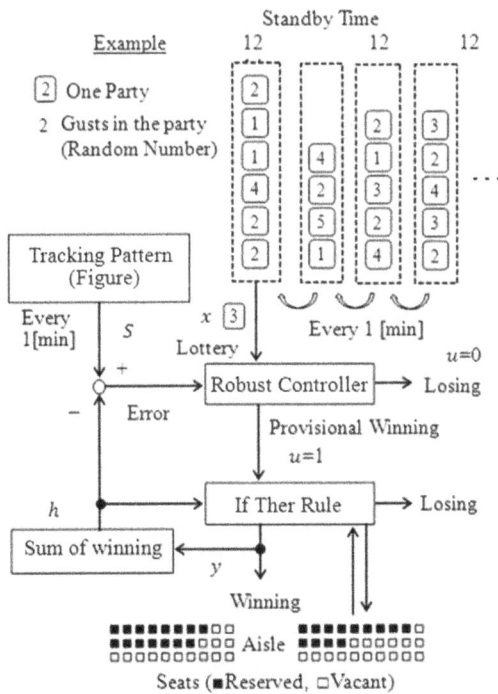

Figure 2. Simulation System (The value in the figure is an example for explanation).

Table 1. Nomenclature.

Symbol	Remarks
t	Simulation time [min] $(0,1,2,3,\cdots)$
k	Control step $(1,2,3,\cdots)$
S_t	Target value (Figure 5) : The total number of seats that should be reserved at a simulation time t
x_k	The number of guests in a party
u_k	Manipulated variable: =1 Winning, =0 Losing
y_k	Output of 'If Then Rule':= $x_k u_k$
h_k	the total number of reserved seats:= $\sum y$

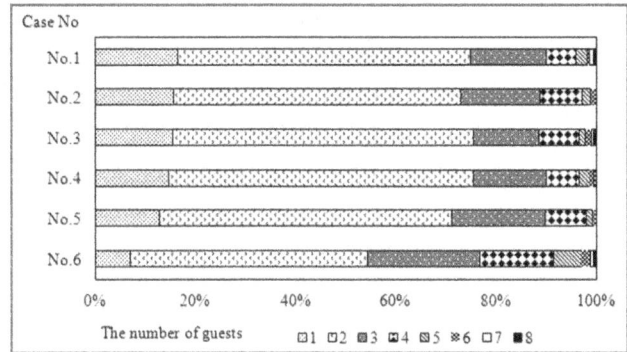

Figure 3. The number of guests in parties: Measurement (1).

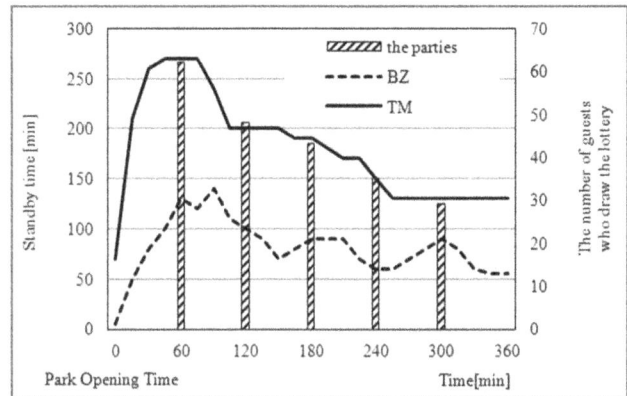

Figure 4. The time-series variations of guests who draw the lottery: Measurement (2) and time-series variations of the standby time (Example BZ: Buzz Lightyear's, TM: Toy Story Mania!).

2.1. Input Data

The number of parties, the number of people in a party, and the time when they draw the lottery are unknown. We guess these as follows;

Measurement (1):

To get the number of guests in a party (x), we counted the number of guests of 3481 parties in total at the different date, time, and place (6 cases in total). Figure 3 shows the results.

Measurement (2):

To obtain the time-series variations of guests who draw the lottery, we counted the number of guests who come in the lottery place during 3700 minutes in total at a different date and time. As a results, we find that the time-series variations of *Measurement* (2) is equal to that of the standby time to enjoy the popular attraction which has a priority admission ticket (Fast-Pass ticket). Figure 4 shows the data of a certain one-day. Then we calculate the number of guests who draw the lottery in 1 minute using the standby time recorded every 15 minutes of the attractions. It becomes the input data of the simulation system.

Measurement (3):

We acquire the standby time of each attraction that the official website shows [9]. Details are as follows.

- Attractions: 12 (Buzz Lightyear's Astro Blasters, Big Thunder Mountain, Star Tours, Tower of Terror, Toy Story Mania!, and so on)
- Period: 01/Jan/2014 - 31/Dec/2014

- Definition of one record: Standby time every 15 minutes of one day at one attraction
- The total number of records: 2658 (Disneyland 1501, DisneySea 1157. We removed the record having an abnormal data or lack data.)

The time-series variations of guests who draw the lottery in Tokyo Disney Park have a variety of choices, for example, there are many guests who come in the evening. These are reproduced by the records of 12 attractions of 365 days.

Process (1): (every 15[min] → every 1[min])

We divide standby time data of every 15 minutes by 15 in straight line interpolation, and obtain the data every 1 minute.

Process (2): (1[min] → guests)

We assume the maximum of the number of guests in a party to be 8 (Refer Figure 3) and partition off the value of *Process* (1) by using a random number between 1 and 8. The division result is the number of parties, and the value of the random number is the number of guests in a party (Refer Figure 2). We do not make the probability distribution of random number (1-8) with a normal distribution, it is based on the composition ratio of *Measurement* (1), and we give the random vibration of 20 [%] (based on Figure 3) as uncertainty of the composition ratio.

2.2. Tracking Pattern

We set the pre-desired tracking pattern [10] (See Figure 5) as a target value to meet *Requirement* (1) and (2). The abscissa of Figure 5 is the simulation time (t), and the ordinate is the total number of seats (S) that should be reversed at a simulation time. The total seating capacity of "Once Upon a Time" is 2500.

2.3. Robust Controller

We apply ON/OFF-Type Discrete Variable Structure Controller [6,7,8] because of the following reasons;

- To follow the tracking pattern (Refer Figure 5), the control logic must be a feedback controller.
- The input data of the simulation have the uncertainties. Therefore, the control logic must have robustness.
- The manipulated variable takes two values of Winning (ON:$u = 1$) or Losing (OFF:$u = 0$).
- The simulation model is a discrete time.

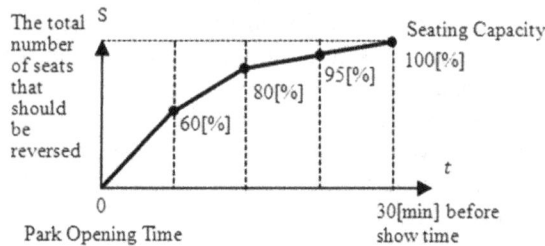

Figure 5. *Tracking Pattern.*

We derive the control logic satisfying Discrete Variable Structure theory condition to guarantee that an error decreases.

First, we define the error as follows,

$$e_k \triangleq S_t - h_k \tag{1}$$

At the next control step, there is the case which the target value changes (See Figure 6).

$$\begin{cases} e_{k+1} = S_t - h_{k+1} \\ e_{k+1} = S_{t+1} - h_{k+1} \end{cases} \tag{2}$$

The total of the reserved seats is calculated,

$$h_{k+1} = h_k + y_k \tag{3}$$

Next, we define the increment seats of the target value

$$\Delta S \triangleq S_{t+1} - S_t \ (\Delta S \geq 0) \tag{4}$$

Substituting Eq. (1), Eq. (3) and Eq. (4) into Eq. (2) leads to

$$\begin{aligned} e_{k+1} &= S_t + \Delta S - h_{k+1} \\ &= S_t + \Delta S - h_k - x_k u_k \\ &= e_k + \Delta S - x_k u_k \ (\Delta S \geq 0) \end{aligned} \tag{5}$$

If Discrete Variable Structure theory condition;

$$\text{When } e_k > 0, must\ be\ e_{k+1} < e_k \tag{6}$$

$$\text{When } e_k \leq 0, \text{must be } e_{k+1} \geq 0 \tag{7}$$

is satisfied, the error decrease. Find the condition of $u_k = 1$ (Winning) to satisfy Eq. (6) and Eq. (7).

About Eq. (6):

$$e_k - e_{k+1} > 0 \tag{8}$$

Substituting Eq. (5) into Eq. (8) leads to

$$e_k - e_k - \Delta S + x_k u_k > 0$$

$$x_k u_k > \Delta S$$

$$\text{As } \Delta S \geq 0, \ \therefore u_k = 1 \text{ when } e_k > 0 \tag{9}$$

About Eq. (7):

$$e_{k+1} = e_k + \Delta S - x_k u_k \geq 0 \tag{10}$$

To satisfy Eq. (10),

$$x_k u_k \leq e_k + \Delta S$$

$$\therefore u_k = 1 \text{ when } x_k \leq e_k + \Delta S$$

$$(x_k > 0, e_k \leq 0, \Delta S \geq 0) \tag{11}$$

Except Eq. (9) and Eq. (11), $u_k = 0$.

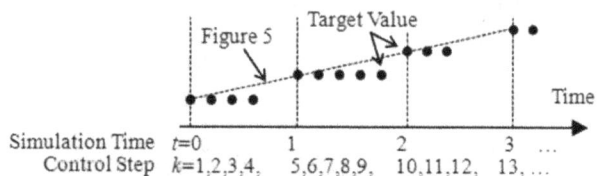

Figure 6. *The details of the target value (The value in the figure is an example for explanation).*

2.4. If Then Rule

However, even if $u_k = 1$, when any of the following condition is satisfied, change $u_k = 0$. We call this 'If Then Rule' (Refer Figure 2).

- To solve *Problem* (1), when $u_k = 1$ (Winning) has chosen continuously, if The number of consecutive winning $\geq$ Random number between 3 and 5, then $u_k = 0$. The random number is generated whenever the control logic operates.
- To meet *Requirement* (3), if this system cannot reserve the seats without sandwiching the aisle between guests in the winner party, then $u_k = 0$.

3. Simulation Result

Table 2 shows the simulation result. Among simulation data of 2658 records (Refer *Measurement* (3)), we obtain an effective result that the each vacant seats of 2598 records are less than 50 seats (50 is 2[%] of the seating capacity 2500). On the other hand, we obtain 25 records that the vacant seats are greater than or equal to 1000 seats.

Figure 7 and Figure 8 show one case of the simulation result. Because there are a few guests who draw the lottery at the park opening time (many guests get Fast-Pass ticket at first), the error is big at an early time (Refer Figure 7). And then it shows the good followability to the tracking pattern. Figure 8 shows that 'If Then Rule' can prevent the continuance of winning.

Table 2. Simulation Result.

Vacant seats	0-49	50-99	100-999	1000-	Total
Records	2598	12	23	25	2658

4. Neural Network Model

In this chapter, we predict the case that the vacant seats increase by using NN (Neural Network Model) [11, 12].

About the bad 25 records in the simulation of Chap.3, which has many vacant seats the obvious reasons for the increase of vacant seats are as follows; foul weather, uncertain weather, or rain is forecasted. However, we cannot say these reasons apply for all bad records. It is difficult to clarify why the vacant seats increases or decreases.

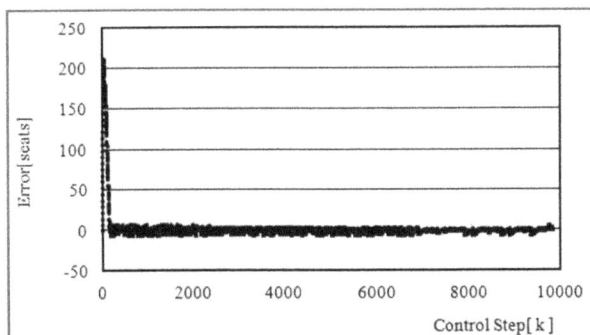

Figure 7. Error (Example, 01/Aug/2014, Attraction: Journey to the Center of the Earth).

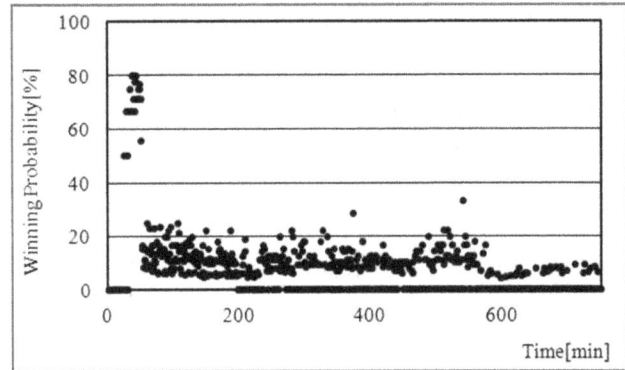

Figure 8. Winning Probability (Example, 01/Aug/2014, Attraction: Journey to the Center of the Earth).

To reduce the vacant seats, we propose that NN predicts the number of vacant seats at the end time of the lottery by using only the morning data. If the bad result is predicted, the staff of the lottery system is able to take the effective measures. For example, they can adjust the tracking pattern (Refer Figure 5) to increase the winners in the early afternoon.

NN has a conventional structure of the layers (input layer, hidden layer, output layer), and the learning method of NN is performed by a conventional error back-propagation algorithm. We calculate NN using R (R is a free software environment for statistical computing and graphics) [13].

Construction:

Input data = the morning data (from Park opening until noon) in the record (2658, 01/Jan/2014 - 31/Dec/2014).

Output data = the number of vacant seats.

Learning data;

Model error = Simulation result of Chap.3 - Output data.

Effect:

Input data = the morning data (from Park opening until noon) in New records (644, 01/Jan/2015 - 30/Apr/2015).

Output data = the number of vacant seats.

Results; Table 3(A), (B)

Table 3(A). Simulation Result (New Records).

Vacant Seats	0-49	50-99	100-999	1000-	Total
Records*1	627	6	5	6*2	644

*1 Simulation method same as Chap. II and Chap. III
*2 Refer Table 3(B): 6 cases

Table 3(B). Simulation Result (Vacant Seats: 1000-).

Simulation Result *3	1229	1781	1850	2047	2142	2176
NN Output*4	1250	null	1058	null	2128	null

*3 Vacant Seats: Table 3(A) 6 records*2
*4 Predicted Vacant Seats
null: NN did not output a solution

In this simulation, including null (NN did not output a solution), NN succeeds in predicting the bad 6*2 records which the vacant seats are greater than or equal to 1000 (Refer Table 3(B)). This result shows the effectiveness of NN. However, as 13 cases that the vacant seats of the simulation result are less than 50, NN predicts that the each vacant seat is more than 1000. As the reason, we think that there are many guests of entering in the evening of the day, or weather is restored in the afternoon, and so on. NN can stand further improvement, which is a future task.

5. Conclusion

We applied the robust feedback controller, ON/OFF-Type Discrete Variable Structure Controller, for the lottery system in Tokyo Disney Park. The main features are as follows:

- This system realizes fine access control characterizing the feedback control to meet *Requirement* (1)-(3).
- The control logic solves *Problem* (2) because it does not use the database, and 'IF Then Rule' solves *Problem* (1).
- Simulation results, based on many real data more than one year and the random vibration of 20[%] as uncertainty, show the effectiveness.
- NN predicts the number of vacant seats by using only the morning data. This method is useful to solve *Problem* (2).

In this paper, we have applied the new lottery system to one show ("Once Upon a Time"), but have confirmed that it is applicable to another shows ("Big Band Beat" and/or "One Man's Dream II"). Furthermore, if a theme park other than Tokyo Disney Park introduces a lottery system, this proposed system is applicable.

References

[1] Tokyo Disney Resort Official Website; http://www.tokyodisneyresort.jp/en/ (at 10/Mar/2015).

[2] XING Yuhong, "The Research and Design of an Applied Electronic Lottery System", Lect Notes Comput Sci, Vol. 7719, pp.653-658, 2013.

[3] William Zame, Jie Xu and Mihaela van der Schaar, "Winning the Lottery: Learning Perfect Coordination With Minimal Feedback", IEEE Journal of Selected Topics in Signal Processing, Vol. 7, No.5, pp. 846-857, 2013.

[4] YANG Jun, LIU Ying, QIN Ping, LIU Antung A, "A review of Beijing's vehicle registration lottery: Short-term effects on vehicle growth and fuel consumption", Energy Policy, Vol. 75 pp. 157-166, 2014.

[5] Title of Invention, "LOTTERY SYSTEM AND LOTTERY INFORMATION PROCESSOR", the name of the invention: Oriental Land Co., Ltd, The publication number assigned to the patent application by the patent office: 2005-266842 (in Japanese).

[6] T. Iwata, M. Yamakita and K. Furuta, "PWM-Type Discrete VSS Controller for On/Off Actuator Systems", Proc. Of the 39th IEEE Conference on Decision and Control, Sydney, Australia, Dec., pp.5131-5136, 2000.

[7] Msc Oscar Ivan Higuera Martinez, Juan Mauricio Salamanca and Hernando Diaz Morales, "Discrete-Time Variable Structure Control for Switching Converters", Transmission and Distribution Conference Conference and Exposition, IEEE/PES, Date 13-15 Aug, 2008

[8] Magnus Hedlund, Janaina G. Oliveira, Hans Bernhoff, "Sliding mode 4-quadrant DCDC converter for a fywheel application", Control Engineering Practice, Vol.21, pp.473-482, 2013.

[9] http://www.deepdisney.com/kimagure2012.aspx, http://s.tokyodisneyresort.jp/tdl/atrc_list.htm, http://s.tokyodisneyresort.jp/tds/atrc_list.htm(at 10/Mar/2015).

[10] N. Sakamoto, "The temperature control of the dryness fireplace by the sliding mode control", SICE Annual Conference 2002, WA10-3, pp.5-7.

[11] Yu-Wei. Chiu, "Machine Learning With R Cookbook - 110 Recipes for Building Powerful Predictive Models with R", Packt Publishing - ebooks Account, March 31, 2015.

[12] Brett Lantz, "Machine Learning with R - Second Edition", Packt Publishing - ebooks Account, Aug 3, 2015.

[13] R: The R Project for Statistical Computing Website; http://www.r-project.org/NeuralNetworkswithR;cran.r-project.org/web/packages/nnet/nnet.pdf (at 10/Mar/2015).

32

Research on Picking Goods in Warehouse Using Grab Picking Robots

Juntao Li[1], Xiaoqing Zhao[2]

[1]School of Information, Beijing Wuzi University, Beijing, China

[2]Graduate Department, Beijing Wuzi University, Beijing, China

Email address:

Lijuntao@bwu.edu.cn (Juntao Li), 2444581373@qq.com (Xiaoqing Zhao)

Abstract: For warehouse picking robots need the help of human vision and hands to quickly identify and get the desired products, this paper describes the procedure to grab picking robot. After receiving an order, grab picking robots use manipulator to grasp the desired products to realize the function of picking goods in warehouse by using technical means such as cameras, image processing.

Keywords: Gab Picking Robot, Picking-Up in Warehouse, Image Processing Technology

1. Introduction

With the development of economy, the users' needs tend to the direction of small batch and many varieties, and the amount and type of goods increase dramatically in a distribution center. In distribution center operations, the proportion of picking is becoming larger and larger. And the picking operation is the most wasted on human costs and time costs. The picking efficiency directly affects the operation efficiency and management benefit, and is also important factors that affect service level in the distribution center. The picking link of traditional distribution center requires a lot of manual operation, and the human account needed occupies over 50% of the total employment in a distribution center, and the intensity of labor by this operation mode is large [1].

With the development and application of grab picking robots (as shown in Fig. 1), the picking speed is improved effectively and manpower is saved effectively. At present, the use of robots in the logistics industry is mainly concentrated in stacking, unstacking, handling and several other parts. With the development of robot vision technology, there have been robot applications on the aspects of picking [2]. With the development of robot visual identification system and the combination between vision system based on two-dimensional plane and three-dimensional laser and

multi-function parallel manipulator, robots can automatically identify the fetching goods parameters such as color, size and location, adopt corresponding manipulator to grasp it, then put it on the collocated carrier or transmission line, and finally realize the function of choosing commodities.

Fig. 1. Grab picking robot.

In November 2015, the Fetch and Harvest executives said that at present their robots could only be used as a mechanical mule, identifying and getting the required products with the aid of human vision and dexterous hands of.

But the Fetch was developing a robot equipped with cameras and grab arm to take goods on the arrival of the shelves by themselves eventually [3].

The operation process of grab picking robot mainly includes the acquisition of goods image, the identification of goods characteristics, the extraction of goods contour, the determination of goods location and grabbing accurately. First of all, grab picking robots collect the images of goods through the camera, and identify the goods color and texture features to determine whether this goods is the desired goods. Secondly, grab picking robots take the contour of the goods to further determine whether this goods is the desired product and to adjust the robots hands according to the contour characteristics of the goods. Thirdly, determine the location where the goods locate according to the existing road signs. Finally, grab picking robots accurately grasp the required goods (as shown in Fig. 2).

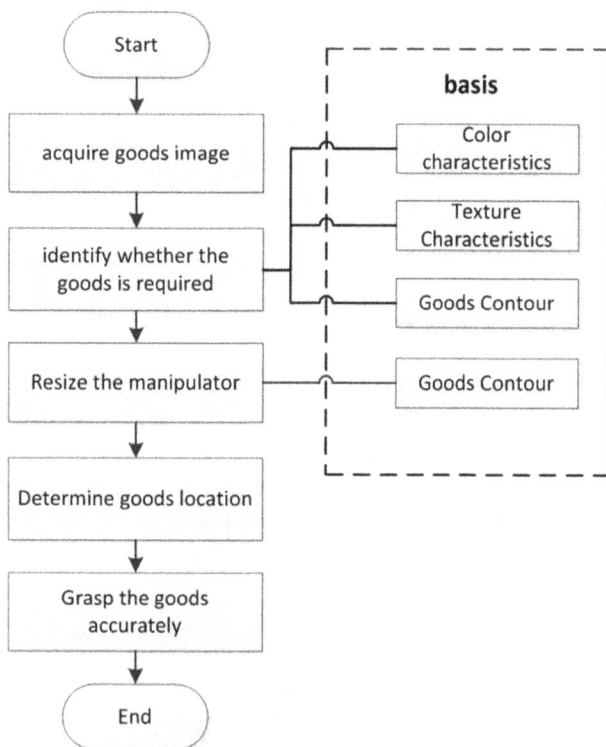

Fig. 2. The operation process of grab picking robots.

2. The Acquisition of Goods Image

The work of acquiring goods image is mainly performed by the camera and image acquisition card. This paper uses the camera instead of color identification device as the "eyes" of the goods identification and sorting system to finish the function of photographing goods image and uses image acquisition card to finish the function of conversing goods image.

The camera has passed CCD (charge-coupled device) image sensor and large scale integrated circuit of digital signal, which can produce images of high quality and high resolution. The image acquisition card uses the six-way composite video input or three-way Y/C input or a

combination of both by the multiplexer switches, chooses one or A set as the current input by software, put the input into digital decoder. And digital decoder will convert the color signals into brightness and color difference signal, output to the A/D converter to convert the modulus. Digital signal after image processing algorithm will be showed by VGA card and stored by computer memory with the use of PCI bus. The image acquisition card's working principle is as shown in Fig. 3. Software development kit provides a rich interface function, being convenient to control image acquisition card to collect image, and set all kinds of working parameters of image acquisition card. According to different functions, the interface function is classify into image acquisition card control function and supported image acquisition function. Image acquisition card control function mainly achieve the function of Setting image acquisition card's parameters, including image acquisition card's data format, source, hue, color saturation, brightness, contrast, display mode, acquisition methods, shielding and so on. When you start work, Image acquisition card control function is usually called to accomplish the image acquisition card initialization work; Supported image acquisition function is responsible for displaying the captured data by sending it to the display card or processing by sending it to the memory. Image acquisition function pushes the image data into memory in the process of program execution, completes the processing of the image by using image processing algorithms in the memory, and shows the result data on the terminal display.

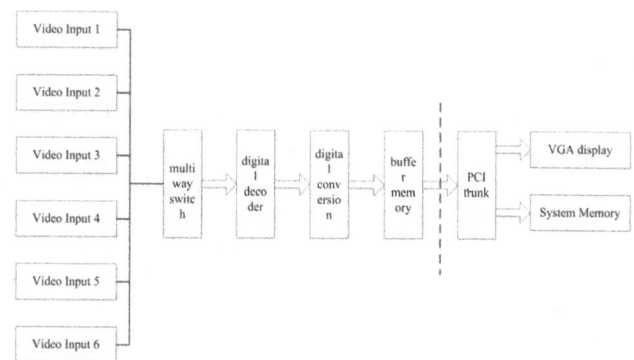

Fig. 3. The working principle of the image acquisition card.

3. The Identification of Goods Characteristics

To judge whether the cargo is the required goods, the goods features need to be taken according to the collected image at first, which mainly contain color features and texture features, etc. Because of the uncertainty of the position and angle of goods, the position and angle of the goods are changing within the image, so the selected goods characteristics must satisfy the rotating invariance and translation invariance. This paper chooses the goods color features and texture features as a basis for the identification of goods.

3.1. The Identification of Color Characteristics

Color is a kind of important visual property of the image, and the definition of color features is clear, and the extraction of color features is easy relatively, so there has been widely in the study of object recognition, target tracking, image retrieval, and other fields [4]. Color features is a global characteristic, is one of the most commonly used in object recognition, and describes the surface properties of scenery corresponding to the images or the image area. The advantage is not affected by changes in the image rotation and translation, and not affected by image size change after further normalization, while the disadvantage is that it does not express the spatial distribution information of color.

3.2. The Identification of Texture Characteristics

Texture feature is also a kind of global features. It describes the surface properties of scenery corresponding to the images or the image area. Unlike color features, texture features is not based on the pixel, and it need to be statistical computed in the area containing multiple pixel. In the pattern matching, the regional characteristic has great superiority, not failing to match due to the local deviation. As a kind of statistical characteristics, texture characteristics often have rotation invariance, and have a stronger ability to resist noise. Texture feature, however, also has its disadvantages that the calculated texture may have larger deviation when the resolution of the image changes.

4. The Extraction of Goods Contour

The cargo's outline is one of the most basic characteristics of the goods image, and it can be extracted by detecting the edge of the goods, which can detect for the required goods and can realize the positioning of goods to control the manipulator movement to grab goods.

4.1. The Process of the Image

The goal of the image processing is to provide the image characteristic value of the discriminated goods type and to output the results of goods identification. Machine vision is based on the theory of digital image processing, and its core content is image processing and recognition [5]. The image processing in machine vision is to transform the captured images by the visual system hardware, so as to achieve the aim of suiting for machine recognition. The original image captured directly often contains all kinds of noise and interference, which requires filtering and increasing the image and improving the quality of the image with all kinds of transformation methods to extract image characteristics and identify. This is the process of image processing.

4.1.1. Graying

The images captured by cameras are colorized. Although the colorized images contain rich information of the goods, the processing needs large amount of calculation. In order to meet the real-time requirement of picking goods, the collected image is conducted the graying process at first in this paper.

The images captured by cameras are RGB mode, that is, each pixel is a combination of R (red), G (green) and B (blue) color values of $0 \sim 255$ and the goal of extracting the color's component is to get the each pixel RGB value information of the input image. R, G, B corresponds each color's value and I refers the grey value of black and white image. The gray scale processing methods mainly include:

(1) The maximum method: grey value is equal to one of the biggest three values.

$$I=Max(R, G, B) \qquad (1)$$

(2) The average method: grey value is equal to the average value of the three.

$$I=(R+G+B)/3 \qquad (2)$$

(3) Weighted average method: grey value is equal to the weighted average value of the three. W_R, W_G, W_B is the corresponding weights of R, G, B according to the indicators of color sensitivity.

$$I=W_RR+W_GG+W_BB \qquad (3)$$

Because people's susceptibility of color is green, red, blue from high to low, a more reasonable gray image will be gained when after continuous study, namely:

$$I=0.3R+0.59G+0.11B \qquad (4)$$

In this algorithm, we choose a charge pal as the goal goods and use the weighted average method shown in the type (4) to process image graying. The unprocessed photo and the processed result are shown in Fig. 4 and Fig. 5.

Fig. 4. *The unprocessed photo.*

Fig. 5. *The grayscale image.*

4.1.2. Denoising

The greyscale image is gained by the inputted RGB mode directly, while not only the effective value of the original image is inputted but the input noise points in the process of obtaining grayscale. In order to extract target image from the gray image effectively and reduce the noise influence, the gray image need to be denoised. So you must choose the appropriate filter, which can effectively reduce the noise interference, and keep the goods location information of the image not destroyed as possible. The median filter is chosen for filtering through a large number of experiments [3].

Median filter is a nonlinear digital filter technique, often used to eliminate the noise in the image or other signals. Its main design thought is to check the input signal sampling and determine whether it represents a signal, and use observation window composed of an odd number of sampling to realize the function. There is observation window's number sorted and the median in the middle of a window as the output. Median filter preserves the image edge details while removing impulse noise, salt and pepper noise. So median filter is used to remove noise in this paper, and specific steps are as follows:

(1) Select a certain size of template and move the template along the row or column direction of image data;

(2) Order by pixel gray value within the window after each move;

(3) Replace the original pixel gray value of window center position by the sorted median.

This can be represented as in mathematics:

$$g(i,j)=\text{median}\{f(i-k,j-l)\},(k,l)\in \omega \qquad (5)$$

Of which i, j is for the current processing pixel coordinates and the scope of k, l depends on the shape and size of the used template.

In this paper, we adopt median filter to process goods images based on the 3 * 3 square template, and the experiment result is shown in Fig. 6.

Fig. 6. The grayscale photo after a median filter.

4.2. The Extraction of Goods Contour

Cargo outline is one of the most basic characteristics of the goods image, and it can be extracted by detecting the edge of the goods, which can not only detect for the required goods

but realize the positioning of goods to control the manipulator movement to fetch goods. Sobel operator is easy to implement on the space, having smooth effect on noise. It can be affected by noise little and can provide more accurate edge information direction. Canny operator can detect the real weak edge by using the non-maximum inhibition and double threshold methods. So the combined method of Sobel operator and Canny operator is used to extract more accurate edge of goods. Specific steps are as follows:

(1) Use Sobel operator to extract the edge of the gray image $A_1(I, J)$ and select double of gradient amplitude mean values as binarization closed, the edge image gained by which is regarded as $A_2(I, J)$.

(2) Project $A_2(I, J)$ along the horizontal and vertical directions and determine the edge location of goods through the wave formed by goods edge.

(3) Get rid of the image outside of the goods and goods location to obtain the image $A_3(I, J)$ of goods support area.

(4) Detect the cargo support area carefully and get the edge image $A_4(I, J)$. First of all, use the gaussian template smoothing images with 9 * 9 size. Secondly, calculate gradient amplitude and direction with a first order partial derivatives of finite difference. Thirdly, inhibit the gradient amplitude non-maximum. Finally, detect the image edge by using closed values, and regard the gradient amplitude size as the image threshold value which make 30% of points for the edge points.

(5) Connect Discontinuity points of $A_3(I, J)$ and search the edge that can be connected within a radius of eight adjacent points of $A_4(I, J)$ when it reaches the edge endpoint until you connect all the clearance of $A_3(I, J)$. The goods edge image gained is written as $A_5(I, J)$.

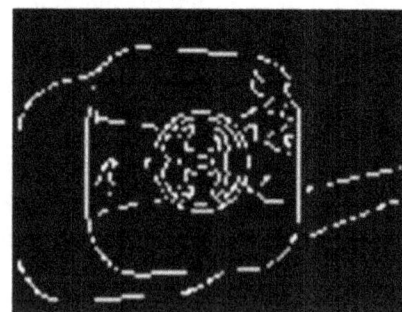

Fig. 7. The edge of goods image.

We have received the edge of goods image (as shown in Fig. 7), but in the actual cases, the detected edge of goods are mostly single pixel edge existing breakpoints. If we do not repair binary image before contour extraction, we do not get a whole goods contour, because the inherent pattern also can produce an edge of goods. So we need to remove isolated points in the image at first, and then use the rectangular probe operator of 3*3 size to conduct expansion operation on the edge image of the goods which has been removed the isolated points. At this point the processed image no longer exists discontinuity points, so the outside contour can be extracted by using edge tracking algorithm. The principle of edge tracking algorithm is arranging discharge point X0, X1... according to certain order, forming the boundary of the object and terminating tracking when the point column forms a closed loop, namely behind is. So the whole algorithm includes three steps. First of all, we scan images along the line to find the edge point, then we look for subsequent points with window operator. Finally we calculate the number of edge pixel points contained in the outline, determine a reasonable threshold, ignore the number of contour pixel points and outlines lower than the threshold, and eliminate the influence of tiny silhouette，therefore gain the complete outer contour of goods (as shown in Fig. 8).

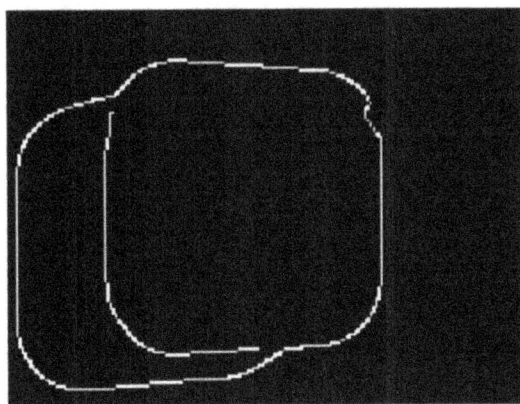

Fig. 8. The complete outer contour of goods.

According to the extracted contour, the weight of the goods can be calculated easily and the location of the goods can be determined, which provides convenience for cargo recognition and grab.

5. The Determination of Goods Location and Grabbing Accurately

The intersecting circle method in the triangulation is used to determine the location of the cherry-pick goods in the warehouse. We establish a right Angle coordinate system with the origin of a corner to determine the absolute coordinates of the two landmarks in warehouse, and the angle of the two way points from the goods to the landmarks, and set for an only circle according to the principle that the inscribed Angle with arc is equal to find out the center of the circle and radius. When the number of landmarks in the image is N and, circles can be determined. Then we take the equations of the two to find the solutions that one is the coordinates the known signpost and the other is the coordinates of the goods.

Grab picking robots determine whether to grab the objects according to the color feature, texture feature and contour features, control the size of the manipulator according to the contour, and move to the cargo location to realize accurate grab eventually.

6. Summary and Prospect

The development and use of grab picking robots fully implement that human vision will be altered to identify goods automatically and that human hands will be altered to grasp goods automatically. Grab picking robots collet the goods image through the camera, and then identify the characteristics of goods such as color characteristics and texture characteristics to determine whether this goods is the desired goods. Besides, grab picking robots do further processing to the collected image to get the goods contour and to further determine whether this goods is the desired goods. Finally, the robot resizes manipulator according to the goods contour to grasp accurately.

The successful development of grab picking robot, is just a link in the storage operation. To fully realize the storage unmanned, intelligent storage technologies such as AGV car are also need to fuse. The successful development of grab picking robot is beneficial to improve the efficiency of the warehouse picking, save labor costs and increase enterprise competitiveness.

Acknowledgements

The study is supported by Beijing Intelligent Logistics System Collaborative Innovation Center and Iintelligent Logistics System Beijing Key Laboratory (NO: BZ0211).

References

[1] Xiaoguang Zhou, Ximei Zhang, Yukun Liu. A flexible picking system based on mobile robot in distribution center [J]. Logistics technology, 2015, 34 (4): 238-240.

[2] Jiantao Tian, Pei Su, Na Wang. Research on visual robot goods identification technology [J]. Journal of Beijing union university, 2014, 28 (4): 12-17.

[3] The industrial control network. Amazon's intelligent warehouse force robot market more popular [EB/OL]. [2016-4-2]. http://www.gkong.com/item/news/2015/11/85518.html.

[4] Qing Li, Yujin Zhang. Image classification method based on the characteristics of the elements and association rules [J]. Electronic journals, 2002, 30 (9): 1262-1265.

[5] Azriel Rosenfeld. Introduction to machine vision [J]. IEEE Control Systems Magazine, 1985, 5: 14-17.

[6] Kuirong Cong, Jie Han, Faliang Chang. Visual robots extract goods contour and position [J]. Journal of Shandong University (engineering science), 2010, 40 (1): 15-18.

[7] Baoxia Cui, Chi Zhang, Tingting Luan. Landmarks extraction method of panoramic vision robot through localization [EB/OL]. 2016 [2016-3-31]. http://www.cnki.net/kcms/detail/21.1189.T.20160302.1642.00 6.html.

[8] Guannan Xu, Qingguan Xia, Zhonghui Zhang. Analysis of visual robots target detection [J]. Mechanical and electronic information, 2015 (30): 128-131.

[9]   Jintian Huang, Huailin Cui. Research on robot visual positioning based on cameras [J]. Journal of Guangdong technical teachers college, 2014, 36 (3): 69-72.

[10] Benfa Zhang, Xiangping Meng. Overview of mobile robot localization method [J]. Shandong industrial technology, 2014 (22): 250.

[11] Zhiqiang Zhang, Zhiyou Niu, Siming Zhao. Overview of Freshwater fish species identification based on machine vision technology [J]. Journal of agricultural engineering, 2011 (11): 388-392.

33

ZnO Nanowire/N719 dye/Polythiophene-SWNT nanocomposite solid state dye sensitized solar cells

S. AbdulAmohsin[1], Sabah Mohammed Mlkat al Mutoki[2], M. Mohamed[1]

[1]Physical Department, Faculty of Science, Thi Qar University , Thi Qar, Iraq.
[2]Electrical Department , Al Furat Al Wast University, Technical Institute of Samawa, Samawa, Iraq.

Email address:

asabah_sh2003@yahoo.com (S. M. M. al Mutoki)

Abstract: We designed and fabricated high efficiency solid state dye sensitized solar cells based on vertical ZnO nanowire arrays by utilizing a mixture paste of LiI, PMII and solid iodine as electrolyte. Poly thiophene -single wall carbon nanotube (PT-SWNT) composites were synthesized on FTO glass by in situ polymerization and employed as counter electrode to replace the conventionally used expensive Pt electrode. The initial results showed the power conversion efficiency of 2.87 % from the device with PPy-SWNT composite coated on FTO glass as counter electrode.

Keywords: Nanocomposite, SWNT, PT, Solid State Dye Sensitized Solar Cell

1. Introduction

The efficiencies of dye-sensitized solar cells (DSCs) recently have been reported up to 12% (Yu et al., 2010) since the DSCs using nanocrystalline structures were published by O'Regan and Grätzel (1991)[1,2]. DSCs have generated excitement because they consist mainly of nontoxic materials and offer a low-cost processing route (such as coating or printing) to thin-film device fabrication. Furthermore, they can be adapted for a variety of indoor and outdoor applications, and achieve high performance with minimal environmental impact. A DSC operates based on the interactions between the cell's anode and cathode, and a film of titanium oxide nanoparticles covered with light-sensitive dye molecules[3,4]. An electrolyte, usually in form of iodide, fills the space between the TiO2 nanoparticles, and helps transfer electrons from the cathode to the dye molecules[5]. The fabrication of DSCs typically requires an electrolyte that enables high charge-collection efficiencies and high open-circuit voltages. The iodide electrolyte is particularly attractive in this regard as its oxidized form, I^{3-}, does not readily accept electrons from the titania surface[6]. This minimizes charge recombination in functioning cell devices. Despite all the benefit and relatively high conversion efficiencies for solar energy, DSCs typically have durability issues associated with the liquid electrolyte, such as electrode corrosion or electrolyte leakage[7-10]. These issues have lead to a significant decrease in conversion efficiency, making these solar cells unsuitable for long-term use. However, the charge recombination in ss-DSCs is 1 to 2 orders of magnitude faster compared to liquid-electrolyte DSCs, which should reduce the achievable open-circuit voltage by 100 to 200 mV.

Currently, the best performing ss-DSCs achieve power conversion efficiencies of up to 7.2%,[11]thus still lagging behind DSCs based on a liquid electrolyte, which are now over 12%.[12] The underperformance of ss-DSC is thought to be associated with three main challenges: i) limited pore-filling of the mesoporous TiO2 with the HTM to ensure the optimum composition of the photoactive layer, 13–16] ii) panchromatic absorption of light, which is currently limited by the maximal film thickness of well-performing ss-DSCs, [17] and iii) efficient charge generation and transport from the excited dye to maximize the current output from the sunlight.[18–25] While issues with efficient charge generation have been extensively addressed (though not resolved) in earlier work, [18–24]the related challenges of effective pore-filling and sufficient panchromatic absorption remains open.

Our purpose is to get high efficiency solid state dye sensitive solar cells base on ZnO NW /N719 dye/PT ,or nanocomposite PT +SWCNTS

I used1- pristine PT by electrochemical method , PT+SWCNTS (10 wt %) ,and Pristine SWCNTS

2. Experimental

2.1. Materials

2.1.1. Materials Synthesis

The vertically aligned ZnO NW arrays were fabricated on FTO glass substrates (SPI Supplies) by a low temperature electrochemical method.[20]They were typically prepared in a two-step process. First, a layer of ZnO thin films was grown on the FTO substrates by using 0.05 M Zinc nitrate (Alpha Aessar) dissolved in a mixture of de-ionized water and methanol (50:50). The growth temperature was kept at 70 °C and the growth time was 5 min under an applied voltage of -2.5 V with two Au electrodes, one of which was connected to the FTO glass substrate as working electrode and the other one was used as a counter electrode. A thin film of about 100-200 nm can be deposited. In the second step, ZnO NW arrays were grown on top of the ZnO thin film using an electrolyte containing 0.01 M Zinc nitrate and 0.01 M hexamethylenetetramine in de-ionized water. The growth was performed at 95 °C with an applied potential of -2.6 V between working and counter electrodes.

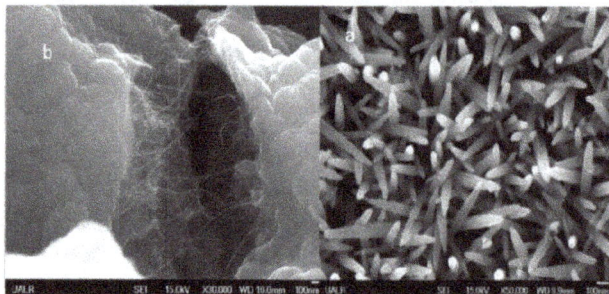

Fig. 1. SEM images of ZnO nanowire arrays (a), Nanocomposite materials PT-SWNTS (b)

As seen from Fig. 1 (a), the nanowires with an average diameter of 100-200 nm are vertically aligned on the substrate show the two components mixed and distributed uniformly in each other (Fig.1 (b)).

Thiophene monomer was distilled twice under reduced pressure before use, and dissolved into a acetonitral to make a 0.1 M solution. The polyThiophene was synthesized with a galvanostatic step method at a constant voltage of 2 V. The working electrode was a commercial ITO-glass with a surface area of 1.5 cm^2. A Platinum wire was used as a counter electrode. The amount of the electrodeposited polythiophene was estimated by weighing the working electrode before and after the electro deposition. Figure 1 displays a schematic diagram of the electrochemical cell for in situ polymerization.

TiO$_2$ nanopartical and PT were brought into contact by sandwiching them to create the solid State Dye Sensitive solar cells . To ensure intimate contact with no leak between the TiO$_2$ and the PT, the two films were sandwiched right after the electrochemical process when the PT was still wet to avoid the oxidation for thin film of polythiophene . A schematic diagram of the fabricated device is illustrated in Figure 3. Current density-voltage (J-V) characteristics have been investigated in the dark and under illumination using an AM1.5 sunlight simulator. Irradiation was achieved from the TiO$_2$ side and from the nanocomposite side explained in figure 4.

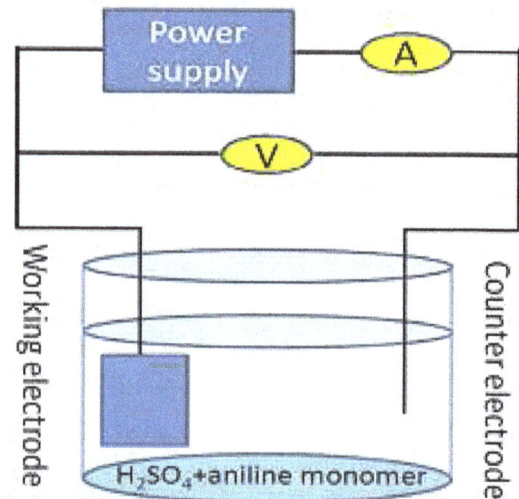

Fig. 2. Schematic diagram of the electrochemical cell used to prepare the PT thin film.

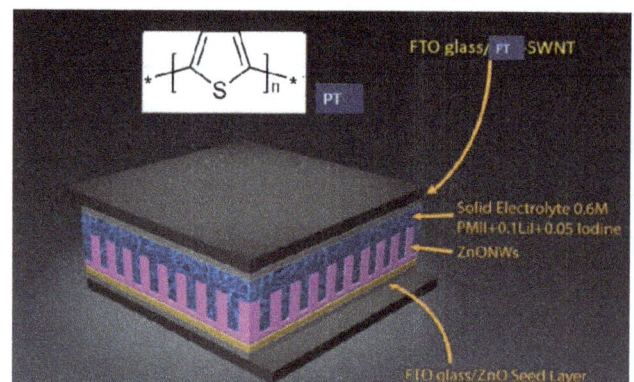

Fig. 3. Schematic diagram of the ZnO nanowire/N719 dye/ Polythiophene-SWNT nanocomposite solid state dye sensitized solar cells.

2.2. Fabrication of ss-DSCS

SWCNTS ,Iodine,1-methyle-3-propyl imidazolium iodine (PMII) ,Zinc nitrate ,Hexamethylentetramine was obtained from cheap tube and Aldrich respectively ,Using airbrushing technique to deposit SWCNTs on FTO glass ,then deposit polythiophene by electrochemical on top of SWCNTS to get good penetration of PT with SWCNTS ,put two drop of organic solvent electrolyte (mixture of 0.1 M LiI,0.6 MPMII,0.005 M I$_2$) after 5 sec the electrolyte will dry and make sandwich them ,finally the Solid State Dye Device ready to measure I-V , We have three device PT ,SWCNTS ,and Nano composite to compare among those Devices to study effect of Nano composite on Solar Cells performance.

2.3. Characterization

Various techniques were employed to characterize the obtained materials in this study, including SEM, UV-Vis absorption, FTIR, and Raman spectroscopy. Raman scattering was collected at room temperature using a Horiba Jobin Yvon LabRam HR800 spectrometer equipped with a charge-coupled detector and a He-Ne laser (633 nm) as excitation sources. Raman shifts were calibrated with a silicon wafer at the 521 cm^{-1} peak. The optical absorption of the composite films was measured by using a Shimadzu double beam spectrophotometer UV-3600. The Fourier transform infrared (FTIR) spectra were collected on a Nicolet MAGNA-IR 550 Series 2 Spectrometer with a resolution of 8 cm^{-1}. The reported spectra were averaged for 32 scans.

Current density-voltage (J-V) characteristics were investigated in the dark and under illumination using a solar simulator at AM1.5 (~100 mW/cm^2) inside a glove box in a nitrogen environment. Illumination was on the TiO$_2$ side. The devices were irradiated in an area of 3×3 mm^2 and data were recorded using a Keithley 2400 source meter.

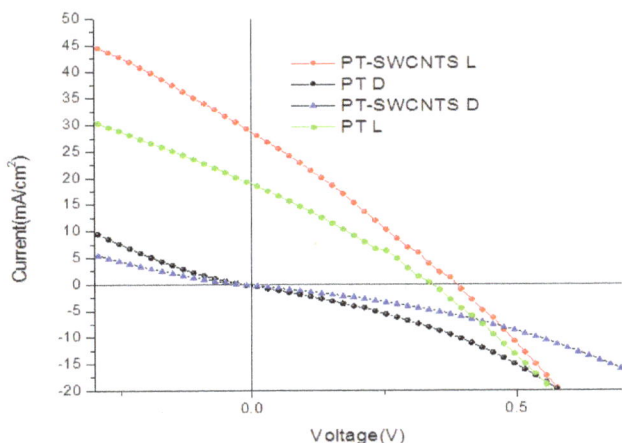

Fig. 4. Photocurrent–voltage characteristics of DSSCs with PT (solid line) and Pt (dash line) counter electrodes under 100 mA/cm^2 light irradiation.

3. Results and Discussion

I study Solid state dye sensitized solar cells base on ZnO NW arrays modified by N719 , and PANI+0.6 M of PMII (1-methyl-3-propyl imidazolium iodide) and 0.1 LiI+0.05 Iodine as a solid electrolyte where The prepare of ZnONW, PANI describe with attached , and immerse the ZnO NW in ethanol solution of N719 dye for 3 days using electrochemical polymerization of PANI on ITO glass then add one drop of 0.6 M of PMII (1-methyl-3-propyl imidazolium iodide) and 0.1 LiI+0.05 Iodine as a solid electrolyte then sandwich them after put separator the efficiency improve from 1.78 to 2.90 %

3.1. Experimental Details

3.1.1. J-V Results
This Figure IV Illumination of Nanocomposite of PT +SWCNTS ,PT,SWCNTS for Solid state dye sensitized solar cells containing ionic liquid and nanocomposite solar cells

As seen in Fig. 4, the performance of the solid state dye-sensitized solar cells was improved by using nanocomposite SWNTs-PT as HTM. Under illumination AM1.5G, the cell with SWNTs has higher output current as compared to the on pristine PT. The short-circuit current density, J$_{SC}$, and the open-circuit voltage, V$_{OC}$ are significantly improved which increased from 19 to 28 mA/cm^2 and 0.33 to 0.39 respectively in the other side the fall factor and efficiency increase from 32% to 42% and 1.78 to 2.9% respectively by using SWNTs. The experimental data demonstrate that the use of SWNTS helps to improve the solar cell performance with an increased conversion efficiency from 1.78 to 2.9. This observation is explained by the significantly increased contact area between the SWNT-PT (HTM) and the paste which facilitates efficient charge transportation in the solar cell. We demonstrated that the PPY-SWNT composites are suitable to improve the stability and efficiency for fabricating high efficiency Solid state dye

Although J_{sc} from ITO side illumination is lower than that from FTO side illumination, the V_{oc} was similar and do not depend on the direction of incident photon. V_{oc} could be considered as a result in the interfacial kinetics between injected electrons with back transfer electrons. Under open-circuit conditions, the rate of injected electrons from dyes is balanced by that of back transfer electrons from TiO$_2$ to spiro-MeOTAD or dyes. Assuming dye regeneration is efficient, recombination from TiO$_2$ to dye is negligible and thus V_{oc} is given by Bisquert et al., 2004[25],

$$V_{oc} = \frac{kT}{q} \ln\left(\frac{J_{sc}\tau_0}{q\alpha n_o}\right)$$

where k is Boltzmann's constant, T is the temperature, q is electronic charge, J_{sc} is the short-circuit current, τ_0 is an electron lifetime, α is the ratio of surface electrons concentration participated in recombination to electrons in the TiO$_2$, and n_o is the electron concentration in the conduction band in the dark. Since the optical process mainly dominates our current output, the J_{sc}/n_o in principle remained constant measurement. From the equation, we believe our identical V_{oc} from different sides illumination is a measure of the electron lifetime.

3.2. Optical Properties of Materials

The photoluminescence spectra of ZnO NW (blue line) ,and N719 grafted ZnO NW (Red line) in methanol ,excited at 325 nm at the same optical density the figure 5 indicates the photoluminescence was quenched in both UV and visible regions when the N719 grafted to the ZnO nanowires to be sensitized .we can conclude the possibility of the dye N719 resonance energy transfer which causes the static quenching due to dye aggregation on to the ZnO NW surface.thus,we attribute this photoluminescence quenching to the efficient photoinduced electron transfer from the dye N719 to the ZnO NW .

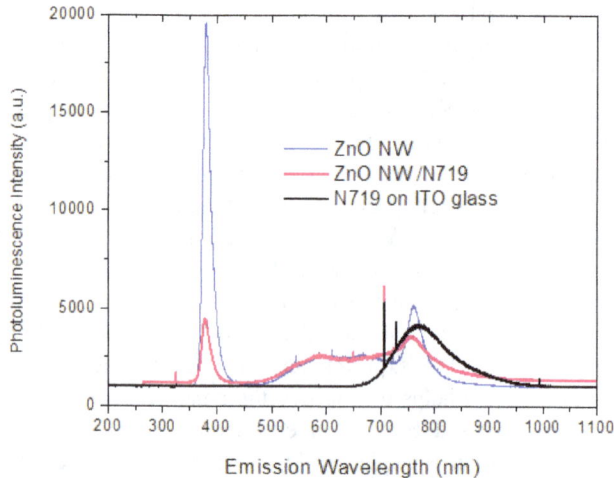

Fig. 5. *The photoluminescence spectra of ZnO nanowires, and ZnO nanowires coated with N719, and N719 on ITO glass.*

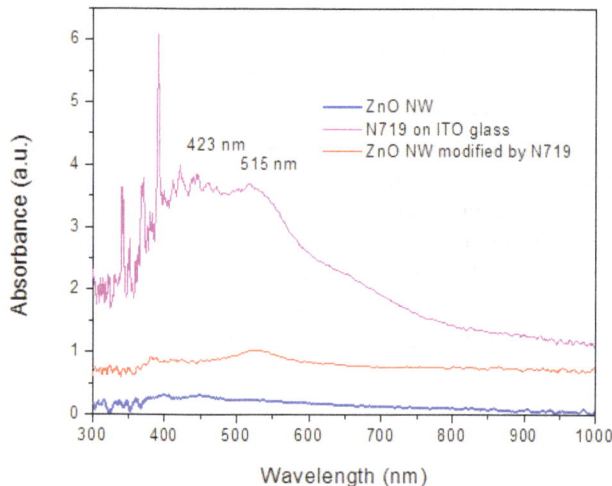

Fig. 7. *The XRD patterns of ITO glass, N719 on ITO glass and N719 on ZnO nanowires.*

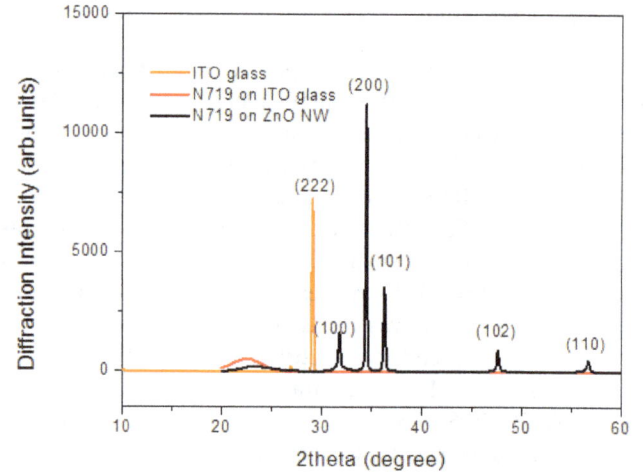

Fig. 6. *The optical absorption spectra of ZnO nanowires, and N719 on ITO glass, and ZnO nanowires modified by N719*

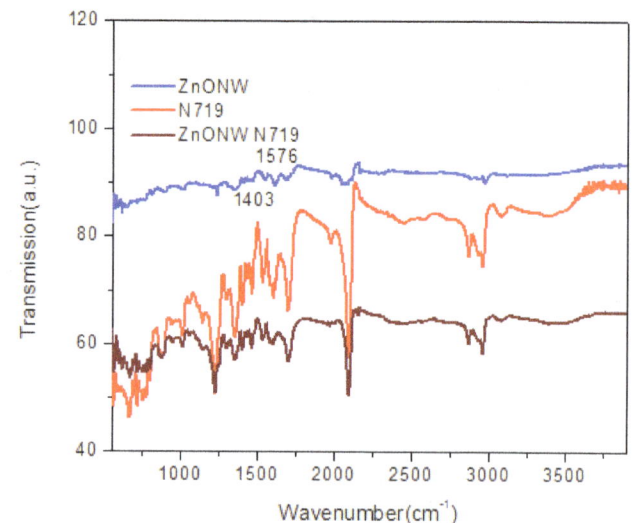

Fig. 8. *FTIR of ZnONW,ZnONW N719,and N719*

In order to understand the reasons behind the increased PCE by N719 dye modification ,where explored their optical properties as seen in the figure 6 where there are 2 main new peaks (423 nm to 515 nm).

The modification of ZnO NW thin film with N719 dye exhibits a strong increase in absorption around 423nm -515 nm ,which is attributed to the additional absorption of the coated porphyrin on ZnO surface .

Crystallography

Figure 7 presents the X-Ray diffraction (XRD) spectra of N719 dye on ZnO NW (black curve),and N719 on ITO glass (Red curve) ,and ITO glass (Pink) the peak at centered at 2 theta 31.11^0 and 36.13^0 are from ITO glass, and the strong peaks 2 theta -34.2^0 is from ZnO NW surface .XRD results explain that the final crystalline structure does not effect after modified by N719 dye and aggregation of N719 dye to ZnO NW surface introduce no change in crystalline structure of ZnO NW

Figure 8 Shows the FTIR spectrum of the ZnO NW synthesized by electrochemical method ,which was aquired in the range 400-4000 cm^{-1} correlated to metal oxide bond (ZnO).from this FTIR for ZnO NW in addition to ZnO NW modified by N719 Dye .the peaks in the range of 1400-1500 cm^{-1} corresponds to C=O bonds .The adsorbed band at 1626 cm^{-1} is assigned O-H bending vibration .the binding between the dye molecules (N719) and the surface of ZnO NW was investigated by FTIR spectra of the dye modified ZnO NW . The solar cells efficiency depend on charge injection process and its highly depending on bonding structure of the dye molecules on the ZnO NW .In addition ,the electron transfer in Solid state dye sensitive solar cells is strongly influenced by electrostatic and chemical interaction between ZnO NW surface and the adsorbed dye molecules[27] Figure 8 shows the FTIR spectra of N719 on ZnO NW compared to the curve of the dye solution .Absorption at 2105 cm^{-1} of N719 dye is attributed to the SCN stretch model of N-bounded SCN ligand [28,27] The IR spectra were observed at 1370 cm^{-1},1610 cm^{-1} is consistent with the bidentate coordinate .The same results were reported that the coordination of N719 dye on ZnO NW occurs mainly by contribution of unidentate and partially by bidentate linkage [26-28]

4. Conclusion

In brief, in order to develop new type of counter electrode to replace the conventional used expensive Pt electrode used in dye sensitized solar cells, we synthesized polythiophene-single wall carbon nanotube (PT-SWNT) composites on FTO glass by in situ electrochemical polymerization. ZnO nanowire arrays were vertically gorwn on ZnO buffered FTO glass and utilized as a 3D photoanode. We designed and fabricated high efficiency solid state dye sensitized solar cells by sandwiching the two parts and utilizing a mixture paste of LiI, PMII and solid iodine as electrolyte. The initial results showed the power conversion efficiency of 2.87% from the device with PT-SWNT composite coated on FTO glass as counter electrode, which is significantly higher that built with PT. The enhanced light conversion is due to improved charge transport in the PT-SWNT composites. The structure of the vertical ZnO array also benefits photon absorption.

References

[1] B. O'regan and M. Grfitzeli, "A low-cost, high-efficiency solar cell based on dye-sensitized," *Nature* 353, 24 (1991).

[2] G. H. Guai, Y. Li, C. M. Ng, C. M. Li, and M. B Chan-Park, "TiO$_2$ composing with pristine, metallic or semiconducting single-walled carbon nanotubes: which gives the Best Performance for a dye-sensitized solar cell," *ChemPhysChem* 13 (10), 2566-2572 (2012).

[3] Y. Xu, C. He, F. Liu, M. Jiao, and S. Yang, "Hybrid hexagonal nanorods of metal nitride clusterfullerene and porphyrin using a supramolecular approach," *Journal of Materials Chemistry* 21 (35), 13538-13545 (2011).

[4] J. H. Zhao, X. C. Yang, M. Cheng, S. F. Li, and L. C. Sun, "New Organic Dyes with a Phenanthrenequinone Derivative as the pi-Conjugated Bridge for Dye-Sensitized Solar Cells," *J. Phys. Chem. C* 117 (25), 12936-12941 (2013).

[5] V. Tjoa, J. Chua, S. S. Pramana, J. Wei, S. G. Mhaisalkar, and N. Mathews, "Facile Photochemical Synthesis of Graphene-Pt Nanoparticle Composite for Counter Electrode in Dye Sensitized Solar Cell," *Acs Applied Materials & Interfaces* 4 (7), 3447-3452 (2012)

[6] G. Yue, J. Wu, Y. Xiao, J. Lin, M. Huang, and Z. Lan, "Application of poly(3,4-ethylenedioxythiophene): polystyrenesulfonate / polypyrrole counter electrode for dye-sensitized solar cells," *The Journal of Physical Chemistry C* 116 (34), 18057-18063 (2012).

[7] J. Zang, C. M. Li, S.-J. Bao, X. Cui, Q. Bao, and C. Q. Sun, "Template-free electrochemical synthesis of superhydrophilic polypyrrole nanofiber network," *Macromolecules* 41 (19), 7053-7057 (2008).

[8] S. Thomas, T.G. Deepak, G.S. Anjusree, T.A. Arun, S. V. Nair, and A. S. Nair, "A review on counter electrode materials in dye-sensitized solar cells," *Journal of Materials Chemistry* A (2014).

[9] S. Siriroj, S. Pimanpang, M. Towannang, W. Maiaugree, S. Phumying, W. Jarernboon, and V. Amornkitbamrung, "High performance dye-sensitized solar cell based on hydrothermally deposited multiwall carbon nanotube counter electrode," *Applied Physics Letters* 100 (24), 243303-243303-243304 (2012).

[10] W. Jarernboon, S. Pimanpang, S. Maensiri, E. Swatsitang, and V. Amornkitbamrung, "Effects of multiwall carbon nanotubes in reducing microcrack formation on electrophoretically deposited TiO$_2$ film," *Journal of Alloys and Compounds* 476 (1), 840-846 (2009).

[11] K. Murakoshi, R. Kogure, Y. Wada, and S. Yanagida, "Fabrication of solid-state dye-sensitized TiO$_2$ solar cells combined with polypyrrole," *Solar Energy Materials and Solar Cells* 55 (1), 113-125 (1998).

[12] T. Kitamura, M. Maitani, M. Matsuda, Y. Wada, and S. Yanagida, "Improved solid-state dye solar cells with polypyrrole using a carbon-based counter electrode," *Chemistry Letters* 30 (10), 1054-1055 (2001).

[13] R. Cervini, Y. Cheng, and G. Simon, "Solid-state Ru-dye solar cells using polypyrrole as a hole conductor," *Journal of Physics D: Applied Physics* 37 (1), 13 (2004).

[14] H. Hlura, T.W. Ebbesen, T. Tanigaki, H. Takahashi, *Chem. Phys. Lett.* 202, 509(1993).

[15] W. A. de Heer, W.S. Bacsa, A. Chatelain, T. Gerfin, R. Humphrey-Baker, L. Forro, and D. Ugarte, "Aligned carbon nanotube films: production and optical and electronic properties," *Science* 268, 845-847 (1995).

[16] J. S. Liu, T. Tanaka, K. Sivula, A. P. Alivisatos, and J. M. J. Frechet, "Employing end-functional polythio-phene to control the morphology of nanocrystal-polymer composites in hybrid solar cells," *Journal of the American Chemical Society* 126 (21), 6550-6551 (2004).

[17] Y.-Ch. Liu, B.-J. Hwang, W.-J. Jian, and R. Santhanam, "In situ cyclic voltammetry-surface-enhanced Raman spectroscopy: studies on the doping–undoping of polypyrrole film," *Thin Solid Films* 374 (1), 85-91 (2000).

[18] J. Duchet, R. Legras, and S. Demoustier-Champagne, "Chemical synthesis of polypyrrole: structure–properties relationship," *Synthetic Met* 98 (2), 113-122, (1998).

[19] A. B. Gonçalves, A. S. Mangrich, and A. J. G. Zarbin, "Polymerization of pyrrole between the layers of α-Tin (IV) Bis (hydrogenphosphate)," *Synthetic Met* 114 (2), 119-124 (2000).

[20] R.H Friend, D.D.C Bradley, and P.D. Townsend, "Photo-excitation in conjugated polymers," *Journal of Physics D: Applied Physics* 20 (11), 1367 (1987).

[21] G Harbeke, D Baeriswyl, H Kiess, and W Kobel, "Polarons and bipolarons in doped polythiophenes," *Physica Scripta* 1986 (T13), 302 (1986).

[22] R. Yang, W. H. Smyrl, D. F. Evans and W. A. Hendrickson, J. Phys. Chem, 96, 1428(1992).

[23] M. J. Antony and M. Jayakannan, J. Phys. Chem. B, 111, 12772(2007).

[24] P. Galář, B. Dzurňák, P. Malý, J. Čermák, A. Kromka, M. Omastová, and B. Rezek, "Chemical Changes and Photoluminescence Properties of UV Modified Polypyrrole," *Int. J. Electrochem. Sci.* 8, 57-70 (2013).

[25] Bisquert.J, and Peter.C .Journal of physical chemistry letters,vo.51,issue1,205-207.

[26] Hwang, K. J.; Yoo, S. J.; Kim, S. S.; Kim, J. M.; Shim, W. G.; Kim, S. I.; Lee, J. W. *J. Nanosci. Nanotechnol.* 2008, *8*, 4976.

[27] Leon, C. P.; Kador, L.; Peng, B.; Thelakkat, M. *J. Phys. Chem. B* 2006, *110*(17), 8723.

[28] Finnie, K. S.; Bartlett, J. R.; Woolfrey, J. L. *Langmuir* 1998, *14*, 2744

34

The Research on Longitudinal Flying Quality of Small Unmanned Aerial Vehicle

Zhian Yan, Hudan Sun

China Electronics Technology Group Corporation Special Mission Aircraft System Engineering Co., Ltd., Chengdu, China

Email address:

yanza@cetca.net.cn (Zhian Yan), sunhd@cetca.net.cn (Hudan Sun)

Abstract: Due to the lack of specific flying quality criteria for UAV (Unmanned Aerial Vehicle), most researchers use the human flying quality criteria to evaluate the UAV flying qualities. This paper, through analyzing the flying performance of the small UAVs before and after the inclusion of the control augmentation systems, has validated the applicability of human flying quality criterion such as CAP and bandwidth criterions to small UAVs and proposed a means of evaluating the flying quality research of UAVs. The research of this paper can provide a reference for the future research of UAV flying quality and help for the design of UAV as well as its flying control system.

Keywords: UAV, Control Augmentation System, Flying Quality, Bandwidth Criterion

1. Introduction

In recent years, more and more UAVs are used in civil and military areas because of their special advantages. In contrast to the big development of UAVs, the research of the UAV flying quality is badly lagged behind. Until now, there is no specific flying quality criterion for UAVs. As we all know, the human flying quality criterion is the design reference and standard for human aircrafts and the control system, and the flying quality criterion can ensure the security and reliability of the aircraft, enhance the efficiency of the aircraft design, and reduce the design period and cost. The mature human flying quality research promoted the quick development of the human aircrafts. Unfortunately, there is no specific UAV flying quality criterion domestic and abroad. All the present researches of UAV flying qualities have been undertaken according to the human flying quality criterion. And more seriously, there is no definitive proof that the human flying quality criterion can be used to evaluate the UAV flying quality, and this is the hot research topic now. The references [1-6], [10-12] analyzed the applicability of human flying quality criterion for UAVs, but the conclusions are different and even contradictory.

As there is no special UAV flying quality criterion, this paper analyzes the flying performance of the small UAV before and after the inclusion of control augmentation system and evaluates the longitudinal flying quality using human flying quality criterion. Through the research of this paper, we conclude some new properties of the flying quality for small UAV and propose a means to evaluate the longitudinal flying quality research of small UAV.

2. Flying Performance Validation with/Without Augment System

The research UAV example is a small UAV about 5Kg, with maximum flying altitude of 2000m, maximum flying speed of 0.06Ma. We choose the design point (h=1000m, Ma=0.05) to evaluate the flying performance of the small UAV. We analyzed the time-domain response performance of the UAV with and without control augment system, as following:

(1) The time-domain response comparison of pitch angle and pitch rate between the UAV with and without control augment system, as "Figure 1" displaying:

pitch angle (theta)

pitch angle (theta)

pitch rate (q)

pitch rate (q)

without control augment system

with control augment system

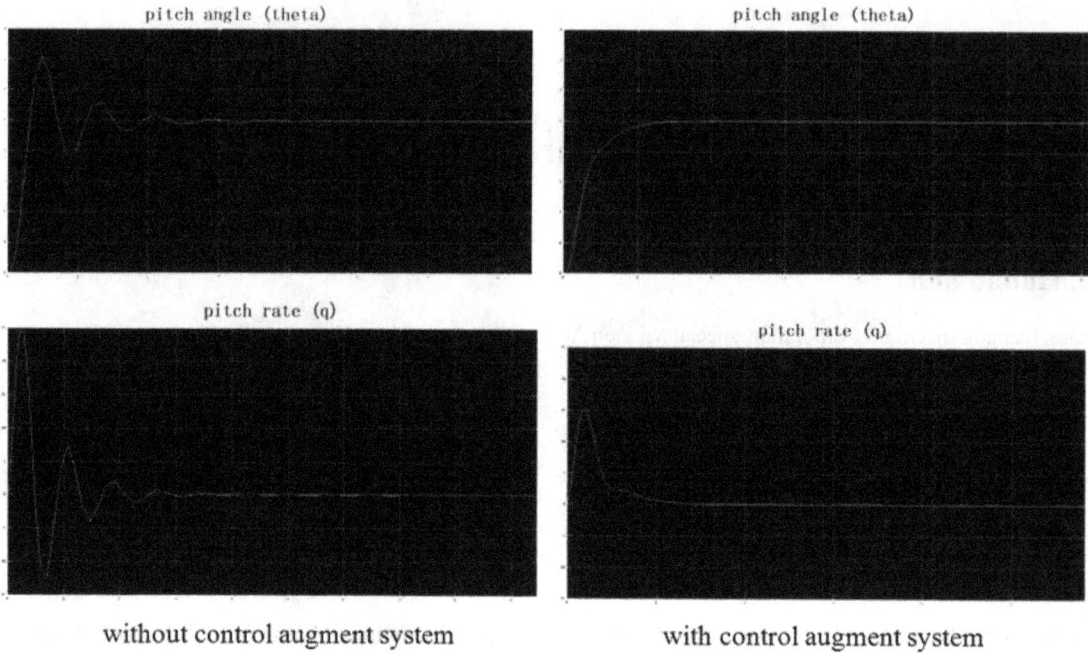

Figure 1. Time-domain response comparison.

(2) The frequency-domain stability comparison between the UAV with and without control augment system, as "Figure 2" displaying:

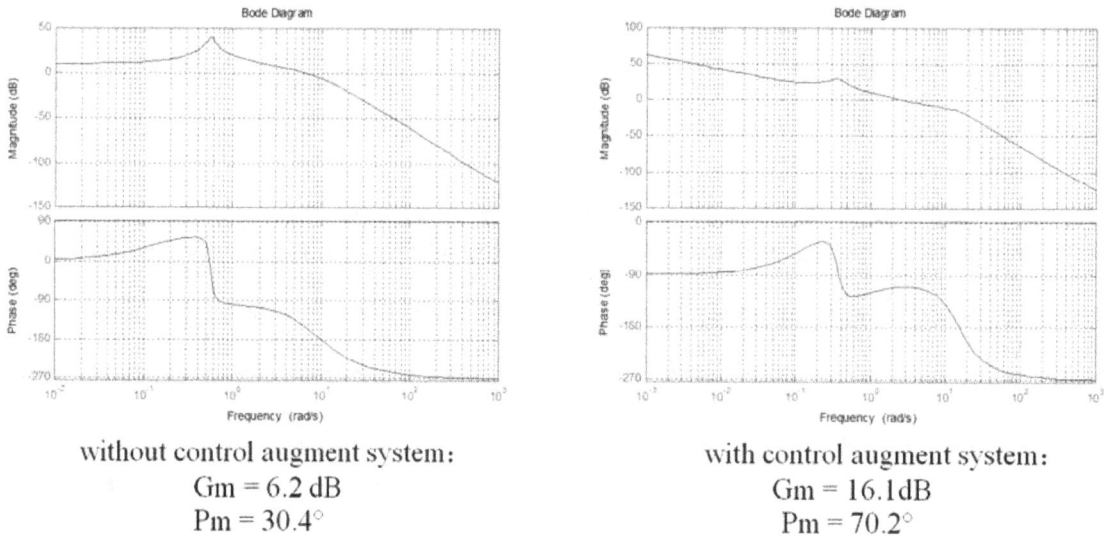

without control augment system:
Gm = 6.2 dB
Pm = 30.4°

with control augment system:
Gm = 16.1dB
Pm = 70.2°

Figure 2. Frequency-domain stability comparison.

From the comparison above, the UAV with the control augment system gets a better the time-domain and frequency-domain performance. So the flying performance of the UAV with the control augment system is much better.

3. Longitudinal Flying Quality Evaluation

We choose the design point (h=1000m,Ma=0.05) to evaluate the flying quality of the example UAV. We evaluate the applicability of human flying quality criterion to the UAV through analyzing the longitudinal flying quality of the

example UAV with and without control augment system by using human flying quality criterion. From the analysis and conclusion of the references [3-6], we evaluate the longitudinal flying quality of the example UAV by using the two typical human flying quality criteria: CAP criterion and bandwidth criterion.

3.1. CAP Criterion Evaluation [4] [8]

The quality level of the CAP criterion is determined by the following three parameters: equivalent short period frequency (ω_{sp}), damping ratio (ζ_{sp}) and stable vertical overload per attack angle (n/a). The equivalent short period frequency and damping ratio can be calculated through equivalent system method [4] [8]. Using the CAP criterion, we can get the longitudinal flying quality results of the example UAV with and without control augment system, as "Table 1" and "Figure 3" displaying.

Table 1. Flying quality results using CAP criterion.

	ω_{sp} /(rad/s)	ζ_{sp}	n/a	level
unaugment	7.7	0.74	8.9	2
augment	10.9	0.95	8.9	3

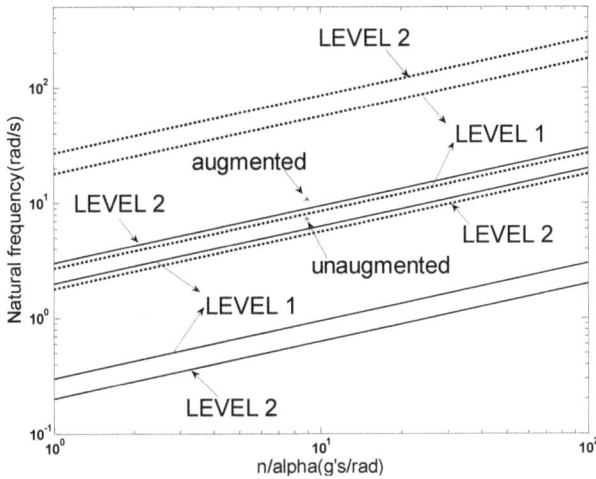

Note:Black line is CAP criterion, and blue line is the revised CAP criterion.

Figure 3. Human flying quality criterion (Phase B).

As we can see from the "Table 1" and "Figure 3", according to the human flying quality CAP criterion, the unaugmented example UAV is in level 2, but the augmented example UAV is in level 3. With the control augment system, the longitudinal flying quality of the example UAV becomes worse. But according to the performance comparison of time and frequency domains above, the flying performance of the augmented example UAV is much better than the unaugmented. So, the flying quality evaluation result based on the human CAP criterion is not consistent with the actual fact.

3.2. Bandwidth Criterion Evaluation

The quality level of the bandwidth criterion is determined by the following three parameters: open loop system bandwidth (ω_{Bw}) and time delay (τ_p). Firstly, bandwidth criterion usually applies to high-order system and does not require the minimum closed-loop pitch attitude bandwidth, so this criterion is applicable to different size aircrafts; Secondly, there are not any restrictions about system response properties, so this criterion can be applicable to the UAV because their responses are always changeful. It is recommended to use bandwidth criterion to evaluate the flying quality of UAV according to references of [3-5] [9]. Through comprehensive analysis, the bandwidth criterion of human flying quality is more applicable to UAV flying quality evaluation. So this paper uses bandwidth to evaluate flying quality of the example UAV.

With reference to the bandwidth criterion, we evaluate longitudinal flying quality of the example UAV at the selected point (h=2000m, Ma=0.05). The evaluation result is in "Table 2".

Table 2. Flying quality results using bandwidth criterion.

	Bandwidth ω_{Bw} /(rad/s)	Time delay τ_p /s	level
unaugment	5.08	0.0372	2
augment	9.08	0.0306	1

From "Table 2", the example UAV with control augment system gets improved flying quality, and this is consistent with the above time and frequency-domain analysis. So using bandwidth criterion to evaluate the UAV flying quality is reasonable and feasible.

4. Flying Quality Results Analysis

From the analysis above, the longitudinal flying quality evaluation results based on CAP criterion and bandwidth criterions are different. Combining the contrastive analysis of the time and frequency-domain, we can conclude that human CAP criterion is not applicable to the small example UAV. The small UAVs always have following features: lightweight, small size, small moment of inertial, and low speed. Refer to the following formula:

$$\omega_{sp} \approx \sqrt{\frac{Z_a M_q}{U_1} - M_a}$$

$$M_a = \frac{Q S_W c_A C_{ma}}{I_{yy}} \tag{1}$$

We can see that short period frequency (ω_{sp}) is related to M_a and velocity U_1, and M_a is inversely proportional to moment of inertial. The short period frequency of small UAVs is always high because of their small moment of inertial and low speed. CAP criterion is not applicable to small UAVs because of their high short period frequency. Given this reason, we propose an improved CAP criterion to apply to small UAVs. The improved CAP criterion adjusts the quality levels of CAP criterion using proportional coefficient N, and the scale relation is as following:

$$\omega_{small-scale} = \omega_{large-scale} \sqrt{N} \tag{2}$$

The coefficient N is the ratio of selected standard big plane wingspan and small UAV wingspan, N=80. Based on the selected coefficient N, we get the improved CAP criterion. According to the improved CAP criterion, the flying quality result of the example UAV is as following "Table 3" and "Figure 3":

Table 3. Flying quality results using improved CAP criterion.

	ω_{sp} /(rad/s)	ζ_{sp}	n/a	level
unaugment	7.7	0.74	8.9	2
augment	10.9	0.95	8.9	1

From "Table 3" and "Figure 3", according to the improved CAP criterion, the flying quality of example UAV without control augment system is in level 2, and in level 1 after being augmented. The example UAV with control augment system gets an improved flying quality, and this is consistent with the above time and frequency-domain analysis. So the improved CAP criterion can be applicable to the small UAV in evaluating longitudinal flying quality.

Based on all the analysis above, we can use bandwidth criterion and improved CAP criterion to evaluate the longitudinal flying quality of small UAVs.

5. Conclusion

This paper validates the applicability of some human flying quality criterions to small UAVs through comparing the flying performance of the example UAV with and without control augment system, and then proposes two longitudinal flying quality criterions of small UAVs: bandwidth criterion and improved CAP criterion. These two criterions present a newest idea and a useful trial to the research of UAV flying quality. And the research of this paper is useful for the design of small UAV as well as its control system design. But both these two criterions proposed by this paper have certain deficiency in analyzing flying quality of small UAVs, such as the definition of criterion boundary, so we still need a large number of fly tests to validate the applicability of these two proposed criterions to UAV flying quality evaluation.

References

[1]   Ning, Dai, Hui, Yang. Flying quality specification research of UAV, Flight Dynamics journal, 2005, 23(4): 13-15.

[2]   Yu jin, Tao, Jian pei, Wang. Flying quality standard research of UAV, Flight Dynamics journal, 2010, 28(1): 13-15.

[3]   Feng, Wang, Sheng jun, Qi. Applicability research of Some typical flying quality to UAV, Flight Dynamics journal, 2013, 31(5): 389-3935.

[4]   Cheng tao, Huang, Li xin, Wang. Longitudinal flying quality evaluation of UAV under remote mode, Flight Dynamics journal, 2013, 39(4): 427-431.

[5]   M. Christopher Cotting. Applicability of Human Flying Qualities Requirements for UAVs, Finding A Way Forward [C]. 27th AIAA Atmospheric Flight Mechanics Conference and Exhibit, 2009: 1-18.

[6]   Holmberg, J., and King, D. J., Flying Qualities Specifications and Design Standards for Unmanned Aerial Vehicles, AIAA-2008-6555.

[7]   Jin yuan, Gao, Lu yu, Li. Aircraft flying quality, 2003.

[8]   Foster, T., M. and Bowman, J. W. Dynamic Stability and Handling Qualities of Small Unmanned Aerial Vehicles, AIAA-2005-1023.

[9]   Giorgio Guglieri, Barbara Pralio, Fulvia Quagliotti. Flight control system design for a micro aerial vehicle [J]. Engineering & Aerospace Technology, 2006,78(2): 87-97.

[10]   Zi quan, Zhou. Flying test engineering, 2010.

[11]   Orkun Simsek, Ozan Tekinalp. System Identification and Handling Quality Analysis of a UAV from Flight Test Data, AIAA-2015-1480.

[12]   Elisa Capello, Giorgio Guglieri. Preliminary assessment of flying and handling qualities for mini-UAVs, Journal of Intelligent & Robotic Systems, 2012.

Permissions

All chapters in this book were first published in AJEP, by Science Publishing Group; hereby published with permission under the Creative Commons Attribution License or equivalent. Every chapter published in this book has been scrutinized by our experts. Their significance has been extensively debated. The topics covered herein carry significant findings which will fuel the growth of the discipline. They may even be implemented as practical applications or may be referred to as a beginning point for another development.

The contributors of this book come from diverse backgrounds, making this book a truly international effort. This book will bring forth new frontiers with its revolutionizing research information and detailed analysis of the nascent developments around the world.

We would like to thank all the contributing authors for lending their expertise to make the book truly unique. They have played a crucial role in the development of this book. Without their invaluable contributions this book wouldn't have been possible. They have made vital efforts to compile up to date information on the varied aspects of this subject to make this book a valuable addition to the collection of many professionals and students.

This book was conceptualized with the vision of imparting up-to-date information and advanced data in this field. To ensure the same, a matchless editorial board was set up. Every individual on the board went through rigorous rounds of assessment to prove their worth. After which they invested a large part of their time researching and compiling the most relevant data for our readers.

The editorial board has been involved in producing this book since its inception. They have spent rigorous hours researching and exploring the diverse topics which have resulted in the successful publishing of this book. They have passed on their knowledge of decades through this book. To expedite this challenging task, the publisher supported the team at every step. A small team of assistant editors was also appointed to further simplify the editing procedure and attain best results for the readers.

Apart from the editorial board, the designing team has also invested a significant amount of their time in understanding the subject and creating the most relevant covers. They scrutinized every image to scout for the most suitable representation of the subject and create an appropriate cover for the book.

The publishing team has been an ardent support to the editorial, designing and production team. Their endless efforts to recruit the best for this project, has resulted in the accomplishment of this book. They are a veteran in the field of academics and their pool of knowledge is as vast as their experience in printing. Their expertise and guidance has proved useful at every step. Their uncompromising quality standards have made this book an exceptional effort. Their encouragement from time to time has been an inspiration for everyone.

The publisher and the editorial board hope that this book will prove to be a valuable piece of knowledge for researchers, students, practitioners and scholars across the globe.

List of Contributors

Ruan Jinhua, Zheng Jingjing and Hu Shouzhong
Fashion Design and Engineering, Shanghai University of Engineering Science, shanghai, China

LI Jun-wei and XIE Yun
Electronic & Information Engineering College, Henan University of Science and Technology, Luoyang Henan, China

Jason Gu
School of Biomedical Engineering, Dalhousie University, Halifax, Canada

Amelia Santoso, Dina Natalia Prayogo and Joniarto Parung
Industrial Engineering Department, University of Surabaya, Surabaya, Indonesia

Chi-Yen Shen, Shuming T. Wang and Rey-Chue Hwang
Electrical Engineering Department, I-Shou University, Kaohsiung City, Taiwan

Yu-Ju Chen
Information Management Department, Cheng-Shiu University, Kaohsiung City, Taiwan

Chuo-Yean Chang
Electrical Engineering Department, Cheng-Shiu University, Kaohsiung City, Taiwan

Asghar Ehsani Fard
Telecommunication, of Non-profit Institution of Higher Education, ABA, Abyek, Qazvin, Iran

Masoud Masomei and Mehdi Hedayeti
Non-profit Institution of Higher Education, ABA, Abyek, Qazvin, Iran

Hamid Chegini
Research Assistant of Non-profit Institution of Higher Education, ABA, Abyek, Qazvin, Iran

Johar Daudi
Department of Aerospace Engineering, School of Engineering, University of Glasgow, Glasgow, UK

Hojatollah Rashidi Alashty
Cybernetics Department of the MATI University – Russian State Technological University named after K. E. Tsiolkovsky and member of research group of the Islamic Azad University, Qaemshahr, Mazandaran, Islamic Republic of Iran (Department of management)

Filimonov Aleksandr Borisovich
Cybernetics Department of the MATI University–Russian State Technological University named after K. E. Tsiolkovsky

Chih-Yung Chen and Ya-Chen Weng
Department of Computer and Communication, Shu-Te University, Kaohsiung City, Taiwan

Yu-Ju Chen
Department of Information Management, Cheng Shiu University, Kaohsiung City, Taiwan

Shen-Whan Chen
Department of Communication Engineering, I-Shou University, Kaohsiung City, Taiwan

Rey-Chue Hwang
Department of Electrical Engineering, I-Shou University, Kaohsiung City, Taiwan

Binbin Fu, Juntao Li and Yiming Wei
School of Information, Beijing Wuzi University, Beijing, China

Wei Lu, Xin Li
Shibei Power Supply Company, Shanghai Municipal Electric Power Company, Shanghai, China

Jamaluddin Mir, Malik Touqir Anwar and M. Yaqoob Wani
SIT Department, the University of Lahore, Islamabad, Pakistan

Majid Mehmood
Computer Science Department, University of Gujrat, Sialkot, Pakistan

H. Bal and N. P. Mahalik
Department of Industrial Technology, Jordan College of Agricultural Sciences and Technology, California State University, Fresno, California, USA

S. K. Mohanty
Colllege of Engineering, Biju Patnaik University of Technology, Bhubaneswar, Odisha, India

B. B. Biswal
National Institute of Technology, Rourkela, India

Tushar Kanti Roy
Department of Electronics & Telecommunication Engineering, Rajshahi University of Engineering & Technology, Rajshahi, Bangladesh

Rachid Sammouda and Hassan Ben Mathkour
Computer Science Department, College of Computer and Information Sciences, King Saud University, Riyadh, Saudi Arabia

He Yunbo, Hu Yongshan, Chen Xin, Gao Jian, Yang Zhijun, Chen Yun, Tang Hui, Ao Yinhui and Zhang Yu
Guangdong Provincial Key Lab. of Computer Integrated Manufacturing Systems, Key Laboratory of Mechanical Equipment Manufacturing and Control Technology of Ministry of Education, School of Electromechanical Engineering, Guangdong University of Technology, Guangzhou, P.R. China

Yue Yang and Zixia Hu
Electrical Engineering, University of Washington, Seattle, Washington, USA

Yanling Yin
Electrical Engineering, Harbin Engineering University, Harbin, USA

De Gu
Key Laboratory of Advanced Process Control for Light Industry (Ministry of Education), Institute of Automation, Jiangnan University, Wuxi, China

Jishuai Wang
Suzhou Institute of Biomedical Engineering and Technology, Chinese Academy of Sciences, Suzhou, China

Ho-Hung Jung
Seoul Metro, Seoul, Korea

Changlong Li and Key-Seo Lee
School of Robotics, Kwangwoon University, Seoul, Korea

Jun-tao Li
School of Information, Beijing Wuzi University, Beijing, China

Hong-jian Liu
Graduate Department, Beijing Wuzi University, Beijing, China

Arnold Adimabua Ojugo
Dept. of Math/Computer, Federal University of Petroleum Resources Effurun, Delta State, Nigeria

Fidelis Obukowho Aghware
Dept. of Computer Science Education, College of Education, Agbor, Delta State, Nigeria

Rume Elizabeth Yoro
Dept. of Computer Sci., Delta State Polytechnic, Ogwashi-Uku, Delta State, Nigeria

Mary Oluwatoyin Yerokun, Andrew Okonji Eboka and Christiana Nneamaka Anujeonye
Dept. of Computer Sci. Education, Federal College of Education (Technical), Asaba, Delta State, Nigeria

Fidelia Ngozi Efozia
Prototype Engineering Development Institute, Fed. Ministry of Science Technology, Osun State, Nigeria

Liyue Chen, Tao Tao, Lizhong Zhang and Bing Lu
Electric power dispatching control center, State Grid Zhejiang Electric Power Company, Hangzhou, China

Zhongling Hang
Department of Automation, Shanghai Jiao Tong University, Shanghai, China

Zhengzheng Cong, Cong Li, Yifeng Shao, Zhize Zhou and Chunmeng Liang
Department of Mechanical Engineering, Guilin University of Aerospace Technology, Guilin, China

Seyedtaha Seyedsadr
Department of Management, Electronic Branch, Islamic Azad University, Tehran, Iran

Mohammadali Afsharkazemi
Department of Management, Tehran Central Branch, Islamic Azad University, Tehran, Iran

Hashem Nikoomaram
Department of Management and Economics, Sciences and Research Branch, Islamic Azad University, Tehran, Iran

Takuya Kamimura and Shinichi Tamura
NBL Technovator Co., Ltd. Shindachimakino, Sennan japan

Yasushi Yagi
ISIR, Osaka University, Mihogaoka, Ibaraki City, Osaka, Japan

Yen-Wei Chen
Ritsumeikan University, Nojihigashi, Kusatsu-shi, Shiga, Japan

Prabhakar Sastri and Andreas Stephanides
Automation and Data Analytics Department, Isa Technologies Pvt. Ltd., Manipal, India

Neama Y and Habil
Medical Technology Institute, Medical Technical University, Baghdad, Iraq

Nguyen Quang Vinh
Department of Radio-Electronic, Institute of Science and Technology, Cau Giay District, Hanoi, Vietnam

Yuan Wang and Keming Tang
College of Information Engineering, Yancheng Teachers University, Yancheng, People's Republic of China

Zhudeng Wang
School of Mathematics and Statistics, Yancheng Teachers University, Jiangsu, People's Republic of China

Abdulkadhum Jaafar Alyasiri , Abdulrahman Saleh Ibrahim and Amel Salih Merzah
President of Foundation of Technical Education (F.T.E), Baghdad, Iraq. Technical College, Baghdad, Iraq

Roman Kaszyński, Adrian Sztandera and Katarzyna Wiechetek
Department of Systems, Signals and Electronic Engineering Faculty of Electrical West Pomeranian University of Technology, Szczecin, Poland

Noriaki Sakamoto
Faculty of Economics, Hosei University, Tokyo, Japan

Juntao Li
School of Information, Beijing Wuzi University, Beijing, China

Xiaoqing Zhao
Graduate Department, Beijing Wuzi University, Beijing, China

S. AbdulAmohsin and M. Mohamed
Physical Department, Faculty of Science, Thi Qar University , Thi Qar, Iraq.

Zhian Yan, Hudan Sun
China Electronics Technology Group Corporation Special Mission Aircraft System Engineering Co., Ltd., Chengdu, China

Index